Genetics and Epigenetics in Complex Diseases

Genetics and Epigenetics in Complex Diseases

Editor

Elixabet Lopez-Lopez

MDPI • Basel • Beijing • Wuhan • Barcelona • Belgrade • Manchester • Tokyo • Cluj • Tianjin

Editor
Elixabet Lopez-Lopez
University of the Basque
Country (UPV/EHU)
Spain

Editorial Office
MDPI
St. Alban-Anlage 66
4052 Basel, Switzerland

This is a reprint of articles from the Special Issue published online in the open access journal *International Journal of Molecular Sciences* (ISSN 1422-0067) (available at: https://www.mdpi.com/books/book/7398).

For citation purposes, cite each article independently as indicated on the article page online and as indicated below:

LastName, A.A.; LastName, B.B.; LastName, C.C. Article Title. *Journal Name* **Year**, *Volume Number*, Page Range.

ISBN 978-3-0365-7857-6 (Hbk)
ISBN 978-3-0365-7856-9 (PDF)

Contents

About the Editor

Elixabet Lopez-Lopez

Dr. Elixabet Lopez-Lopez is an assistant professor at the Department of Biochemistry and Molecular Biology at the University of the Basque Country (UPV/EHU), and she is affiliated with the Biocruces Bizkaia Health Research Institute. She has dedicated her career so far to the search for pharmacogenomics markers to improve the treatment of pediatric oncology patients. During her PhD training and subsequent postdoctoral experience at the UPV/EHU (Spain), her main focus was on genetic variants associated with adverse drug reactions and response in pediatric acute lymphoblastic leukemia and osteosarcoma patients. She was able to find SNPs associated with drug toxicity by studying candidate genes in the methotrexate and vincristine pathways. Additionally, she was the first author in the first study to show that polymorphisms in miRNA processing genes are associated with toxicity after treatment in cancer patients. Later, in the context of her work as a postdoctoral fellow at St. Jude Children's Research Hospital (USA), she was able to analyze genome-wide expression and regulation (miRNAs, methylation, SNPs, and copy number alterations) in association with methotrexate response in acute lymphoblastic leukemia patients. At her current position as part of the group "Genetics and Epigenetics of Complex Diseases", she is in charge of the research line of pharmacogenetics in childhood ALL. vii

Preface to "Genetics and Epigenetics in Complex Diseases"

Many of the most common diseases are influenced by a combination of multiple factors, which include environmental effectors as well as genetic and epigenetic variations. Therefore, these diseases are grouped under the term "complex" diseases because, from the point of view of genetics, they cannot be explained by simple Mendelian inheritance.

The aim of this compilation of articles was to identify genetic and epigenetic factors involved in such diseases in order to improve not only the knowledge of risk factors for those diseases, which could be of help for prevention, but also the understanding and characterization of each disease and to optimize and personalize the treatment.

Elixabet Lopez-Lopez
Editor

International Journal of
Molecular Sciences

Editorial

Genetics and Epigenetics in Complex Diseases

Elixabet Lopez-Lopez [1,2]

[1] Department of Biochemistry and Molecular Biology, University of the Basque Country, UPV/EHU, 48940 Leioa, Spain; elixabet.lopez@ehu.eus
[2] Pediatric Oncology Group, BioCruces Bizkaia Health Research Institute, 48903 Barakaldo, Spain

Many of the most common diseases are influenced by a combination of multiple factors, which include environmental effectors, as well as genetic and epigenetic variations. Therefore, these diseases are grouped under the term "complex" diseases because, from the point of view of genetics, they cannot be explained by simple Mendelian inheritance.

The aim of this Special Issue was to identify genetic and epigenetic factors involved in such diseases, in order to improve not only the knowledge of risk factors for those diseases, which could be of help for prevention, but also to improve the understanding and characterization of each disease, and to optimize and personalize their treatment.

On the one hand, seven research articles identified genetic and epigenetic variations of relevance for pediatric and adult malignancies, coronary artery disease, body shape and metabolic traits, and Alzheimer's Disease.

In particular, two articles focused on pediatric malignancies. While Schedel et al. identified a germline variant in *RAD21* that can predispose to childhood lymphoblastic leukemia or lymphoma without displaying a Cornelia-de-Lange syndrome phenotype [1], Michler et al. identified a germline variant in *POT1* in a child with acute myeloid leukemia and showed a connection between this variant and POT expression and telomeric dysregulation [2]. Regarding adult malignancies, Pacholewska et al. identified local decreases in methylation levels in chronic lymphocytic leukemia patients harboring mutations at *SF3B1*, mostly in proximity to telomeric regions, enriched in cancer-related signaling genes [3], and Campos Gudiño et al. observed that SKP1, CUL1, and F-box protein complex member genes were frequently altered at the genetic and epigenetic levels in many cancer types, which might contribute to the development and progression of these malignancies [4].

Regarding other disease entities, Chou et al. proposed that combining resistin and sST2 levels with weighted genetic risk scores of *RETN* and *IL1RL1* could be helpful for the prediction of outcome in coronary artery disease [5]. Moreover, Wu et al. showed that genetic and epigenetic variations of *KLF14* were associated with body shape indices, metabolic traits, insulin resistance, and metabolically healthy status, effects that were mediated by age, sex and obesity [6]. Finally, Tortora et al. hypothesized that a polymorphism in *CD33* could be a risk factor for Alzheimer's disease, through the binding of sialic acid, acting as an enhancer of the CD33 inhibitory effects on amyloid plaque degradation, based on in silico analyses [7].

On the other hand, four articles reviewed the literature on the role of genetic variants in vitamin D-binding protein-related diseases, psoriatic disease, vasovagal syncope, and virus-induced epigenetic changes that lead to carcinogenesis.

First, Rozmus et al. reviewed the relationship between polymorphisms in the VDBP gene which might lead to vitamin D deficiencies and diseases such as diabetes, polycystic ovarian syndrome, metabolic syndrome, or Parkinson's disease [8]. Secondly, Queiro et al. discussed the association of genetic variants in the NF-κB pathway with the risk of suffering psoriatic disease, as well as with the comorbidities that frequently accompany it, and their relevance for improving treatment selection [9]. Then, Matveeva et al. summarized data on the genetics of vasovagal syncope, describing the inheritance pattern of the disorder,

Citation: Lopez-Lopez, E. Genetics and Epigenetics in Complex Diseases. *Int. J. Mol. Sci.* **2023**, *24*, 8186. https://doi.org/10.3390/ijms24098186

Received: 13 February 2023
Accepted: 18 April 2023
Published: 3 May 2023

candidate gene association studies and genome-wide studies [10]. Lastly, Pietropaolo et al. reviewed the role of epigenetic changes that take place in the host cells in virus-induced cancers [11].

As a whole, this Special Issue covers different aspects of genetic and epigenetic variation with a role in different complex diseases, sheds light on possible pathways that lead to their involvement in these diseases, and suggests possible applications of such knowledge.

Conflicts of Interest: The author declares no conflict of interest.

References

1. Schedel, A.; Friedrich, U.A.; Morcos, M.N.; Wagener, R.; Mehtonen, J.; Watrin, T.; Saitta, C.; Brozou, T.; Michler, P.; Walter, C.; et al. Recurrent Germline Variant in *RAD21* Predisposes Children to Lymphoblastic Leukemia or Lymphoma. *Int. J. Mol. Sci.* **2022**, *23*, 5174. [CrossRef] [PubMed]
2. Michler, P.; Schedel, A.; Witschas, M.; Friedrich, U.A.; Wagener, R.; Mehtonen, J.; Brozou, T.; Menzel, M.; Walter, C.; Nabi, D.; et al. Germline POT1 Deregulation Can Predispose to Myeloid Malignancies in Childhood. *Int. J. Mol. Sci.* **2021**, *22*, 11572. [CrossRef] [PubMed]
3. Pacholewska, A.; Grimm, C.; Herling, C.D.; Lienhard, M.; Königs, A.; Timmermann, B.; Altmüller, J.; Mücke, O.; Reinhardt, H.C.; Plass, C.; et al. Altered DNA Methylation Profiles in SF3B1 Mutated CLL Patients. *Int. J. Mol. Sci.* **2021**, *22*, 9337. [CrossRef] [PubMed]
4. Campos, G.R.; Farrell, A.C.; Neudorf, N.M.; McManus, K.J. A Comprehensive Assessment of Genetic and Epigenetic Alterations Identifies Frequent Variations Impacting Six Prototypic SCF Complex Members. *Int. J. Mol. Sci.* **2021**, *23*, 84. [CrossRef] [PubMed]
5. Chou, H.-H.; Hsu, L.-A.; Juang, J.-M.J.; Chiang, F.-T.; Teng, M.-S.; Wu, S.; Ko, Y.-L. Synergistic Effects of Weighted Genetic Risk Scores and Resistin and sST2 Levels on the Prognostication of Long-Term Outcomes in Patients with Coronary Artery Disease. *Int. J. Mol. Sci.* **2022**, *23*, 4292. [CrossRef] [PubMed]
6. Wu, S.; Hsu, L.-A.; Teng, M.-S.; Chou, H.-H.; Ko, Y.-L. Differential Genetic and Epigenetic Effects of the *KLF14* Gene on Body Shape Indices and Metabolic Traits. *Int. J. Mol. Sci.* **2022**, *23*, 4165. [CrossRef] [PubMed]
7. Tortora, F.; Rendina, A.; Angiolillo, A.; Di Costanzo, A.; Aniello, F.; Donizetti, A.; Febbraio, F.; Vitale, E. CD33 rs2455069 SNP: Correlation with Alzheimer's Disease and Hypothesis of Functional Role. *Int. J. Mol. Sci.* **2022**, *23*, 3629. [CrossRef] [PubMed]
8. Rozmus, D.; Plominski, J.; Augustyn, K.; Cieslinska, A. rs7041 and rs4588 Polymorphisms in Vitamin D Binding Protein Gene (VDBP) and the Risk of Diseases. *Int. J. Mol. Sci.* **2022**, *23*, 933. [CrossRef] [PubMed]
9. Queiro, R.; Coto, P.; González-Lara, L.; Coto, E. Genetic Variants of the NF-kappaB Pathway: Unraveling the Genetic Architecture of Psoriatic Disease. *Int. J. Mol. Sci.* **2021**, *22*, 13004. [CrossRef] [PubMed]
10. Matveeva, N.; Titov, B.; Bazyleva, E.; Pevzner, A.; Favorova, O. Towards Understanding the Genetic Nature of Vasovagal Syncope. *Int. J. Mol. Sci.* **2021**, *22*, 10316. [CrossRef] [PubMed]
11. Pietropaolo, V.; Prezioso, C.; Moens, U. Role of Virus-Induced Host Cell Epigenetic Changes in Cancer. *Int. J. Mol. Sci.* **2021**, *22*, 8346. [CrossRef] [PubMed]

International Journal of
Molecular Sciences

MDPI

Article

Recurrent Germline Variant in *RAD21* Predisposes Children to Lymphoblastic Leukemia or Lymphoma

Anne Schedel [1], Ulrike Anne Friedrich [1], Mina N. F. Morcos [2], Rabea Wagener [3], Juha Mehtonen [4], Titus Watrin [3], Claudia Saitta [5], Triantafyllia Brozou [3], Pia Michler [1], Carolin Walter [6], Asta Försti [7,8], Arka Baksi [9], Maria Menzel [1], Peter Horak [10], Nagarajan Paramasivam [11], Grazia Fazio [5], Robert J Autry [7,8], Stefan Fröhling [10], Meinolf Suttorp [1], Christoph Gertzen [12], Holger Gohlke [12,13], Sanil Bhatia [3], Karin Wadt [14], Kjeld Schmiegelow [15], Martin Dugas [6,16], Daniela Richter [17,18], Hanno Glimm [17,18,19], Merja Heinäniemi [4], Rolf Jessberger [9], Gianni Cazzaniga [5,20], Arndt Borkhardt [3], Julia Hauer [2,21,*] and Franziska Auer [2]

1 Pediatric Hematology and Oncology, Department of Pediatrics, University Hospital Carl Gustav Carus, TU Dresden, 01307 Dresden, Germany; anne.schedel@uniklinikum-dresden.de (A.S.); ulrikeanne.friedrich@uniklinikum-dresden.de (U.A.F.); pia.michler@uniklinikum-dresden.de (P.M.); maria.menzel@uniklinikum-dresden.de (M.M.); meinolf.suttorp@uniklinikum-dresden.de (M.S.)
2 Department of Pediatrics, School of Medicine, Technical University of Munich; 80804 Munich, Germany; mina.morcos@tum.de (M.N.F.M.); f.auer@tum.de (F.A.)
3 Department of Pediatric Oncology, Hematology and Clinical Immunology, Heinrich-Heine University Duesseldorf, Medical Faculty, 40225 Duesseldorf, Germany; rabea.wagener@med.uni-duesseldorf.de (R.W.); titus.watrin@med.uni-duesseldorf.de (T.W.); triantafyllia.brozou@med.uni-duesseldorf.de (T.B.); sanil.bhatia@med.uni-duesseldorf.de (S.B.); arndt.borkhardt@med.uni-duesseldorf.de (A.B.)
4 Institute of Biomedicine, School of Medicine, University of Eastern Finland, Yliopistonranta 1, FI-70211 Kuopio, Finland; juha.mehtonen@uef.fi (J.M.); merja.heinaniemi@uef.fi (M.H.)
5 Tettamanti Research Center, Pediatrics, University of Milan Bicocca, Fondazione MBBM/San Gerardo Hospital, 20900 Monza, Italy; c.saitta2@campus.unimib.it (C.S.); g.fazio@hsgerardo.org (G.F.); gianni.cazzaniga@asst-monza.it (G.C.)
6 Institute of Medical Informatics, University of Muenster, 48149 Muenster, Germany; carolin.walter@uni-muenster.de (C.W.); martin.dugas@med.uni-heidelberg.de (M.D.)
7 Division of Pediatric Neurooncology, German Cancer Research Center (DKFZ), German Cancer Consortium (DKTK), 69120 Heidelberg, Germany; a.foersti@kitz-heidelberg.de (A.F.); robert.autry@kitz-heidelberg.de (R.J.A.)
8 Hopp Children's Cancer Center Heidelberg (KiTZ), 69120 Heidelberg, Germany
9 Institute of Physiological Chemistry, Medical Faculty Carl Gustav Carus, TU Dresden, 01307 Dresden, Germany; arka.baksi@tu-dresden.de (A.B.); rolf.jessberger@tu-dresden.de (R.J.)
10 Division of Translational Medical Oncology, National Center for Tumor Diseases (NCT) and German Cancer Research Center (DKFZ), 69120 Heidelberg, Germany; peter.horak@nct-heidelberg.de (P.H.); stefan.froehling@nct-heidelberg.de (S.F.)
11 Computational Oncology, Molecular Diagnostics Program, National Center for Tumor Diseases (NCT), 69120 Heidelberg, Germany; n.paramasivam@dkfz-heidelberg.de
12 Institute for Pharmaceutical and Medicinal Chemistry, Heinrich-Heine-Universität Duesseldorf, Universitätsstraße 1, 40225 Duesseldorf, Germany; christoph.gertzen@hhu.de (C.G.); gohlke@uni-duesseldorf.de (H.G.)
13 John von Neumann Institute for Computing (NIC), Jülich Supercomputing Centre (JSC), Institute of Biological Information Processing (IBI-7: Structural Biochemistry), Forschungszentrum Jülich GmbH, 52425 Jülich, Germany
14 Department of Clinical Genetics, University Hospital of Copenhagen, Faculty of health and Medical Sciences, University of Copenhagen, 2100 Copenhagen, Denmark; karin.wadt@regionh.dk
15 Department of Paediatrics and Adolescent Medicine, Copenhagen University Hospital Rigshospitalet, 2100 Copenhagen, Denmark; kjeld.schmiegelow@regionh.dk
16 Institute of Medical Informatics, Heidelberg University Hospital, 69120 Heidelberg, Germany
17 Department of Translational Medical Oncology, National Center for Tumor Diseases (NCT) Dresden, 01307 Dresden, Germany; daniela.richter@nct-dresden.de (D.R.); hanno.glimm@uniklinikum-dresden.de (H.G.)
18 German Cancer Consortium (DKTK), 01307 Dresden, Germany
19 Translational Functional Cancer Genomics, National Center for Tumor Diseases (NCT) and German Cancer Research Center (DKFZ), 69120 Heidelberg, Germany
20 Medical Genetics, Department of Medicine and Surgery, University of Milan Bicocca, 20900 Monza, Italy
21 German Cancer Consortium (DKTK), 81675 Munich, Germany

Citation: Schedel, A.; Friedrich, U.A.; Morcos, M.N.F.; Wagener, R.; Mehtonen, J.; Watrin, T.; Saitta, C.; Brozou, T.; Michler, P.; Walter, C.; et al. Recurrent Germline Variant in *RAD21* Predisposes Children to Lymphoblastic Leukemia or Lymphoma. *Int. J. Mol. Sci.* **2022**, *23*, 5174. https://doi.org/10.3390/ijms23095174

Academic Editor: Elixabet Lopez-Lopez

Received: 12 April 2022
Accepted: 2 May 2022
Published: 5 May 2022

* Correspondence: julia.hauer@mri.tum.de; Tel.: +49-(89)-3068-3940

Abstract: Somatic loss of function mutations in cohesin genes are frequently associated with various cancer types, while cohesin disruption in the germline causes cohesinopathies such as Cornelia-de-Lange syndrome (CdLS). Here, we present the discovery of a recurrent heterozygous *RAD21* germline aberration at amino acid position 298 (p.P298S/A) identified in three children with lymphoblastic leukemia or lymphoma in a total dataset of 482 pediatric cancer patients. While *RAD21* p.P298S/A did not disrupt the formation of the cohesin complex, it altered *RAD21* gene expression, DNA damage response and primary patient fibroblasts showed increased G2/M arrest after irradiation and Mitomycin-C treatment. Subsequent single-cell RNA-sequencing analysis of healthy human bone marrow confirmed the upregulation of distinct cohesin gene patterns during hematopoiesis, highlighting the importance of *RAD21* expression within proliferating B- and T-cells. Our clinical and functional data therefore suggest that *RAD21* germline variants can predispose to childhood lymphoblastic leukemia or lymphoma without displaying a CdLS phenotype.

Keywords: acute lymphoblastic leukemia; trio sequencing; germline cancer predisposition; *RAD21*; cohesin complex

1. Introduction

The cohesin complex is one of the most essential keepers of genome stability, ensuring proper cell development and proliferation. Cohesin complex genes are ubiquitously expressed and are indispensable for cell survival [1]. Its central element is a highly conserved protein complex, formed as a ring-like structure by the helical proteins SMC1 and SMC3, which are in turn connected by RAD21 [2] and STAG 1/2 (also known as SA 1/2) [3,4] (Figure 1A). The co-factor WAPL is important for the cleavage in early phases of mitosis [5–7] and PDS5B can act both as maintenance and as a cohesin releasing factor [8]. Cohesin genes are first and foremost known for their involvement in chromatid aggregation and organized segregation in anaphase [9–11] with RAD21 cleavage marking the onset of anaphase [12]. Additionally, the complex participates in DNA double-strand break (DSB) repair, by holding the chromatids together during homologous recombination [13,14]. More recently, the cohesin complex has been implicated to govern the structure and function of chromatin. In this regard, the complex is involved in gene transcription through chromatid folding and RNA recruitment together with the CCCTC- binding factor (CTCF) [15,16], and has been shown to take part in the formation of topologically associated domains (TADs) [17].

RAD21-inactivating heterozygous somatic mutations are a well-established correlate of various human cancers, such as acute myeloid leukemia (AML) [18]. Furthermore, two cases with somatic truncating mutations in *RAD21* were recently identified in a study of pediatric precursor B-cell ALL (BCP-ALL) with very early relapse [19] and somatic cohesin mutations have been reported in pediatric high hyperdiploid leukemia [20]. Germline aberrations in cohesin complex genes are rare, but if present, cause syndromal disorders termed cohesinopathies. Cornelia-de-Lange syndrome (CdLS) is one of the best described examples, which exerts a condition of variable penetrance and expressivity presenting with neuro-developmental delays and abnormalities of the limbs [21]. While this syndrome is not typically known to confer cancer predisposition, an index case of a child with simultaneous occurrence of acute lymphoblastic leukemia (ALL) and CdLS caused by a *NIPBL* frameshift mutation has recently been reported [22]. Nevertheless, a possible link between additional germline cohesin complex gene mutations and childhood leukemia as well as cancer in general is still lacking. We find this quite surprising, given the established role of cohesins in various cancer types. Here, we describe a recurrent and functionally relevant mutated position within *RAD21* in three children with lymphatic malignancies originating from three different independent cancer cohorts.

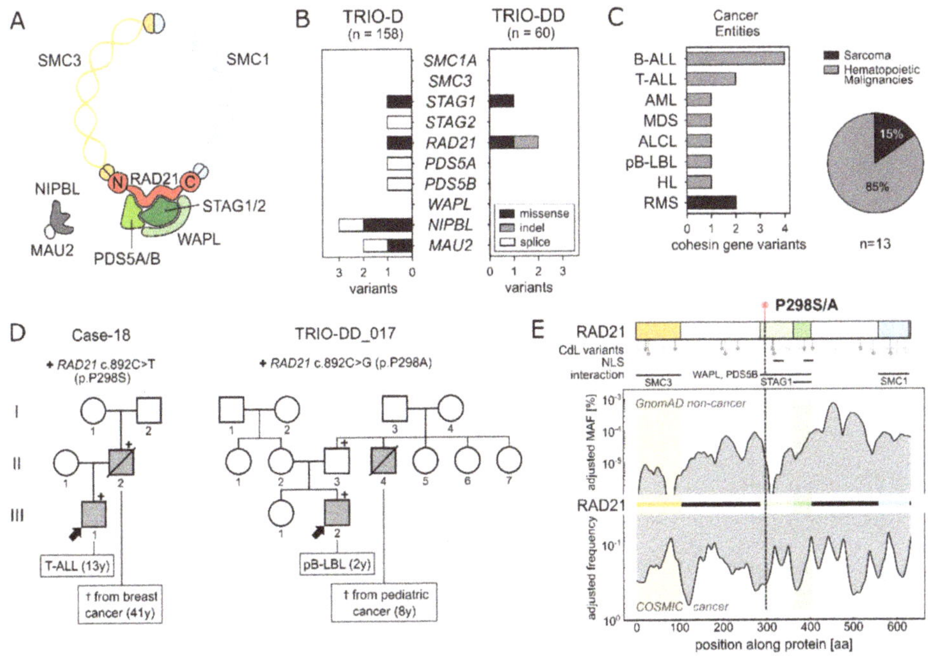

Figure 1. (**A**): The cohesin complex is formed by the 4 main core units SMC1 and SMC3 connected by RAD21 and STAG1 or STAG2. WAPL and PDS5 as co-factors and NIPBL and MAU2 as loaders are depicted. (**B**): Two patient cohorts (TRIO-D: n = 158 and TRIO-DD n = 60) were analyzed for germline variants within cohesin genes as depicted in Supplementary Table S1. Only non-synonymous variants with a MAF < 0.1% (gnomAD non-cancer population) were included. (**C**): Tumor entities of patients carrying a coding variant in one of the cohesin genes as shown in (**B**) (both cohorts combined, n = 13). Hematological malignancies account for 84.6% of cancers in the patients with germline cohesin variants. Further cohesin variants were identified in 2 patients with rhabdomyosarcoma. ALL: Acute lymphoblastic leukemia, AML: Acute myeloid leukemia, MDS: Myelodysplastic syndrome, ALCL: Anaplastic large-cell lymphoma, pB-LBL: precursor B-cell lymphoblastic lymphoma, HL: Hodgkin lymphoma, RMS: Rhabdomyosarcoma. (**D**): Family pedigrees of patients carrying the heterozygous germline *RAD21* variant p.P298S/A. Index patients are marked with an arrow. Family members affected by cancer are highlighted in grey. Variant carriers are marked with "+". (**E**): Upper: RAD21 protein structure displaying the interaction domains with SMC3 (1-103 amino acids (AA)), WAPL and PDS5B (287-403AA), STAG1/STAG2 (362-403AA) and SMC1 (558-628AA, available online: http://genesdev.cshlp.org/content/23/18/2224.long accessed on 10 April 2022). Lollipops below depict the positions of variants known in Cornelia de Lange (CdL) syndrome patients, adapted from Krab et al. 2020, with light gray representing missense variants and in-frame deletions and darker gray representing protein truncations. Lower: Distribution of variant frequencies along RAD21, based on two databases: The top shows the adjusted MAF (%) of *RAD21* germline variants in the gnomAD non-cancer database, while the bottom shows the adjusted frequency of variants in the COSMIC (somatic cancer mutations) database.

2. Results

2.1. Identification of a Recurrent RAD21 Germline Alteration (p.P298S/A)

To add a novel piece to the understanding of cohesins in cancer predisposition, we analyzed whole exome sequencing data of an unselected German parent–child cohort of children with cancer (n = 60, TRIO-DD), as well as a recently published parent–child pediatric cancer cohort (n = 158, TRIO-D) [23] for germline variants in cohesin complex genes (Supplementary Table S1). Overall, in both childhood cancer cohorts, 13 variants (Minor allele frequency (MAF) < 0.1%; gnomAD non-cancer database) in seven different cohesin genes were identified (Figure 1B). All were transmitted from one of the parents, were mutually exclusive and significantly enriched in leukemia (lymphoid origin = 6, myeloid origin = 2) and lymphoma (n = 3) patients as compared to patients with solid tumors within the cohorts (Fisher's exact test; p = 0.0081) (Figure 1C and Figure S1). Thereof, CdLS phenotypes were observed in one AML patient carrying *NIPBL* p.(G998E) (Case-92) and in one BCP-ALL patient harboring *MAU2* p.(N410S) (Case-74) (Supplementary Table S2). Nonetheless, none of the two patients presented with a definitive diagnosis of CdLS.

Interestingly, among all cohesin complex variants, one recurrently mutated nucleotide leading to an amino acid (AA) exchange at position 298 of *RAD21* (rs148308569) was identified in two families (one per cohort), in the absence of otherwise known-pathogenic variants (ClinVar) (Figures S2 and S3, Supplementary Tables S3 and S4). While the affected pediatric cancer patients carrying the recurrent *RAD21* variation did not show signs of CdLS, both three-generation pedigrees displayed a remarkable family history of early-in-life cancer (Figure 1D). In family I (Case-18), the heterozygous *RAD21* p.P298S (c.892C>T) variant was identified in a 13-year-old boy with T-ALL. His father, who transmitted *RAD21* p.P298S to his son, had died from breast cancer at the age of 41. Family II (TRIO-DD_017) displayed an alternative AA substitution at the same protein position (*RAD21* p.P298A; c.892C>G), which was detected in a 2-year-old patient with precursor B-cell lymphoblastic lymphoma (pB-LBL). Here, the variant was inherited from the healthy father, whose brother had died during childhood from cancer of unknown subtype (8y).

RAD21 p.P298 is evolutionarily conserved across species (GERP-score 5.61, phastCons = 1), located within the WAPL/PDS5B binding domain, and has not yet been reported in individuals with CdLS [24] (Figure 1E, Supplementary Table S5). While a low MAF at *RAD21* p.P298 and its surrounding AA indicates that these positions are rarely mutated in the germline of the non-cancer population (gnomAD database n = 118,479; MAF *RAD21* p.P298S < 10^{-6} and p.P298A < 10^{-5}), high somatic variation frequencies (COSMIC database n = 37,221) are observed at the end of the SMC3 interaction domain and the start of the WAPL/PDS5B interacting domain, where the variants are located (Figure 1E). Furthermore, the CADD scores indicate potential deleterious effects with values of 22.3 and 22.5 for *RAD21* p.P298S and *RAD21* p.P298A, respectively. To assess the structural impact of *RAD21* p.P298S/A, we aimed to generate a computational model of the 50 adjacent residues on each side. However, several approaches failed to generate a secondary structure for this region, reflecting the substitution site as part of a very flexible and intrinsically disordered region (predicted disorder content of *RAD21*: 51.7%) (Figure S4).

2.2. *RAD21 p.P298S/A Alters Cell Cycle and DNA Damage Responses*

Given that *RAD21* p.P298S/A is located in a hyper-flexible domain, we next aimed to investigate its interaction with cohesin complex partners. Therefore, the identified *RAD21* variants were cloned and transfected into HEK293T cells (R32-hRAD21). In analogy to *RAD21* WT, neither protein expression nor nuclear localization were affected by the variants *RAD21* p.P298S/A (Figure S5). Immunoprecipitation assays of the nuclear fraction showed binding of RAD21 with WAPL and PDS5B for the WT, as well as for both mutant proteins RAD21 p.298S/A, respectively (Figure 2A). Furthermore, the interaction of RAD21 WT and RAD21 p.P298S/A to SMC1 and STAG2 were comparable (Figure S6), suggesting that RAD21 p.P298S/A does not perturb the formation of the cohesin complex.

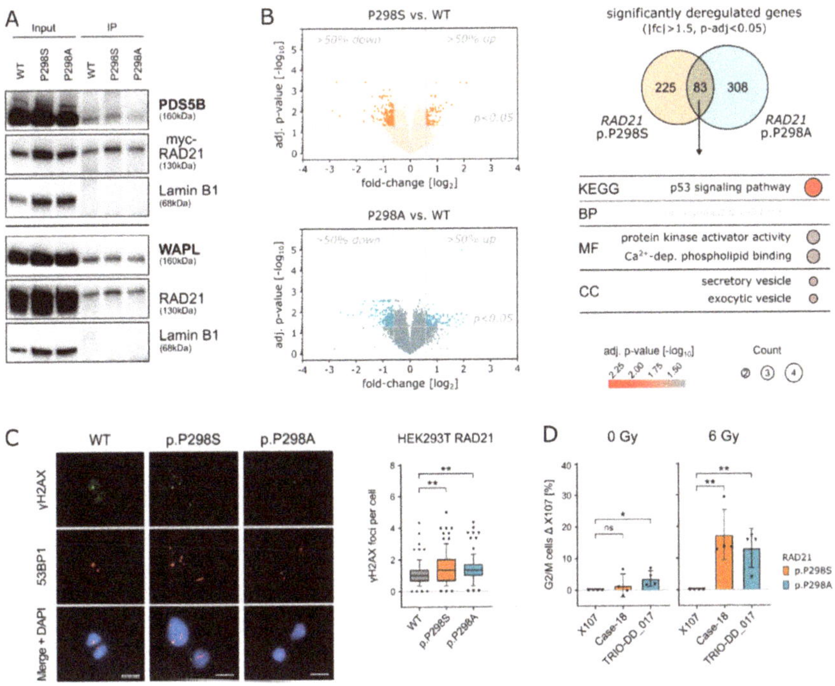

Figure 2. (**A**): Immunoprecipitation was performed on HEK293T cells overexpressing cMyc-tagged RAD21 WT, RAD21 p.P298S or RAD21 p.P298A. Cells were FCS-deprived and after 24 h arrested with colchicine (0.5 μg/mL) for 2 h, and the nuclear fraction was used for immunoprecipitation with the cMyc-tag. While the upper and lower panel represent one immunoprecipitation assay, they were run on two independent immunoblots and therefore presented as two panels. (**B**): Volcano plot of average gene expression based on microarray data. Fold-change and adjusted p-values are calculated by comparing *RAD21* p.P298S to WT (orange, top panel) and *RAD21* p.P298A to WT (blue, bottom panel). Probes with > 50% up- or downregulation and an adjusted p-value < 0.05 are considered as differentially expressed (DE) and highlighted in dark orange (*RAD21* p.P298S, top panel) or dark blue (*RAD21* p. P298A, bottom panel). DE genes are compared between *RAD21* p.P298S vs. WT and *RAD21* p.P298A vs. WT and show an overlap >20%. GO-term analysis of shared DE genes from the previous analysis identified enriched GO-terms. All GO-terms that exceed the significance (Benjamini–Hochberg FDR < 0.05) are represented. (**C**): Left: representative images of γH2AX (green) and 53BP1 (red) foci. DAPI (blue) was used for DNA labelling. Scale bar: 10 μm. Right: quantification of γH2AX foci per cell in HEK293T *RAD21* WT, p.P298A and p.P298S cells. Experiments were performed as 3 independent replicates. Values are expressed in boxplots with whiskers from percentile 10–90. For the statistical analysis, Student's t-test was performed (** = $p \leq 0.01$). (**D**): X107 (healthy control, *RAD21* WT), Case-18 (*RAD21* p.P298S), and TRIO-DD_017 (*RAD21* p.P298A) primary fibroblasts were subjected to irradiation with 6 Gy (n = 4) and the cell cycle analyzed using propidium iodide staining. For indicated p-values, Student's t-testing was performed (* = $p \leq 0.05$; ** = $p \leq 0.01$). Case-18 and TRIO-DD_017 were adjusted to X107 as a baseline response.

Since one additional function of the complex is the control of transcriptional regulation through genome-wide chromatin organization [25,26], we next tested the effect of *RAD21* p.P298S/A on gene expression by microarray analysis in the cell line system described above. Hierarchical clustering of differentially expressed genes (|fc| > 1.5, adj. p-value < 0.05) showed a clear clustering of replicates and a separation of each condition (Figure S7). In total, 308 and 391 genes were differentially regulated (|fc| > 1.5, adj. p-value < 0.05) in cells carrying the *RAD21* variants p.P298S/A, respectively. A total of 83 genes were significantly up-/down-regulated in both *RAD21* cell line models (Figures 2B and S8, Supplementary Table S6). GO term analysis of these genes identified "p53 signaling pathway" as the most prominent among enriched deregulated signaling pathways (Figure 2B). In line with these observations, HEK293T cells carrying *RAD21* p.P298S/A showed an increased number of γH2AX and 53BP1 co-localized foci indicating the extent of DNA double-strand breaks resulting from the mutated RAD21 protein compared to the WT (** = $p \leq 0.01$; Student's t-test) (Figure 2C).

Based on these results, we questioned whether patients carrying *RAD21* p.P298S/A would also display DNA damage signaling abnormalities during normal and cellular stress conditions. Therefore, primary patient fibroblasts carrying the respective *RAD21* p.P298S/A variants in comparison to *RAD21* WT control fibroblasts were challenged through irradiation to induce DNA damage and their response assessed via cell-cycle analysis. Both fibroblastic cell lines carrying *RAD21* p.P298A and *RAD21* p.P298S displayed a significant G2/M cell-cycle arrest compared to a WT control after ionizing irradiation (Figures 2D and S9). Likewise, upon treatment with the DNA cross-linking agent Mitomycin-C (MMC), *RAD21* p.P298S fibroblasts arrested more cells at the S/G2/M cell-cycle stage ($p = 0.0033$; Student's t-test) (Figure S10). Therefore, the observed G2/M cell cycle arrest is a potential phenotype of the increased DNA damage occurring in cells carrying *RAD21* p.P298S/A upon exposure to stress conditions and further underlines the increased risk of malignant transformation for predisposed patients.

2.3. *Amino Acid Replacements (S/A) at Position 298 of RAD21 Lead to Altered RAD21 Expression Levels*

To elucidate the molecular mechanism of *RAD21* dysregulation mediated through both variants, we employed an additional variant specific model by generating a HEK293T cell line with doxycycline-inducible expression of siRNA targeting the endogenous *RAD21* and concomitant expression of EGFP-tagged pRTS-1-*RAD21* WT, p.P298A or p.P298S [27]. Three days after doxycycline induction, cells of each condition were EGFP-sorted and subjected to RNA-Sequencing (Figure S11A). In parallel, endogenous RAD21 downregulation and its replacement by EGFP-tagged RAD21 was verified by Western Blot analysis (Figure S11B), while the presence of the respective *RAD21* variants was additionally validated by Sanger Sequencing (Figure S11C). In total, the RNA-Sequencing yielded only 50 commonly deregulated genes between both variants and RAD21 WT (Figure S12, Supplementary Table S7) (adj. p-value < 0.05). These results are in line with published data confirming only modest gene expression changes with mostly weak effects observed immediately upon cohesin loss [28]. Nevertheless, *RAD21* itself ranked as the top downregulated gene for both, the *RAD21* p.P298A and the *RAD21* p.P298S variant conditions, compared to the WT *RAD21* cells (Figure 3A,B). Therefore, these data provide evidence that the here identified amino acid replacements at position 298 of RAD21 confer a functional effect in hampering proper *RAD21* transcription levels.

Thus, to identify vulnerable populations during hematopoietic differentiation, which are dependent on high *RAD21* expression and would be potentially susceptible to *RAD21* p.P298S/A, single-cell RNA-Sequencing (scRNA-Seq) data of healthy human bone marrow from the Human Cell Atlas were analyzed for cohesin complex gene expression. In line with its essential role in mitosis, *RAD21* expression was primarily up-regulated in actively dividing cells within the G2/M or S-phase compared to cells in G1 ($p < 2.2 \times 10^{-16}$, Wilcoxon test) (Figures 3C and S13). Particularly high *RAD21* transcript levels clustered with *SMC3* and *PTTG1* transcripts and were detected in cycling pre- and pro-B-cells, while *RAD21* expression in common lymphoid progenitors (CLPs) and hematopoietic stem and progenitor cells (HS/PCs) was significantly lower ($p < 2.2 \times 10^{-16}$, Wilcoxon test) (Figures 3D and S14). These data are in line with the expression pattern of *RAD21* in human leukemias, as observed in gene and protein expression data across various hematological malignancies (Figure S15).

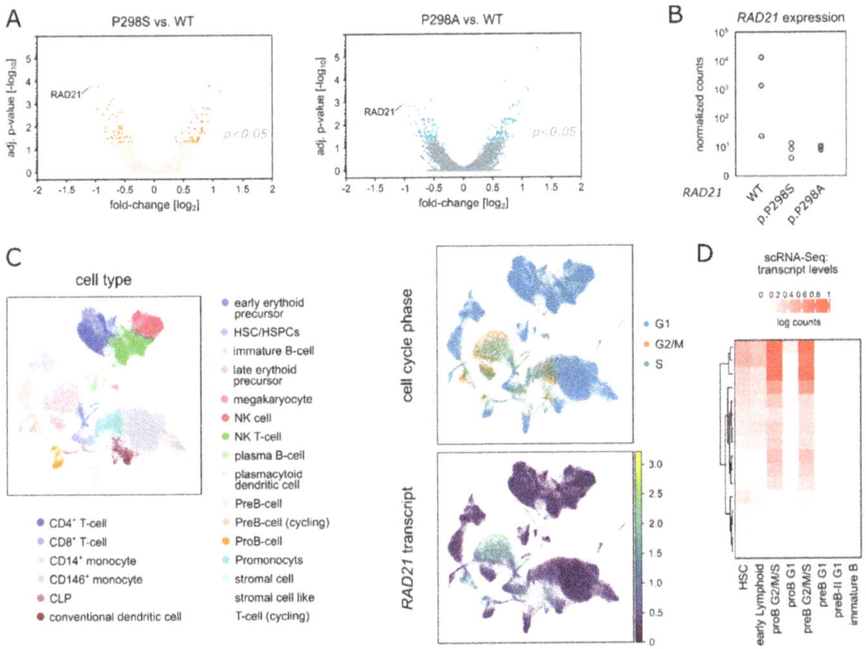

Figure 3. (**A**): Volcano plot of average gene expression based on bulk RNA-Sequencing data. Fold-change and adjusted *p*-values are calculated by comparing *RAD21* p.P298S to WT (orange, left panel) and *RAD21* p.P298A to WT (blue, right panel). Genes with an adjusted *p*-value < 0.05 are considered as differentially expressed and highlighted in dark orange (*RAD21* p.P298S, left panel) or dark blue (*RAD21* p.P298A, right panel). (**B**): Expression of *RAD21*, as the top down-regulated gene in both *RAD21* variants, is separately indicated for *RAD21* WT, p.P298S and p.P298A (three biological replicates each, bulk RNA-Sequencing). (**C**): Left: UMAP-visualization of the healthy human bone marrow scRNA-seq data. Right: Cell cycle stages colored on the UMAP-visualization (upper) and *RAD21* gene expression colored on the UMAP-visualization (lower). (**D**): Heat map indicating the cohesin complex genes' expression levels in cells of the different stages of B-cell differentiation.

2.4. *RAD21 p.P298S/A Is Recurrently Found in Pediatric Lymphoblastic Leukemia/Lymphoma*

To confirm a correlation between germline *RAD21* p.P298S/A and pediatric leukemia, we analyzed an additional unpublished pediatric cancer cohort of 150 children with relapsed ALL (Italian IntReALL standard risk study; R-ALL) for *RAD21* p.P298S/A. Here, we identified a third case with *RAD21* p.P298A in a boy who was diagnosed with B-cell precursor ALL (BCP-ALL) at 12 years old and had a combined bone marrow/CNS relapse 5 years later (Table 1). In a fourth cohort including 114 children and adolescents with therapy refractory leukemia and lymphoma (INFORM), no germline indels or missense variants affecting *RAD21* were identified, suggesting no enrichment in the relapsed or therapy refractory patients. To further cross-validate *RAD21* p.P298S/A in a non-pediatric cancer setting, a cohort of 2300 young adults (<51 years) with cancer was mined (MASTER program). In this extensive sample collection, only one patient harboring *RAD21* p.P298A with a solid tumor was identified (Table 1). Therefore, amongst all cohorts, *RAD21* p.P298S/A was found to be enriched in pediatric vs. adult cancers (3/482 vs. 1/2300; Fisher's exact test; $p = 0.018$). Overall, we did not observe an enrichment in the relapsed or therapy refractory patient cohorts suggesting that *RAD21* p.P298S/A predisposes to lymphoid precursor malignancies with no influence on therapy response.

Table 1. Cohort descriptions and identified RAD21 variants analyzed in context of clinical phenotypic and pathogenic findings. HR = High risk, SR = Standard risk, N/A = not applicable, pB-LBL = B-cell lymphoblastic lymphoma, T-ALL = T-cell acute lymphoblastic leukemia, BCP-ALL = precursor B-cell acute lymphoblastic leukemia, MPNST = Malignant peripheral nerve sheath tumor.

		TRIO-DD	TRIO-D	R-ALL	INFORM	MASTER
Cohort	Number of patients	**n = 60** pediatric	**n = 158** pediatric	**n = 150** pediatric	**n = 114** pediatric	**n = 2300** adult
	% Hematopoietic malignancies	38.3%	51.3%	100%	100%	3.7%
	Inclusion criteria	Primary diagnosis	Primary diagnosis	IntReALL SR	Therapy refractory	Young adults < 51 y
Patient	Sex	Male	Male	Male	-	Female
	Age	2	13	12	-	53
	Tumor	pB-LBL	T-ALL	BCP-ALL	-	MPNST
	Risk group	SR	HR	SR	-	N/A
RAD21 variant p.P298	Protein exchange	ENSP00000297338.2 p.P298A	ENSP00000297338.2 p.P298S	ENSP00000297338.2 p.P298A	-	ENSP00000297338.2 p.P298A
	Base exchange	ENST00000297338.2 c.892 C>G	ENST00000297338.2 c.892 C>T	ENST00000297338.2 c.892 C>G	-	ENST00000297338.2 c.892 C>G
	SNP ID	rs148308569	rs148308569	rs148308569	-	rs148308569
	MAF GnomAD	10^{-5}	10^{-6}	10^{-5}	-	10^{-5}
	MAF within the cohort	1.7×10^{-2}	6.5×10^{-3}	6.7×10^{-3}	-	4.3×10^{-4}
Genetic history	Genetic counselling *	+	+	unknown	-	unknown
	Family history	+	+	unknown	-	unknown
2nd Hit	Somatic Mutations	unknown	*KRAS* p.Q61R	*KRAS* p.G12C	-	*PTCH2* p.A68V

SR = standard risk, HR = high risk, * based on criteria from Jongmans et al. Eur J Med Genet 59 (2016) 116-125 und Ripperger et al., Am J Med Genet A. (2017).

3. Discussion

The cohesin complex is a cogwheel of ordered chromosome alignment and segregation during cell division, homologous-recombination-driven DNA repair and regulation of gene expression [5,29,30]. RAD21 is essential for this machinery as it connects the SMC1 and SMC3 cohesin subunits and thereby generates the functional ring-like structure of cohesin

Overall, within all analyzed datasets, comprising in total 482 pediatric cancer patients and 2300 adult cancers as controls, we present three children with lymphoblastic leukemia/lymphoma all carrying a recurrent *RAD21* germline variation at position 298. None of the patients displayed a CdLS phenotype, which is in line with previous reports, showing that *RAD21* variants are known to display reduced CdLS phenotype expressivity [24]. Furthermore, as with other *RAD21* missense variants in cancer [31], the here identified *RAD21* p.P298S/A alterations are heterozygous and mutually exclusive to other variants in cohesin complex genes.

The observed familial cancer history in two of the patients demonstrates an increased cancer risk across generations. Nevertheless, due to the incomplete penetrance and the tumor variance, additional factors such as synergizing germline mutations or environmental influences to drive tumor evolution need to be taken into account. Interestingly, in two patients carrying *RAD21* p.P298S/A we identified a known pathogenic KRAS hot-spot mutation as a common somatic denominator in the respective tumors, which is in line with a recently published association between cohesin complex mutations and RAS signaling in cancer progression [32].

Functionally, the described alterations at position 298 did not disturb the formation of the cohesin complex, which is also rarely seen in variants without detrimental gene disruption [33]. Mechanistically, we could show that the described variants caused deregulations of proper *RAD21* transcript levels, which in the long-term affected p53 signaling. By applying irradiation and MMC as external stressors this effect was further enhanced as seen by increased cell cycle arrest in primary patient cells carrying *RAD21* p.P298S/A. Likewise, *RAD21* variants have been previously described in radiosensitive cancer patients [34] and CdLS patients displaying increased DNA damage sensitivity [35,36]. Furthermore, embryonic stem cells of *RAD21* heterozygous mice show significantly reduced survival after treatment with MMC [30]. Thus, the increased G2/M arrest in germline cells carrying *RAD21* p.P298S/A emphasizes the crucial role of properly functioning cohesins to avoid chromosomal instabilities during the repair of both interstrand MMC-DNA cross-links [37] and irradiation-induced DNA DSB [14,38].

Although cohesin complex genes are supposed to be ubiquitously expressed owing to their inevitability for basic cellular processes, we utilized scRNA-Seq to newly demonstrate that cohesin complex partners are differentially regulated during B-cell lineage specification in human bone marrow. Even though HS/PCs require cohesin, *Rad21* haploinsufficiency in mice was postulated to display distinct hematopoietic phenotypes in comparison to other cohesin subunit knockout models [39], further supporting the here described cohesin gene specific expression patterns during early B-cell differentiation. Interestingly, high expression of WAPL was identified particularly in HS/PCs, pointing towards a so far unrecognized role of WAPL within the stem cell compartment. *STAG2, RAD21, SMC3* and *SMC1* loss of function is known to induce stemness potential such as enhanced self-renewal and differentiation arrest in human and mouse HS/PCs [33,40]. Along these lines, it was also shown that cohesin facilitates V(D)J recombination in pro-B cells [41] and T-cell receptor α locus rearrangement [42].

Moreover, cohesins and their associated proteins are being recognized to act as master transcriptional regulators of hematopoietic genes [43]. Therefore, their deregulation can be regarded as a critical first step in the evolution of hematopoietic malignancies [40,44]. Intriguingly, the here identified patients harboring *RAD21* p.P298S/A all suffered from precursor lymphoblastic malignancies, which suggests either stem and progenitor cells or early lymphoid precursors as the origins of the disease.

Taken together, in addition to *RAD21* germline and somatic loss-of-function variants that result in cohesinopathies and predominantly myeloid cancers, respectively, our data propose a third category of *RAD21* variants that mediate germline predisposition to lymphoblastic malignancies in childhood. Understanding the influence of *RAD21* germline variants may offer new treatment options such as their potential sensitivity to PARPP inhibitors which are already included in clinical trials in leukemias with somatically mutated cohesin [45].

4. Materials and Methods

4.1. Patients

Patients ≤ 19 years of age were unselectively recruited at the Pediatric Oncology Department, Dresden (years 2019–2020), or as previously described [23,46,47]. Consent of the families was obtained according to the Ethical Vote EK 181042019 (Dresden) and in line with the Declaration of Helsinki. For the IntReALL cohort, patients' parents or their legal guardians gave informed consent to genetic analyses in the context of add-on studies linked to the clinical protocol to which patients were enrolled.

4.2. Whole Exome Sequencing (WES)

Germline DNA was extracted from the patient's fibroblasts using AllPrep DNA/RNA Mini Kit (Qiagen, Venlo, Netherlands) and from PBMCs of the parents and the remaining patient's using the QIAamp DNA Blood Mini Kit (Qiagen). Sequenceable next-generation libraries for WES were generated with the SureSelect Human All Exon V7 kit (Agilent Technologies, Santa Clara, California, USA). The libraries were sequenced on a NovaSeq 6000 platform (Illumina, San Diego, CA, USA) in paired-end mode (2 × 150bp) and with final on-target coverage of ≥100×. Processing of the WES data was performed as previously described [23].

4.3. Cell Culture

Primary fibroblasts were initially cultivated in BIO-AMF™-2 Medium (Biological-Industries, Kibbutz Beit Haemek, Israel) up to a passage of 5. For experimental analysis, fibroblasts were cultured in Dulbecco's Modified Eagle Medium (DMEM; GIBCO/Thermo Fisher Scientific, Waltham, Massachusetts, USA) with 20% fetal calf serum (FCS; GIBCO), 1% Penicillin/Streptomycin (P/S; 10,000 units/mL; GIBCO) and 1% MEM Non-essential Amino Acids (NEAA; GIBCO) up to a passage of 13.

HEK293T cells transfected with R32-hRAD21 were cultured in DMEM with 10% FCS, 1% P/S and 1% NEAA. All cells were kept at 37 °C and 5% CO_2.

4.4. Cloning

The inducible RAD21 system (pRTS-1-RAD21) was gifted from Kerstin Wendt and Olaf Stemman [27]. Mutated cDNAs for *RAD21* p.P298A and p.P298S were created by site directed mutagenesis by PCR and cloned into the pMC3.Hygro (=R32-hRAD21) and the pRTS-1 (=pRTS-1-RAD21) plasmid via MluI/SpeI and SwaI/XhoI restriction sites, respectively, utilizing the following primer pairs (Table 2):

Table 2. Primer sequences for cloning.

Name	Sequence (5′ → 3′)
hRad21_MluI_F	GGCGCacgcgtgccaccATGTTCTACGCACATTTTGTTCTC
hRad21_SpeI_R	CCTCGactagtTATAATATGGAACCTTGGTCCAGGTGTTGC
hRad21_SwaI_F	GGCGCATTTAAATCATGTTCTACGCAC
hRad21_XhoI_R	CCTCGCTCGAGTCCATATAATATGGAACC
hRad21_P298S_F	GATCAAACAACACTTGTTtCAAATGAGGAAGAAGCATTTGC
hRad21_P298S_R	GCAAATGCTTCTTCCTCATTTGaAACAAGTGTTGTTTGATC
hRad21_P298A_F	GATCAAACAACACTTGTTgCAAATGAGGAAGAAGCATTTGC
hRad21_P298A_R	GCAAATGCTTCTTCCTCATTTGcAACAAGTGTTGTTTGATC

4.5. HEK293T Cell Transfection

R32-hRAD21

HEK293T cells were seeded at a density of 4x10^5 cells and stably transfected with 4 µg of Vector [48] (R32-hRAD21 or R32-hRAD21 p.P298S or R32-hRAD21 p.P298A using Lipofectamine2000 (Invitrogen) and selected with Hygromycin (Invitrogen/Thermo Fisher Scientific, Waltham, MA, USA) at a concentration of 200 µg/mL for 7 days. Continuous culturing was performed with Hygromycin concentration altering between 100 µg/mL and 200 µg/mL, put freshly 3 times a week.

pRTS-1-RAD21

HEK293T cells were seeded at a density of 5x10^5 cells and stably transfected with 4 µg of Vector [27] (pRTS-1-RAD21, pRTS-1-RAD21 p.P298S or pRTS-1-RAD21 p.P298A using Lipofectamine2000 (Invitrogen) and selected with Hygromycin (Invitrogen) at a concentration of 400 µg/mL for 7 days. Continuous culturing was performed with Hygromycin at concentrations altering between 200 µg/mL and 400 µg/mL, put freshly 3 times a week.

4.6. Microarray (R32-hRAD21)

Stably transfected HEK293T cells overexpressing R32-hRAD21 with either WT, p.P298S or p.P298A conditions were seeded onto 10 cm plates in a density of 2 × 10^6 cells in quadruplicates. After 48 h, control cells were harvested and 6 × 10^6 cells were pelleted and stored at −80 °C for later RNA extraction. RNA was extracted using the RNeasy Mini Kit (Qiagen #74106) with 350 µL of RLT Buffer+ BME using QIAshredder (#79656) and RNAse-Free DNase Set (Qiagen #79254). RNA was stored at −80 °C.

RNA samples were sent to Macrogen Europe B.V. (Amsterdam, Netherlands) for gene expression analysis using the SurePrint G3 Human Gene Expression 8 × 60K v3 microarray (Agilent, Inc., Santa Clara, CA, USA). Put briefly, Cy3-labeled cRNA was prepared from 1~5 µg total RNA (Quick Amp Labeling Kit, Agilent), subsequently fragmented and (1.65 µg) hybridized to the microarray. Scanning was performed by the SureScan Microarray Scanner System G4900DA (Agilent).

For analysis, raw data were extracted using the software provided by Agilent Feature Extraction Software (v11.0.1.1). The raw data for the same probe was summarized automatically in the Agilent feature extraction protocol to provide expression data for each gene probed on the array. Flag A-tagged probes were filtered out and the remaining gProcessedSignal values were log transformed and quantile normalized.

Furthermore, all technical replicates (n = 4) of one sample were combined and samples were compared pairwise by fold-change values: *RAD21* p.P298A vs. WT, *RAD21* p.P298S vs. WT and *RAD21* p.P298A vs. *RAD21* p.P298S. The p-value calculated with an independent Student's t-test was corrected for multiple testing and used to define the significance of these pairwise comparisons. Genes with an absolute fold-change of 1.5 or more and an adjusted p-value below 0.05 were considered as significantly up- or down-regulated. These data (n = 995 probes) were used to perform a two-dimensional hierarchical clustering using Euclidean distance and complete linkage. Results were represented as heat map (seaborn.clustermap v.0.10.1 with prior optimal leaf ordering, Python v.3.6). The same analysis was performed for a smaller set (n = 83 probes), which were differentially expressed in both mutants *RAD21* p.P298A and p.P298S vs. WT was similarly analyzed and represented.

4.7. Quantitative Real-Time (qRT)-PCR Analysis

RNA was extracted from primary fibroblasts (TRIO_DD_018; TRIO_DD_025; 2.0–3.0 × 10^6 cells) using the RNaeasy Mini Kit (Qiagen #74106) with 350 µL of RLT Buffer+ beta-ME using QIAshredder (#79656) and RNAse-Free DNase Set (Qiagen #79254). A total of 3 independent RNA extractions were performed, and 1 µg of RNA was reverse transcribed into cDNA using the QuantiTect Reverse Transcription Kit (Quiagen #205311) following manufacturer's instructions. The qRT-PCR was performed using TaqMan Universal Master Mix II following manufacturer's instructions (Thermo Fisher Scientific, Waltham, MA, USA, #PN4428173) for 20 µL reaction with 1.5 µL of cDNA. The following TaqMan assays were used: TBP (Hs00427620_m1), HPRT1 (Hs02800695) and POT1 (Hs00209984_m1). Expression of mRNA was analysed by the comparative $\Delta\Delta$-C_T method and plotted in relation to the control sample.

4.8. GO-Term Analysis

Gene Ontology (GO) term analysis was performed using the web server EnrichR (https://maayanlab.cloud/Enrichr/; accessed on 13 April 2021) [49]. GO terms of the categories "Molecular Function", "Biological Pathway", "Cellular Component" and "KEGG" were analyzed and results with an adjusted p-value < 0.05 are represented.

4.9. Cell Sorting and RNA-Sequencing (pRTS-1-RAD21)

HEK293T pRTS-1-RAD21 cells stably selected with Hygromycin, were induced with Doxycycline at a concentration of 2 µg/mL for 72 h. All cells were trypsinized, and washed with cold PBS. Cells were diluted in cold FACS Buffer (PBS + 2 µM EDTA) and kept on ice until sorting. Cell sort for high EGFP was performed on an FACSAria II (BD).

RNA Extraction was performed using the RNA Micro Kit (Qiagen) following manufacturer's instruction. RNA quality analysis was performed on an Agilent 2100 bioanalyzer, with all samples showing RIN values of 10. RNA libraries were prepared by mRNA enrichment by poly-dT pull down using the NEBNext Poly(A) kit based on manufacturer's recommendations (New England Biologies, Ipswitch, MA, USA). Sequencing was carried out as 2 × 50 bp reads and read depths of 30–50 million on an Illumina NovaSeq 6000.

FastQC (v.0.11.9; http://www.bioinformatics.babraham.ac.uk/, accessed on 10 April 2022) was used to perform a basic quality control of the resulting sequencing data. Fragments were aligned to the human reference genome hg38 with support of the Ensembl 104 splice sites using the aligner gsnap (v2020-12-16) [50]. Counts per gene and sample were obtained based on the overlap of the uniquely mapped fragments with the same Ensembl annotation using featureCounts (v2.0.1) [51]. The normalization of raw fragments based on library size and testing for differential expression between the different cell types/treatments was performed with the DESeq R package (v1.30.1) [52]. Sample to sample Euclidean distance, Pearson and Spearman correlation coefficients (r) and PCA based upon the top 500 genes showing highest variance were computed to explore correlation between biological replicates and different libraries. To identify differentially expressed genes, counts were fitted to the negative binomial distribution and genes were tested between conditions using the Wald test of DESeq2. Resulting p-values were corrected for multiple testing with the Independent Hypothesis Weighting package (IHW 1.12.0) [53]. Genes with a maximum of 5% false discovery rate (padj $\leq$ 0.05) were considered as significantly differentially expressed.

4.10. Statistical Analyses

For statistical analysis, the two-tailed Student's unpaired t-test was performed. Differences with a p value < 0.05 were considered to be significant, ns = $p > 0.05$, * = $p \leq 0.05$, ** = $p \leq 0.01$, *** = $p \leq 0.001$, **** = $p \leq 0.0001$.

Supplementary Materials: The following are available online at https://www.mdpi.com/article/10.3390/ijms23095174/s1.

Author Contributions: Conceptualization, J.H. and F.A.; Data curation, U.A.F., R.W., C.S., T.B., A.F., M.M., P.H., N.P., G.F., R.J.A., S.F., M.S., K.W., K.S., D.R., H.G. (Hanno Glimm), M.H., G.C., A.B. (Arndt Borkhardt) and F.A.; Formal analysis, A.S.; Funding acquisition, J.H.; Investigation, A.S., U.A.F., C.G. and F.A.; Methodology, A.S., U.A.F., M.N.F.M., R.W., J.M., T.W., P.M., C.W., A.B. (Arka Baksi), C.G., H.G. (Holger Gohlke) and S.B.; Resources, R.W., C.S., T.B., C.W., A.F., P.H., N.P., G.F., R.J.A., S.F., M.S., H.G. (Holger Gohlke), K.W., K.S., M.D., D.R., H.G. (Hanno Glimm), M.H., R.J., G.C., A.B. (Arndt Borkhardt) and J.H.; Software, U.A.F., R.W., J.M., C.W., M.D., D.R., H.G. (Hanno Glimm) and M.H.; Supervision, J.H. and F.A.; Validation, A.S., M.N.F.M., T.W. and P.M.; Visualization, A.S., U.A.F., J.M. and C.G.; Writing—original draft, F.A.; Writing—review and editing, A.S., U.A.F., M.N.F.M., R.W., R.J., J.H. and F.A. All authors have read and agreed to the published version of the manuscript.

Funding: J.H. is supported by ERC Stg 85222 "PreventALL", ERA Per Med.JTC 2018 "GEPARD", Deutsche Krebshilfe (DKH) Excellenz Förderprorgamm 70114539 and Sonnenstrahl e.V. A.S. is supported by Deutsche José Carreras Leukämie-Stiftung e.V. together with Gesellschaft für Paediatrische Onkologie und Haematologie e.V. G.C. is supported by the Italian Association for Cancer Research (AIRC), grant IG2018 n.21999; C.S., G.F. and G.C. are partly supported by the Monza parents committee 'Comitato Maria Letizia Verga'. M.H. and J.M. is supported by ERA Per Med.JTC 2018 "GEPARD". R.J. is supported by the Deutsche Forschungsgemeinschaft (DFG, JE150/31-1). R.W. acknowledges the financial support of the Forschungskommission (2020-28) Heinrich Heine University Duesseldorf. This article/publication is based upon work from COST Action LEukaemia GENe Discovery by data sharing, mining and collaboration (CA16223), supported by COST (European Cooperation in Science and Technology; www.cost.eu accessed on 10 April 2022).

Institutional Review Board Statement: The study was conducted according to the guidelines of the Declaration of Helsinki and approved by the Ethics Committee of University Clinic Carl Gustav Carus Dresden (EK 181042019).

Informed Consent Statement: Informed consent was obtained from all subjects involved in the study.

Data Availability Statement: The *RAD21* variant was submitted to ClinVar (https://www.ncbi.nlm.nih.gov/clinvar/, accessed on 4 May 2022). The datasets generated during and/or analysed during the current study are available from the corresponding author on reasonable request.

Acknowledgments: The authors would like to thank all members of our groups for useful discussions and for their critical reading of the manuscript. Particular thanks goes to Helena Jambor, Glen Pearce, Dalileh Nabi, Evelin Schröck, Arne Jahn, Silke Furlan, Carolin Melzig and Rajanya Ghosh.

Conflicts of Interest: The authors declare no conflict of interest.

References

1. Waldman, T. Emerging themes in cohesin cancer biology. *Nat. Rev. Cancer* **2020**, *20*, 504–515. [CrossRef]
2. Haering, C.H.; Lowe, J.; Hochwagen, A.; Nasmyth, K. Molecular architecture of SMC proteins and the yeast cohesin complex. *Mol. Cell* **2002**, *9*, 773–788. [CrossRef]
3. Losada, A.; Hirano, M.; Hirano, T. Identification of Xenopus SMC protein complexes required for sister chromatid cohesion. *Genes Dev.* **1998**, *12*, 1986–1997. [CrossRef] [PubMed]
4. Michaelis, C.; Ciosk, R.; Nasmyth, K. Cohesins: Chromosomal proteins that prevent premature separation of sister chromatids. *Cell* **1997**, *91*, 35–45. [CrossRef]
5. Busslinger, G.A.; Stocsits, R.R.; van der Lelij, P.; Axelsson, E.; Tedeschi, A.; Galjart, N.; Peters, J.M. Cohesin is positioned in mammalian genomes by transcription, CTCF and Wapl. *Nature* **2017**, *544*, 503–507. [CrossRef]
6. Chan, K.L.; Roig, M.B.; Hu, B.; Beckouet, F.; Metson, J.; Nasmyth, K. Cohesin's DNA exit gate is distinct from its entrance gate and is regulated by acetylation. *Cell* **2012**, *150*, 961–974. [CrossRef]
7. Tedeschi, A.; Wutz, G.; Huet, S.; Jaritz, M.; Wuensche, A.; Schirghuber, E.; Davidson, I.F.; Tang, W.; Cisneros, D.A.; Bhaskara, V.; et al. Wapl is an essential regulator of chromatin structure and chromosome segregation. *Nature* **2013**, *501*, 564–568. [CrossRef]
8. Losada, A.; Yokochi, T.; Hirano, T. Functional contribution of Pds5 to cohesin-mediated cohesion in human cells and Xenopus egg extracts. *J. Cell Sci.* **2005**, *118*, 2133–2141. [CrossRef]
9. Guacci, V.; Koshland, D.; Strunnikov, A. A direct link between sister chromatid cohesion and chromosome condensation revealed through the analysis of MCD1 in S. cerevisiae. *Cell* **1997**, *91*, 47–57. [CrossRef]
10. Nasmyth, K.; Peters, J.M.; Uhlmann, F. Splitting the chromosome: Cutting the ties that bind sister chromatids. *Science* **2000**, *288*, 1379–1385. [CrossRef]
11. Tanaka, T.; Fuchs, J.; Loidl, J.; Nasmyth, K. Cohesin ensures bipolar attachment of microtubules to sister centromeres and resists their precocious separation. *Nat. Cell Biol.* **2000**, *2*, 492–499. [CrossRef] [PubMed]
12. Uhlmann, F.; Lottspeich, F.; Nasmyth, K. Sister-chromatid separation at anaphase onset is promoted by cleavage of the cohesin subunit Scc1. *Nature* **1999**, *400*, 37–42. [CrossRef]
13. Heidinger-Pauli, J.M.; Mert, O.; Davenport, C.; Guacci, V.; Koshland, D. Systematic reduction of cohesin differentially affects chromosome segregation, condensation, and DNA repair. *Curr. Biol.* **2010**, *20*, 957–963. [CrossRef] [PubMed]
14. Sjogren, C.; Nasmyth, K. Sister chromatid cohesion is required for postreplicative double-strand break repair in Saccharomyces cerevisiae. *Curr. Biol.* **2001**, *11*, 991–995. [CrossRef]
15. Parelho, V.; Hadjur, S.; Spivakov, M.; Leleu, M.; Sauer, S.; Gregson, H.C.; Jarmuz, A.; Canzonetta, C.; Webster, Z.; Nesterova, T.; et al. Cohesins functionally associate with CTCF on mammalian chromosome arms. *Cell* **2008**, *132*, 422–433. [CrossRef]

16. Wendt, K.S.; Yoshida, K.; Itoh, T.; Bando, M.; Koch, B.; Schirghuber, E.; Tsutsumi, S.; Nagae, G.; Ishihara, K.; Mishiro, T.; et al. Cohesin mediates transcriptional insulation by CCCTC-binding factor. *Nature* **2008**, *451*, 796–801. [CrossRef]
17. Canela, A.; Maman, Y.; Jung, S.; Wong, N.; Callen, E.; Day, A.; Kieffer-Kwon, K.R.; Pekowska, A.; Zhang, H.; Rao, S.S.P.; et al. Genome Organization Drives Chromosome Fragility. *Cell* **2017**, *170*, 507–521 e518. [CrossRef]
18. Kon, A.; Shih, L.Y.; Minamino, M.; Sanada, M.; Shiraishi, Y.; Nagata, Y.; Yoshida, K.; Okuno, Y.; Bando, M.; Nakato, R.; et al. Recurrent mutations in multiple components of the cohesin complex in myeloid neoplasms. *Nat. Genet.* **2013**, *45*, 1232–1237. [CrossRef]
19. Antic, Z.; Yu, J.; Bornhauser, B.C.; Lelieveld, S.H.; van der Ham, C.G.; van Reijmersdal, S.V.; Morgado, L.; Elitzur, S.; Bourquin, J.P.; Cazzaniga, G.; et al. Clonal dynamics in pediatric B-cell precursor acute lymphoblastic leukemia with very early relapse. *Pediatr. Blood Cancer* **2022**, *69*, e29361. [CrossRef]
20. Moura-Castro, L.H.; Pena-Martinez, P.; Castor, A.; Galeev, R.; Larsson, J.; Jaras, M.; Yang, M.; Paulsson, K. Sister chromatid cohesion defects are associated with chromosomal copy number heterogeneity in high hyperdiploid childhood acute lymphoblastic leukemia. *Genes Chromosomes Cancer* **2021**, *60*, 410–417. [CrossRef]
21. Horsfield, J.A.; Print, C.G.; Monnich, M. Diverse developmental disorders from the one ring: Distinct molecular pathways underlie the cohesinopathies. *Front. Genet.* **2012**, *3*, 171. [CrossRef] [PubMed]
22. Fazio, G.; Massa, V.; Grioni, A.; Bystry, V.; Rigamonti, S.; Saitta, C.; Galbiati, M.; Rizzari, C.; Consarino, C.; Biondi, A.; et al. First evidence of a paediatric patient with Cornelia de Lange syndrome with acute lymphoblastic leukaemia. *J. Clin. Pathol.* **2019**, *72*, 558–561. [CrossRef]
23. Wagener, R.; Taeubner, J.; Walter, C.; Yasin, L.; Alzoubi, D.; Bartenhagen, C.; Attarbaschi, A.; Classen, C.F.; Kontny, U.; Hauer, J.; et al. Comprehensive germline-genomic and clinical profiling in 160 unselected children and adolescents with cancer. *Eur. J. Hum. Genet.* **2021**. [CrossRef]
24. Krab, L.C.; Marcos-Alcalde, I.; Assaf, M.; Balasubramanian, M.; Andersen, J.B.; Bisgaard, A.M.; Fitzpatrick, D.R.; Gudmundsson, S.; Huisman, S.A.; Kalayci, T.; et al. Delineation of phenotypes and genotypes related to cohesin structural protein RAD21. *Hum. Genet.* **2020**, *139*, 575–592. [CrossRef] [PubMed]
25. Yan, J.; Enge, M.; Whitington, T.; Dave, K.; Liu, J.; Sur, I.; Schmierer, B.; Jolma, A.; Kivioja, T.; Taipale, M.; et al. Transcription factor binding in human cells occurs in dense clusters formed around cohesin anchor sites. *Cell* **2013**, *154*, 801–813. [CrossRef] [PubMed]
26. Merkenschlager, M.; Nora, E.P. CTCF and Cohesin in Genome Folding and Transcriptional Gene Regulation. *Annu. Rev. Genom. Hum. Genet.* **2016**, *17*, 17–43. [CrossRef] [PubMed]
27. Zuin, J.; Dixon, J.R.; van der Reijden, M.I.; Ye, Z.; Kolovos, P.; Brouwer, R.W.; van de Corput, M.P.; van de Werken, H.J.; Knoch, T.A.; van IJcken, W.F.; et al. Cohesin and CTCF differentially affect chromatin architecture and gene expression in human cells. *Proc. Natl. Acad. Sci. USA* **2014**, *111*, 996–1001. [CrossRef]
28. Rao, S.S.P.; Huang, S.C.; Glenn St Hilaire, B.; Engreitz, J.M.; Perez, E.M.; Kieffer-Kwon, K.R.; Sanborn, A.L.; Johnstone, S.E.; Bascom, G.D.; Bochkov, I.D.; et al. Cohesin Loss Eliminates All Loop Domains. *Cell* **2017**, *171*, 305-320.e324. [CrossRef]
29. Uhlmann, F. SMC complexes: From DNA to chromosomes. *Nat. Rev. Mol. Cell Biol.* **2016**, *17*, 399–412. [CrossRef]
30. Xu, H.; Balakrishnan, K.; Malaterre, J.; Beasley, M.; Yan, Y.; Essers, J.; Appeldoorn, E.; Tomaszewski, J.M.; Vazquez, M.; Verschoor, S.; et al. Rad21-cohesin haploinsufficiency impedes DNA repair and enhances gastrointestinal radiosensitivity in mice. *PLoS ONE* **2010**, *5*, e12112. [CrossRef]
31. De Koninck, M.; Losada, A. Cohesin Mutations in Cancer. *Cold Spring Harb. Perspect. Med.* **2016**, *6*, a026476. [CrossRef] [PubMed]
32. Martin-Izquierdo, M.; Abaigar, M.; Hernandez-Sanchez, J.M.; Tamborero, D.; Lopez-Cadenas, F.; Ramos, F.; Lumbreras, E.; Madinaveitia-Ochoa, A.; Megido, M.; Labrador, J.; et al. Co-occurrence of cohesin complex and Ras signaling mutations during progression from myelodysplastic syndromes to secondary acute myeloid leukemia. *Haematologica* **2020**, *106*, 2215–2223. [CrossRef] [PubMed]
33. Mullenders, J.; Aranda-Orgilles, B.; Lhoumaud, P.; Keller, M.; Pae, J.; Wang, K.; Kayembe, C.; Rocha, P.P.; Raviram, R.; Gong, Y.; et al. Cohesin loss alters adult hematopoietic stem cell homeostasis, leading to myeloproliferative neoplasms. *J. Exp. Med.* **2015**, *212*, 1833–1850. [CrossRef] [PubMed]
34. Severin, D.M.; Leong, T.; Cassidy, B.; Elsaleh, H.; Peters, L.; Venter, D.; Southey, M.; McKay, M. Novel DNA sequence variants in the hHR21 DNA repair gene in radiosensitive cancer patients. *Int. J. Radiat. Oncol. Biol. Phys.* **2001**, *50*, 1323–1331. [CrossRef]
35. Revenkova, E.; Focarelli, M.L.; Susani, L.; Paulis, M.; Bassi, M.T.; Mannini, L.; Frattini, A.; Delia, D.; Krantz, I.; Vezzoni, P.; et al. Cornelia de Lange syndrome mutations in SMC1A or SMC3 affect binding to DNA. *Hum. Mol. Genet.* **2009**, *18*, 418–427. [CrossRef]
36. Vrouwe, M.G.; Elghalbzouri-Maghrani, E.; Meijers, M.; Schouten, P.; Godthelp, B.C.; Bhuiyan, Z.A.; Redeker, E.J.; Mannens, M.M.; Mullenders, L.H.; Pastink, A.; et al. Increased DNA damage sensitivity of Cornelia de Lange syndrome cells: Evidence for impaired recombinational repair. *Hum. Mol. Genet.* **2007**, *16*, 1478–1487. [CrossRef]
37. Boamah, E.K.; White, D.E.; Talbott, K.E.; Arva, N.C.; Berman, D.; Tomasz, M.; Bargonetti, J. Mitomycin-DNA adducts induce p53-dependent and p53-independent cell death pathways. *ACS Chem. Biol.* **2007**, *2*, 399–407. [CrossRef]
38. Sanders, J.T.; Freeman, T.F.; Xu, Y.; Golloshi, R.; Stallard, M.A.; Hill, A.M.; San Martin, R.; Balajee, A.S.; McCord, R.P. Radiation-induced DNA damage and repair effects on 3D genome organization. *Nat. Commun.* **2020**, *11*, 6178. [CrossRef]
39. Cheng, H.; Zhang, N.; Pati, D. Cohesin subunit RAD21: From biology to disease. *Gene* **2020**, *758*, 144966. [CrossRef]
40. Mazumdar, C.; Shen, Y.; Xavy, S.; Zhao, F.; Reinisch, A.; Li, R.; Corces, M.R.; Flynn, R.A.; Buenrostro, J.D.; Chan, S.M.; et al. Leukemia-Associated Cohesin Mutants Dominantly Enforce Stem Cell Programs and Impair Human Hematopoietic Progenitor Differentiation. *Cell Stem Cell* **2015**, *17*, 675–688. [CrossRef]

41. Degner, S.C.; Verma-Gaur, J.; Wong, T.P.; Bossen, C.; Iverson, G.M.; Torkamani, A.; Vettermann, C.; Lin, Y.C.; Ju, Z.; Schulz, D.; et al. CCCTC-binding factor (CTCF) and cohesin influence the genomic architecture of the Igh locus and antisense transcription in pro-B cells. *Proc. Natl. Acad. Sci. USA* **2011**, *108*, 9566–9571. [CrossRef] [PubMed]
42. Seitan, V.C.; Hao, B.; Tachibana-Konwalski, K.; Lavagnolli, T.; Mira-Bontenbal, H.; Brown, K.E.; Teng, G.; Carroll, T.; Terry, A.; Horan, K.; et al. A role for cohesin in T-cell-receptor rearrangement and thymocyte differentiation. *Nature* **2011**, *476*, 467–471. [CrossRef] [PubMed]
43. Panigrahi, A.K.; Pati, D. Road to the crossroads of life and death: Linking sister chromatid cohesion and separation to aneuploidy, apoptosis and cancer. *Crit. Rev. Oncol. Hematol.* **2009**, *72*, 181–193. [CrossRef] [PubMed]
44. Rivas, M.A.; Meydan, C.; Chin, C.R.; Challman, M.F.; Kim, D.; Bhinder, B.; Kloetgen, A.; Viny, A.D.; Teater, M.R.; McNally, D.R.; et al. Smc3 dosage regulates B cell transit through germinal centers and restricts their malignant transformation. *Nat. Immunol.* **2021**, *22*, 240–253. [CrossRef] [PubMed]
45. Padella, A.; Ghelli Luserna Di Rora, A.; Marconi, G.; Ghetti, M.; Martinelli, G.; Simonetti, G. Targeting PARP proteins in acute leukemia: DNA damage response inhibition and therapeutic strategies. *J. Hematol. Oncol.* **2022**, *15*, 10. [CrossRef] [PubMed]
46. Worst, B.C.; van Tilburg, C.M.; Balasubramanian, G.P.; Fiesel, P.; Witt, R.; Freitag, A.; Boudalil, M.; Previti, C.; Wolf, S.; Schmidt, S.; et al. Next-generation personalised medicine for high-risk paediatric cancer patients—The INFORM pilot study. *Eur. J. Cancer* **2016**, *65*, 91–101. [CrossRef]
47. Horak, P.; Klink, B.; Heining, C.; Groschel, S.; Hutter, B.; Frohlich, M.; Uhrig, S.; Hubschmann, D.; Schlesner, M.; Eils, R.; et al. Precision oncology based on omics data: The NCT Heidelberg experience. *Int. J. Cancer* **2017**, *141*, 877–886. [CrossRef]
48. Linka, R.M.; Risse, S.L.; Bienemann, K.; Werner, M.; Linka, Y.; Krux, F.; Synaeve, C.; Deenen, R.; Ginzel, S.; Dvorsky, R.; et al. Loss-of-function mutations within the IL-2 inducible kinase ITK in patients with EBV-associated lymphoproliferative diseases. *Leukemia* **2012**, *26*, 963–971. [CrossRef]
49. Xie, Z.; Bailey, A.; Kuleshov, M.V.; Clarke, D.J.B.; Evangelista, J.E.; Jenkins, S.L.; Lachmann, A.; Wojciechowicz, M.L.; Kropiwnicki, E.; Jagodnik, K.M.; et al. Gene Set Knowledge Discovery with Enrichr. *Curr. Protoc.* **2021**, *1*, e90. [CrossRef]
50. Wu, T.D.; Nacu, S. Fast and SNP-tolerant detection of complex variants and splicing in short reads. *Bioinformatics* **2010**, *26*, 873–881. [CrossRef]
51. Liao, Y.; Smyth, G.K.; Shi, W. featureCounts: An efficient general purpose program for assigning sequence reads to genomic features. *Bioinformatics* **2014**, *30*, 923–930. [CrossRef] [PubMed]
52. Love, M.I.; Huber, W.; Anders, S. Moderated estimation of fold change and dispersion for RNA-seq data with DESeq2. *Genome Biol.* **2014**, *15*, 550. [CrossRef] [PubMed]
53. Ignatiadis, N.; Huber, W. Covariate powered cross-weighted multiple testing. *J. R. Stat. Soc. Ser. B* **2021**, *83*, 720–751. [CrossRef]

International Journal of *Molecular Sciences*

MDPI

Article

Germline POT1 Deregulation Can Predispose to Myeloid Malignancies in Childhood

Pia Michler [1], Anne Schedel [1], Martha Witschas [1], Ulrike Anne Friedrich [1], Rabea Wagener [2], Juha Mehtonen [3], Triantafyllia Brozou [2], Maria Menzel [1], Carolin Walter [4], Dalileh Nabi [5], Glen Pearce [6], Miriam Erlacher [7,8,9], Gudrun Göhring [10], Martin Dugas [11], Merja Heinäniemi [3], Arndt Borkhardt [2], Friedrich Stölzel [12], Julia Hauer [1,13,*] and Franziska Auer [13]

1 Pediatric Hematology and Oncology, Department of Pediatrics, University Hospital "Carl Gustav Carus", TU Dresden, 01307 Dresden, Germany; Pia.Michler@uniklinikum-dresden.de (P.M.); Anne.Schedel@uniklinikum-dresden.de (A.S.); Martha.Witschas@uniklinikum-dresden.de (M.W.); UlrikeAnne.Friedrich@uniklinikum-dresden.de (U.A.F.); Maria.Menzel@uniklinikum-dresden.de (M.M.)
2 Department of Pediatric Oncology, Hematology and Clinical Immunology, Medical Faculty, Heinrich-Heine University Duesseldorf, 40225 Duesseldorf, Germany; Rabea.Wagener@med.uni-duesseldorf.de (R.W.); Triantafyllia.Brozou@med.uni-duesseldorf.de (T.B.); Arndt.Borkhardt@med.uni-duesseldorf.de (A.B.)
3 Institute of Biomedicine, School of Medicine, University of Eastern Finland, Yliopistonranta 1, FI-70211 Kuopio, Finland; juha.mehtonen@uef.fi (J.M.); merja.heinaniemi@uef.fi (M.H.)
4 Institute of Medical Informatics, University of Muenster, 48149 Muenster, Germany; carolin.walter@uni-muenster.de
5 Department of Neuropediatrics Charité-Universitätsmedizin Berlin, Corporate Member of Freie Universität Berlin, Humboldt-Universität zu Berlin, and Berlin Institute of Health, 10117 Berlin, Germany; Dalileh.nabi@charite.de
6 Institute of Physiological Chemistry, Medical Faculty "Carl Gustav Carus", TU Dresden, 01307 Dresden, Germany; glen.pearce@tu-dresden.de
7 Department of Pediatrics and Adolescent Medicine, Division of Pediatric Hematology and Oncology, Faculty of Medicine, University Medical Center Freiburg, 79106 Freiburg, Germany; miriam.erlacher@uniklinik-freiburg.de
8 German Cancer Consortium (DKTK), 79106 Freiburg, Germany
9 German Cancer Research Center (DKFZ), 69120 Heidelberg, Germany
10 Department of Human Genetics, Hannover Medical School, 30625 Hannover, Germany; Goehring.Gudrun@mh-hannover.de
11 Institute of Medical Informatics, Heidelberg University Hospital, 69120 Heidelberg, Germany; martin.dugas@med.uni-heidelberg.de
12 Hematology and Oncology, University Hospital "Carl Gustav Carus", TU Dresden, 01307 Dresden, Germany; Friedrich.Stoelzel@uniklinikum-dresden.de
13 National Center for Tumor Diseases (NCT), 01307 Dresden, Germany; Franziska.Auer@uniklinikum-dresden.de
* Correspondence: Juliahauer@hotmail.com; Tel.: +49-351-458-3522

Citation: Michler, P.; Schedel, A.; Witschas, M.; Friedrich, U.A.; Wagener, R.; Mehtonen, J.; Brozou, T.; Menzel, M.; Walter, C.; Nabi, D.; et al. Germline POT1 Deregulation Can Predispose to Myeloid Malignancies in Childhood. *Int. J. Mol. Sci.* **2021**, *21*, 11572. https://doi.org/10.3390/ijms222111572

Academic Editor: Elixabet Lopez-Lopez

Received: 1 October 2021
Accepted: 23 October 2021
Published: 26 October 2021

Abstract: While the shelterin complex guards and coordinates the mechanism of telomere regulation, deregulation of this process is tightly linked to malignant transformation and cancer. Here, we present the novel finding of a germline stop-gain variant (p.Q199*) in the shelterin complex gene *POT1*, which was identified in a child with acute myeloid leukemia. We show that the cells overexpressing the mutated *POT1* display increased DNA damage and chromosomal instabilities compared to the wildtype counterpart. Protein and mRNA expression analyses in the primary patient cells further confirm that, physiologically, the variant leads to a nonfunctional *POT1* allele in the patient. Subsequent telomere length measurements in the primary cells carrying heterozygous *POT1* p.Q199* as well as *POT1* knockdown AML cells revealed telomeric elongation as the main functional effect. These results show a connection between *POT1* p.Q199* and telomeric dysregulation and highlight *POT1* germline deficiency as a predisposition to myeloid malignancies in childhood.

Keywords: acute myeloid leukemia; pediatric; trio sequencing; germline cancer predisposition; POT1; shelterin complex

1. Introduction

Telomeres play an essential role in preserving our genetic material by protecting chromosomal ends from degradation, while at the same time avoiding unwanted DNA damage responses [1]. The shelterin complex, which consists of six different protein subunits—TRF1, TRF2, RAP1, TIN2, TPP1 and POT1 (Figure 1A)—is responsible for safeguarding and properly maintaining telomeric DNA [2]. Shelterin deregulation leads to uncapped telomeres, which subsequently risks irreversible cellular changes, including genome instabilities, cellular aging or senescence and malignant transformation [3].

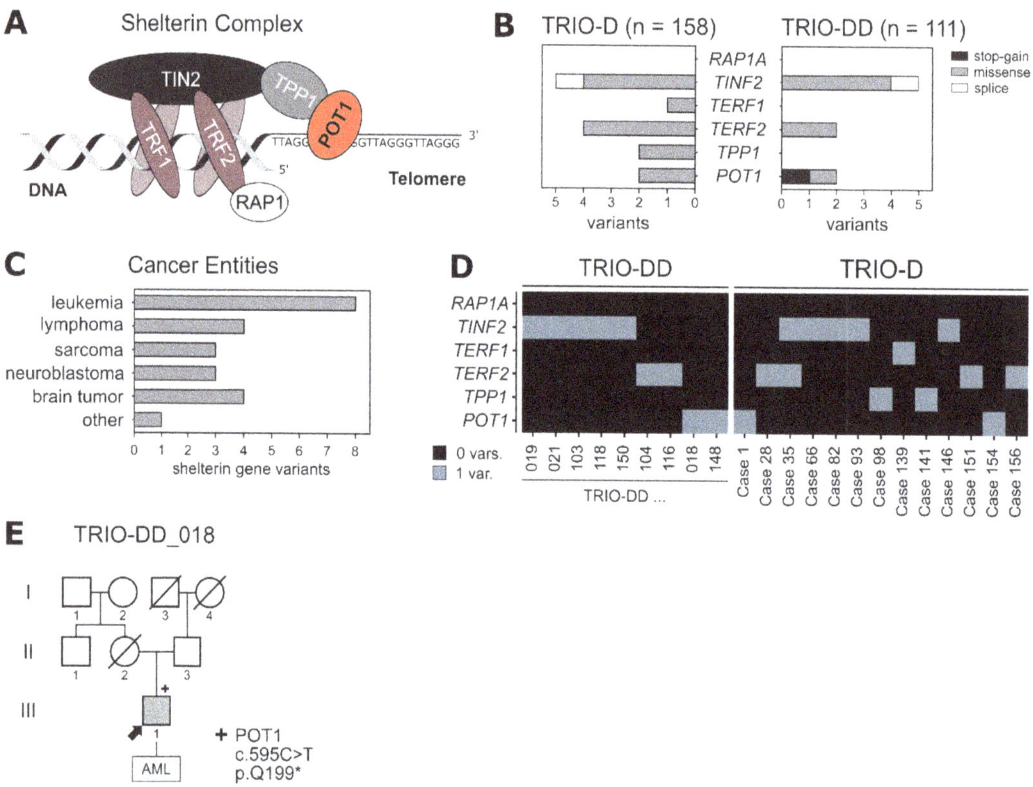

Figure 1. (**A**) Schematic drawing of the shelterin complex. POT1 (orange) is associated with the core complex via TPP1. (**B**) Two pediatric patient cohorts (TRIO-D, *n* = 158; TRIO-DD, *n* = 111) were analysed for germline variants within the shelterin complex genes. Only protein-altering variants with a MAF < 0.2% (healthy population) were included. (**C**) Tumor entities of the patients harboring the shelterin complex variants identified in (**B**). (**D**) Distribution of the protein-altering variants from (**B**) among the patients. (**E**) Family pedigree of the patient carrying the heterozygous germline *POT1* variant p.Q199*. The index patient is marked with an arrow. Variant carrier is marked with "+". Half-siblings were not included.

Protection of telomeres 1 (POT1), located on chromosome 7q, is an indispensable part of the shelterin complex [4,5]. In humans, POT1 uniquely recognises and binds telomeric single-stranded (ss) DNA via its N-terminal oligonucleotide/oligosaccharide-binding (OB) domains (OB1 and OB2) [6]. C-terminally, POT1 connects to the other shelterin complex proteins by interacting with TPP1 [7–9]. While POT1 exerts a multitude of functions, the most prominent comprise regulation of the telomerase-dependent telomere length [5,6], as well as repression of ATR-mediated DNA damage responses [10,11]. Therefore, it is not surprising that POT1 malfunction has been linked to the development of various types of human cancer [12]. While somatic *POT1* mutations are most prevalent in angiosarcoma [13], monoallelic *POT1* germline mutations are associated with a wide range of cancers, including familial melanoma, chronic lymphocytic leukemia, angiosarcoma and glioma [12,14]. Recently, germline *POT1* variants were found to predispose to myeloid and lymphoid neoplasms in adults [15]. Nevertheless, whether *POT1* germline aberrations can likewise confer susceptibilities to childhood cancer is unclear. Here, we present a novel loss of function stop-gain mutation in *POT1* (p.Q199*) in a boy with acute myeloid leukemia (AML). We further show how the resulting *POT1* haploinsufficiency confers telomere elongation and genomic instability, thereby generating a susceptible environment for malignant transformation.

2. Results

2.1. Identification of a Novel Germline POT1 Stop-Gain Mutation (p.Q199*) in a Child with AML

To elucidate whether shelterin complex mutations can predispose to the development of pediatric cancer, we analysed whole exome sequencing data of two independent parent–child cohorts of pediatric cancer patients (TRIO-D, $n = 158$ [16]; TRIO-DD, $n = 111$) for rare germline variants (minor allele frequency (MAF) < 0.2%) in the shelterin complex genes (Supplementary Table S1). Overall, 23 variants were identified in 269 pediatric cancer patients (Figure 1B) across various tumor entities (Figure 1C), with missense mutations being the most prominent. While one patient (case 35) presented with both a *TERF2* variant and a *TINF2* variant, all the other variants were mutually exclusive (Figure 1D).

Interestingly, a novel heterozygous *POT1* stop-gain mutation was found in an 8-year-old boy with AML. This germline variant (ENST00000357628.8:c.595C>T) is not yet described in public databases and causes the substitution of a glutamine to a stop codon at position 199 (p.Q199*) of the POT1 protein. Sanger sequencing validated the presence of the variant in the germline of the affected boy (variant allele frequency (VAF) of 0.5), while the variant was absent in his healthy father (Figure 1E, Supplementary Figure S1A). Since his mother's death at the age of 43 was not cancer-related, it is not clear whether *POT1* p.Q199* arose de novo or was maternally transmitted. Apart from the *POT1* variant, no other pathogenic or likely pathogenic variants according to the ACMG guidelines were identified in this patient (Supplementary Figure S1B). Accordingly, the *POT1* germline variant might play a role in the AML onset in this child. Clinically, the boy itself presented with pancytopenia, aplastic bone marrow and 70% myeloid blast cells in the bone marrow at the age of 8. Furthermore, the somatic molecular genetic makeup of the myeloid blast cells displayed an atypical monosomy 7q- with derivative chromosome 7, potentially leading to a loss of the remaining *POT1* wild-type (WT) allele (located on chromosome 7q) in the tumor (Supplementary Tables S2 and S3). Since the derivative chromosome 7 was still present in the blast cells, loss of heterozygosity could not be unambiguously verified in the tumor sample (detected VAF of 0.5). Due to an insufficient treatment response, the patient underwent hematopoietic stem cell transplantation with his haploidentical half-brother as the donor. Subsequently, the patient had a fast and sustained engraftment with a complete donor chimerism of 100% and is currently tapered from immunosuppression without any signs of graft-versus-host disease and relapse-free on day 526 after haploidentical cell transplantation.

2.2. POT1 p.Q199* Leads to a Loss of POT1 Expression from the Respective Allele

POT1 p.Q199* is located within the oligonucleotide-/oligosaccharide-binding 2 (OB2) domain, subsequently leading to a loss of around 2/3 of the full-length POT1 protein, including its complete TPP1-binding domain. Furthermore, the novel *POT1* p.Q199* variant was not found in previous studies on germline *POT1* variants in adults with myeloid malignancies (lollipops in Figure 2A). The low frequency of *POT1* germline variants within the healthy population (gnomAD database V2.1, $n = 118{,}479$) and a constraint score of 0.18 indicating high intolerance of the protein towards loss-of-function variants (Figure 2A) strongly suggest a functionally relevant role of POT1 in cancer susceptibility.

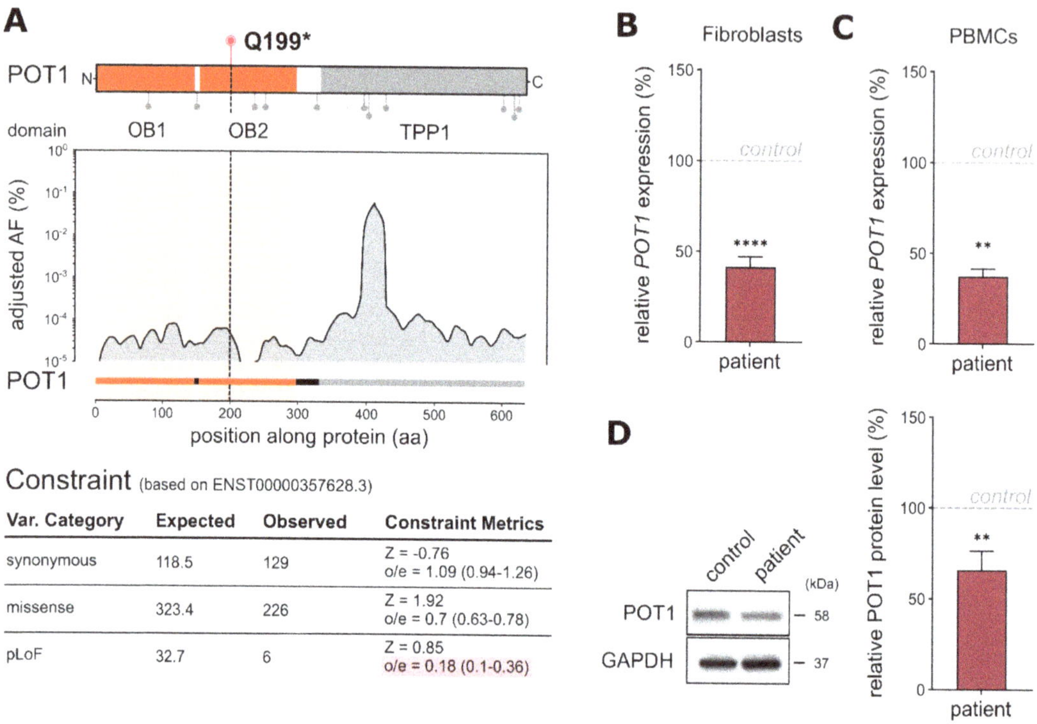

Figure 2. (**A**) Top: POT1 protein structure displaying the interaction domains with single-stranded DNA (1–299) and TPP1 (320–634). Lollipops below depict the positions of variants found in adult AML with predicted loss-of-function variants being displayed in light-grey (adapted from Lim et al., 2021). The variant p.Q199* lies within the interaction domain with single-stranded DNA. Bottom: distribution of the adjusted variant frequencies (AF (%)) along POT1 based on germline variants in the gnomAD noncancer database with the respective constraint metrics (based on Ensembl canonical transcript ENST00000357628.3). (**B**) QRT-PCR analyses of the primary fibroblast samples carrying heterozygous *POT1* p.Q199* compared to the *POT1* WT fibroblasts from an unrelated child. TaqMan probe binds to *POT1* exon 6–7 on chromosome 7. The assays were performed as three independent experiments, each with three technical replicates. (**C**) QRT-PCR analysis of PBMCs of the boy harboring heterozygous *POT1* p.Q199* compared to his father (carrying WT *POT1*). The assays were performed as two independent experiments, each with three technical replicates. (**D**) Representative Western blot and quantitative POT1 protein level analysis of the patient's fibroblasts harboring *POT1* p.Q199* compared to a control fibroblast sample (WT *POT1*). Three independent Western blots from each genotype were performed. The data represent the means ± SEM. The two-tailed Student's unpaired *t*-test was performed for the statistical analysis. ** = $p \leq 0.01$, **** = $p \leq 0.0001$.

To assess the immediate effect of *POT1* p.Q199* on gene expression, we performed quantitative real-time (qRT)-PCR analysis. Thereby, we could confirm haploinsufficiency of *POT1* in the patient's fibroblast cells harboring heterozygous *POT1* p.Q199* compared to a *POT1* WT fibroblast control ($p \leq 0.0001$, Student's *t*-test; Figure 2B), as well as in the patient's nonleukemic peripheral blood mononuclear cells referenced to blood from his father ($p = 0.024$, Student's *t*-test; Figure 2C). Additionally, Western blot analyses of the patient's primary fibroblasts revealed reduced POT1 levels, further validating the loss of one intact *POT1* allele in the patient (Figure 2D).

2.3. *POT1 p.Q199* Overexpression Confers Increased DNA Damage Response Induction*

Depletion of POT1 is known to cause de-repression of ATR-dependent DNA damage responses [10]. To test a potential functional influence of the truncated POT1 p.Q199* protein with regard to DNA damage signaling, N-terminally c-Myc-tagged *POT1* p.Q199* was cloned, transfected into HEK293T cells and compared to its WT counterpart (Supplementary Figure S2A). Western Blot analysis of the c-Myc-tag confirmed overexpression of the WT POT1 as well as a truncated protein corresponding to p.Q199* in the respective cell line system (Supplementary Figure S2B). Subsequent DNA damage analyses showed that POT1 p.Q199* overexpression led to an increase of DNA double-strand breaks in transfected HEK293T cells determined by gammaH2AX and 53BP1 immunofluorescence assays ($p \leq 0.001$, Student's *t*-test) (Figure 3A), which is in line with a deregulated DNA damage response and inappropriate repair by nonhomologous end joining [17]. These results could further be enhanced during stress conditions (irradiation with 3Gy, $p \leq 0.001$, Student's *t*-test) (Figure 3B). Interestingly, compared to the overexpression system, the primary fibroblasts of the patient carrying heterozygous *POT1* Q199* did not corroborate this deregulated DNA damage phenotype (Figure 3C,D), suggesting another mode of action to enable malignant transformation in the physiological setting.

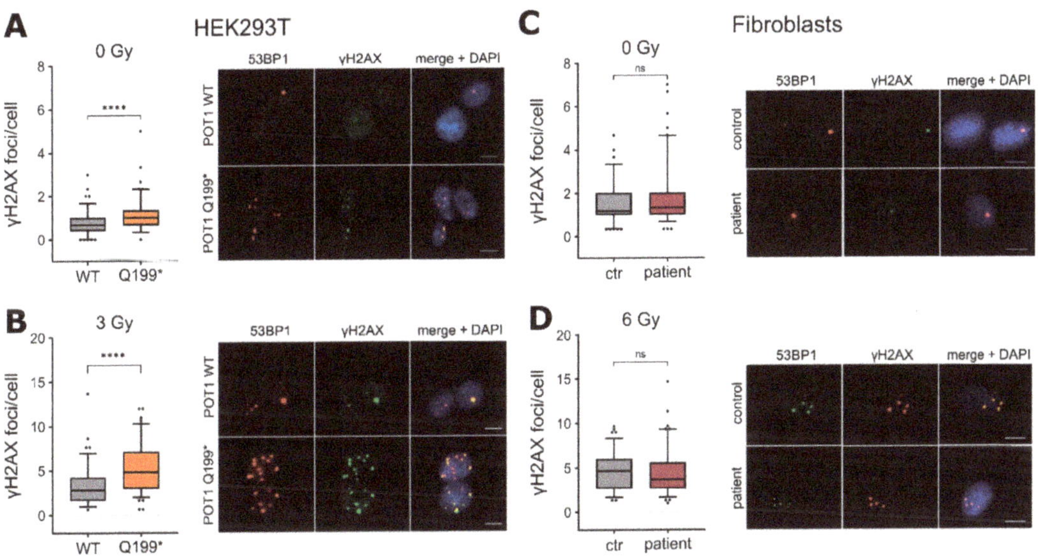

Figure 3. (**A**) Left: Quantification of the yH2AX foci per cell in the HEK293T *POT1* WT and *POT1* p.Q199* cells. Right: Representative images of the yH2AX (green) and 53BP1 (red) foci. DAPI (blue) was used for DNA labeling. Scale bar: 10 µm. (**B**) Left: quantification of yH2AX foci per cell in HEK293T *POT1* WT and *POT1* p.Q199* cells. The cells were exposed to 3 Gy ionising radiation. Right: Representative images analogous to (**A**). (**C**) Immunostaining analogous to (**A**) in the primary fibroblasts of the patient. (**D**) Immunostaining analogous to (**A**) in the primary fibroblasts of the patient with the cells exposed to 6 Gy ionising radiation. The experiments were performed as three biological replicates. The values are expressed in boxplots with whiskers from percentile 5–95. For the statistical analysis, two-tailed Student's unpaired *t*-test was performed. ns = $p > 0.05$, **** = $p \leq 0.0001$. ctr = control.

2.4. *POT1 p.Q199* Leads to Telomere Elongation and Chromosomal Instability*

Instead of a deregulated DNA damage phenotype, the patient's fibroblasts showed dysregulation of another main POT1 function, namely telomere length maintenance. Here, relative telomere length measurements by means of qRT-PCR indicated significant telomere elongation in primary fibroblast cells harboring *POT1* p.Q199* compared to WT *POT1* control fibroblasts from an age-matched child ($p = 0.019$, Student's *t*-test) (Figure 4A). This phenotype became even more apparent when the culture time was prolonged and the telomere length was assessed at a high passage number (passage 20: $p \leq 0.0001$,

Student's t-test) (Figure 4A). The same trend of longer telomeres was observed in the HEK293T overexpression system, as well as in the patient's blood cells (Supplementary Figure S3A,B). Elongated telomeres in the boy's peripheral blood (granulocytes) could further be validated by clinical diagnostics (ΔΔCt value of 1.42). Corroborating these results, the second telomere length analysis of the patient's blood after stem cell transplantation (haploidentical half-brother) confirmed reversion of the elongated telomere phenotype (Supplementary Figure S3C).

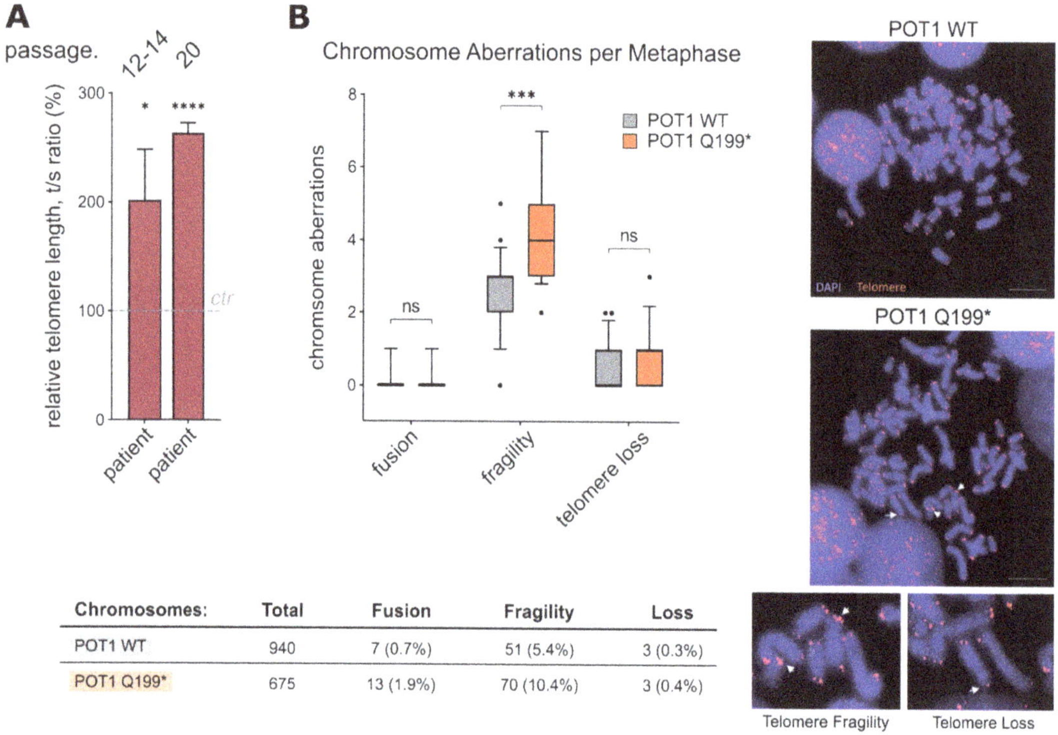

Chromosomes:	Total	Fusion	Fragility	Loss
POT1 WT	940	7 (0.7%)	51 (5.4%)	3 (0.3%)
POT1 Q199*	675	13 (1.9%)	70 (10.4%)	3 (0.4%)

Figure 4. (**A**) Relative telomere length (rTL) analysis by qRT-PCR (comparative ΔΔCt method) with DNA isolated from fibroblast samples (the cells were grown for 12–14 passages or 20 passages). Three biological replicates, each with three technical replicates, were performed. The values are expressed as the means ± SEM. (**B**) Left: Telomere fluorescence in situ hybridisation analysis on metaphase chromosomes of the stably transfected WT and p.Q199* POT1 HEK293T cells. Chromosomal aberrations are categorised in telomere fusion, fragility and loss. Right: Representative images of metaphase chromosomes. Red fluorescence shows telomere signals, and chromosomal DNA was stained with DAPI (blue). White arrows mark the respective chromosomal aberration. Scale bar: 10 μm. ns = $p > 0.05$, * $p = \leq 0.05$, *** = $p \leq 0.001$, **** = $p \leq 0.0001$.

Additionally, to analyse chromosomal stability, telomere FISH assays on metaphase chromosomes were carried out. In line with our results on telomere deregulation, the *POT1* p.Q199* HEK293T cells showed a significant increase in telomere fragility compared to the cells overexpressing WT POT1 ($p = 0.0002$, Student's t-test) (Figure 4B).

2.5. Reduced POT1 Levels in Myeloid Cells Confer Telomere Elongation

Within the hematopoietic system, POT1 shows the highest expression levels in hematopoietic stem and progenitor cells, as visualised by single-cell RNA sequencing data from the Human Cell Atlas (Figure 5A), while Western blot analyses show varying levels of POT1 in the ALL and AML cell lines (Supplementary Figure S4A). The clinical phenotype of the boy harboring the germline stop-gain *POT1* variant suggests a link between *POT1* p.Q199* and susceptibility to myeloid malignancies with 7q loss. To test this hypothesis, we generated a shRNA-mediated POT1 knockdown model of the AML cell line HL-60 (*POT1* mutation status = WT). Therefore, we used two different shRNAs—shRNA1 with a predicted knockdown level of 56% and shRNA2 with a predicted knockdown level of 92%. Successful transfection was validated via Sanger sequencing (Supplementary Figure S4B) and POT1 haploinsufficiency in both shRNA cell line models was confirmed by qRT-PCR (Figure 5B). Subsequent telomere length measurements confirmed significant telomeric elongation in the cells carrying the stronger knockdown with shRNA 2 (Figure 5C), thereby corroborating our previous results seen in the *POT1* p.Q199* overexpression model as well as in the primary patient cells.

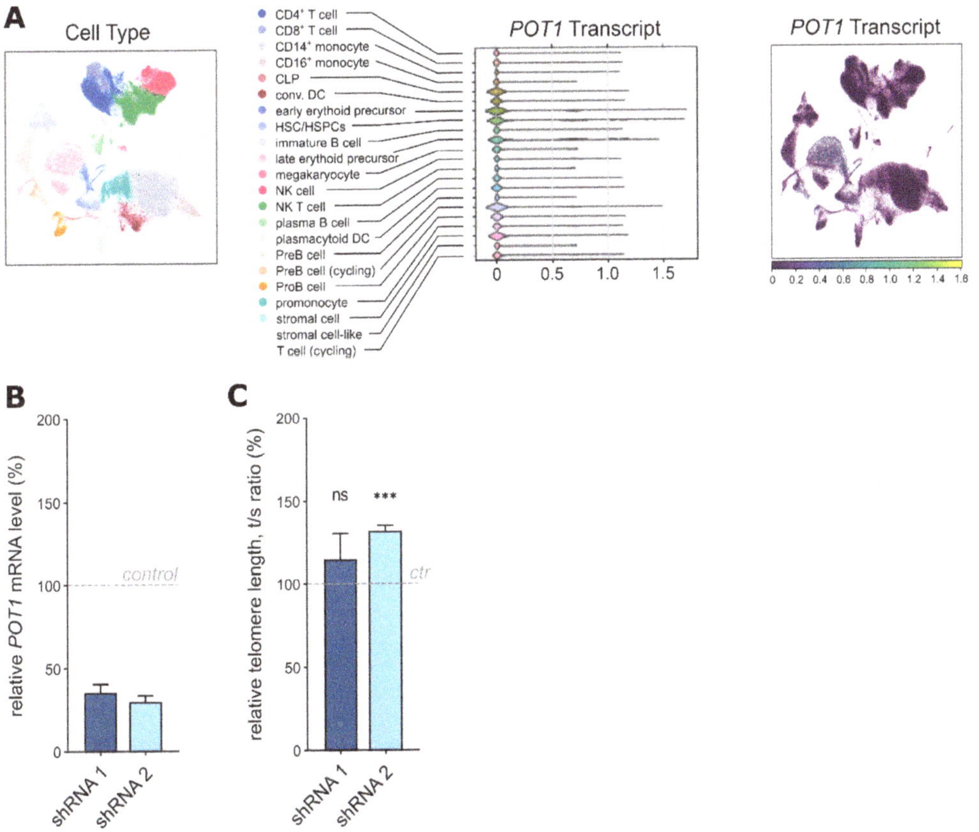

Figure 5. (**A**) Principal component analyses of the single-cell RNA sequencing data displaying POT1 expression within the healthy human bone marrow. CLP: common lymphoid progenitor, DC: dendritic cell, HSC: hematopoietic stem cell, HSPC: hematopoietic stem and progenitor cell, NK: natural killer cell. (**B**) QRT-PCR analysis confirming downregulation of POT1 in the HL-60 cells carrying shRNA 1 and 2 compared to the nontargeting shRNA control. (**C**) Relative telomere length (rTL) analysis by qRT-PCR (comparative ΔΔCt method) with DNA isolated from the HL-60 samples; shRNA 2 showed significant telomere elongation compared to the nontargeting shRNA control (ctr). ns = $p > 0.05$, *** = $p \leq 0.001$.

3. Discussion

Telomeres act as "molecular clocks" by defining the proper lifespan of each cell. Accordingly, deregulation of the shelterin complex which guards and maintains telomeric DNA is closely linked to cancer progression [2,18]. Here, we present a pediatric patient with AML who harbors a novel stop-gain variant, conferring germline haploinsufficiency of the shelterin complex gene *POT1*.

We show that the HEK293T cells overexpressing POT1 p.Q199* display an increased number of gammaH2AX foci and chromosomal fragility compared to WT POT1 overexpression, which points towards a loss of function phenotype. This also corroborates the previous findings of the induction of DNA damage signaling and telomere fragility in *Pot1a* deficient murine models [19,20]. Compared to the overexpression model, the primary fibroblasts harboring heterozygous *POT1* p.Q199* displayed strong telomeric elongation. This is in line with published data from human embryonic stem cells (hESC), in which cancer-associated *POT1* mutations did not trigger DNA damage responses but led to longer telomeres [21]. This telomere elongation is further confirmed by numerous in vitro POT1 knockout and overexpression models [5,21–23].

Causal relationships between longer telomeres and an increased cancer risk have been reported [24]. Interestingly, even though AML is commonly associated with short telomere syndromes (STS) [25], our data support an opposite scenario of telomere elongation mediated by *POT1* p.Q199* in our patient. Although counterintuitive, this observation is in line with the absence of classical STS phenotypes [3] in our patient and adult AML patients harboring *POT1* variants [15]. Moreover, previous reports on *POT1* germline variants in other cancers connect *POT1* deregulation with telomere elongation [26,27]. Therefore, our findings suggest that germline *POT1* haploinsufficiency causes abnormally long telomeres, which might generate a susceptible cell population with extended proliferative capacity and the potential to acquire additional mutations required for malignant transformation [21]. Likewise, long telomeres are more fragile and pose an increased risk for genomic instabilities, favoring cancer progression [18]. *POT1* is mostly expressed in stem and progenitor cells as depicted by scRNA-Seq data and human hematopoietic stem cells (HSCs) harboring CRISPR-Cas9-induced heterozygous *POT1* stop-gain mutations do not display fitness disadvantages in vivo [21]. Therefore, the precursor cells of the hematopoietic system could be directly affected by *POT1* aberrations, which is in line with the clinical phenotype observed in the patient.

Taken together, our data highlight a potential role of *POT1* germline deregulation in the context of predisposition to myeloid malignancies in childhood, which is mediated through telomere elongation. Although still in its infancy [28], antitelomerase therapy should be considered as a potential anticancer strategy to counteract malignant transformation through longer telomeres in *POT1*-deficient tumors.

4. Materials and Methods

4.1. Patients

Patients $\leq$ 19 years of age were unselectively recruited at the Pediatric Oncology Department, Dresden (years 2019–2021), or as previously described [16]. Consent of the families was obtained according to Ethical Vote EK 181,042,019 (Dresden) and in line with the Declaration of Helsinki.

4.2. Whole Exome Sequencing (WES)

Germline DNA was extracted from the patients' fibroblasts using an AllPrep DNA/RNA Mini Kit (Qiagen, Venlo, Netherlands) and from PBMCs of the parents and the remaining patients using a QIAamp DNA Blood Mini Kit (Qiagen). Next-generation libraries were generated with a SureSelect Human All Exon V7 kit (Agilent Technologies, Santa Clara, California). The libraries were sequenced on a NovaSeq 6000 platform (Illumina) in the paired-end mode (2 $\times$ 150 bp) and with the final target coverage of $\geq$100$\times$. Read files in the fastq format were generated with bcl2fastq v2.19.0, and trimmomatic v0.33 was used to remove adapter and low-quality sequences [29]. The alignment to the human reference genome GRCh37 was performed using BWA-MEM v0.7.12 [30] and Samtools v1.2 [31]. The tool Peddy 0.4.6 [32] performed gender and relatedness analyses to validate the correct sample assignment and the expected relationship of the patient's data with the corresponding parents' data. Single nucleotide variants (SNVs) and insertion/deletions (indels) were called using GATK v4.1.4.1 and VarScan2 v2.3.9 [33], applying the trio mode.

Initial variant interpretation was carried out with the CPSR pipeline [34], which classified the variants as pathogenic, likely pathogenic, variant of unknown significance (VUS), likely benign or benign. Additional variant interpretation was manually performed (e.g., by taking the CADD scores into account [35] as well as by utilising an extended cancer gene list).

4.3. Sanger Sequencing Validation

POT1 p.Q199* was validated via PCR and subsequent Sanger sequencing using the forward primer ACTCTACTCTCTTATGGCAGGT and the reverse primer CATCACCTTCAGAGATCTTGCC (5′ → 3′).

4.4. POT1 Variation Analysis

Allele frequencies of all the coding germline variants present in *POT1* (Ensembl transcript ID ENST00000297338) in a global healthy population taken from the gnomAD non-cancer exome r.2.1.1 dataset (n = 118,479) were summed up codon-wise. The variants must have been VEP-annotated to one of the following consequences for inclusion: start_lost, missense_variant, inframe_insertion, inframe_deletion, stop_gain, frameshift_variant, coding_sequence_variant, stop_lost, incomplete_terminal_codon_variant, transcript_ablation, transcript_amplification, protein_altering_variant. The collected dataset was smoothed using the LOWESS algorithm (fraction: 0.06, iterations: 3) prior to plotting.

4.5. Cloning

The coding sequence of human WT POT1 was obtained from plasmid pLPC mycPOT1 (a gift from Titia de Lange (Addgene plasmid No. 12387; http://www.addgene.org/12387/) (accessed on 25 October 2021); RRID: Addgene 12387) [5]. The mutant sequence for POT1 p.Q199* was created using site-directed mutagenesis by PCR. After digestion of the PCR product and vector, ligation of the vector DNA and the insert DNA was performed using a Quick Ligation™ Kit (New England Biologies (NEB) No. M2200S, Ipswitch, Massachusetts). Recombinant DNA was introduced into 10-beta competent *E. coli* (NEB No. C3019H) by transformation. The organisms containing vector sequences were selected and validated by Sanger sequencing. The plasmids were amplified in maxi cultures and POT1 WT/Q199* DNA was purified with a NucleoBond® Xtra Maxi EF Kit (Qiagen, Venlo, Netherlands) according to the manufacturer's instructions.

4.6. Cell Culture

The primary fibroblasts were cultivated in the BIO-AMF™-2 Medium (Biological-Industries, Kibbutz Beit Haemek, Israel) for up to five passages. For experimental analyses, the fibroblasts were cultured in Dulbecco's Modified Eagle's Medium (DMEM, obtained from GIBCO/Thermo Fisher Scientific, Waltham, Massachusetts) with 20% fetal calf serum (FCS; GIBCO), 1% penicillin/streptomycin (P/S; 10,000 units/mL; GIBCO) and 1% MEM Non-essential Amino Acids (NEAA; GIBCO).

HEK293T cells were cultured in DMEM with 10% FCS, 1% P/S and 1% NEAA. HL-60 cells and the additionally tested leukemia/lymphoma cell lines were cultivated in the RPMI 1640 (GIBCO) medium with 20% FCS and 1% P/S. All the cells were kept at 37 °C and 5% CO_2.

4.7. HEK293T Cell Transfection

HEK293T cells were seeded at a density of 4×10^5 cells and transfected with POT1 WT and POT1 Q199* plasmids using Lipofectamine 2000 (Invitrogen/Thermo Fisher Scientific, Waltham, Massachusetts) according to the manufacturer's instructions. The cells were selected with 2 µg/mL puromycin (Invitrogen), reduced to 1 µg/mL after one week.

Validation of successful transfection was done by Sanger sequencing of reverse-transcribed mRNA.

4.8. POT1 Knockdown in HL-60 Cells

For lentivirus production, 15 × 10^6 HEK293T cells were seeded on 15 cm dishes and transfected with VSV-G, pCMVdr8.2dvpr and shRNA plasmids containing specific POT1 knockdown sequences or non-coding shRNA. The transfection was performed using Lipofectamine 2000 according to the manufacturer's instructions. The following plasmids were obtained from Sigma-Aldrich, St. Louis, Missouri: control (SHC016), shRNA1 (TRCN0000039804; predicted knockdown level, 0.56), shRNA2 (TRCN0000010352; predicted knockdown level, 0.92). The plasmids were amplified in maxi cultures and POT1 WT/Q199* DNA was purified using a NucleoBond® Xtra Maxi EF kit (Qiagen).

HL-60 cells were transduced via spinfection on Retronectin (TakaraBio, Saint-Germain-en-Laye, France) coated 6-well plates. The successfully transduced HL-60 cells were selected with 1 µg/mL puromycin (Invitrogen), which was reduced to 0.5 µg/mL after 1 week. Knockdown of POT1 in HL-60 cells was quantified by qRT-PCR using POT1 TaqMan assays (Hs01565611_m1).

4.9. Quantitative Real-Time (qRT)-PCR Analysis

RNA was extracted from the primary fibroblasts (TRIO_DD_018; TRIO_DD_025; 2.0–3.0 × 10^6 cells) using an RNaeasy Mini Kit (Qiagen No. 74106) with 350 µL of the RLT Buffer+ beta-ME using a QIAshredder (Qiagen, No. 79656) and an RNAse-Free DNase Set (Qiagen No. 79254). Three independent RNA extractions were performed. One µg of RNA was reverse-transcribed into cDNA using a QuantiTect Reverse Transcription Kit (Qiagen No. 205311) following the manufacturer's instructions. Quantitative RT-PCR was performed using a TaqMan Universal Master Mix II following the manufacturer's instructions (Thermo Fisher Scientific No. PN4428173, Waltham, Massachusetts) for a 20 µL reaction with 1.5 µL of cDNA. The following TaqMan assays (Thermo Fisher Scientific, Waltham, Massachusetts) were used: TBP (Hs00427620_m1), HPRT1 (Hs02800695_m1) and POT1 (Hs01565611_m1; Chr.7: 124,822,386–124,929,983). Expression of mRNA was analysed by means of the comparative $\Delta\Delta C_T$ method and plotted in relation to the control sample.

4.10. Western Blot

For whole cell lysates, 2–2.5 × 10^6 (fibroblast samples) and 5 × 10^6 (HEK293T cells stably overexpressing POT1 WT or p.Q199*) were lysed in a RIPA buffer (50 mM Tris, 150 mM NaCl, 0.5% sodium deoxycholate, 1% Triton and 0.1% SDS 20%, with 10× PhosSTOP (PS, Roche) and 25× PIC (Protease Inhibitor Cocktail, Roche, Basel, Switzerland) for 30 min on ice. The protein concentration was measured with the Bradford protein assay (Roti-Quant, Roth) by determining OD_{595nm}. Twenty µg (HEK293T cells protein) or 15 µg (fibroblast cells protein) were heated for 10 min at 95 °C while shaking at 350 rpm and loaded accordingly onto a BIORAD Mini-Protean TGX Gel 4–20% (Bio-Rad Laboratories, Hercules, California). The blot was run for 3 h at 80 V. The transfer was performed using a Trans-Blot Turbo 1× Transfer System (high-molecular-weight, BIO-RAD). The immunoblot was blocked in 5% milk at room temperature for 1 h. After three washes with 1× TBS-T, the blot was incubated overnight at 4 °C with the following antibodies: a c-myc-Taq antibody (Invitrogen No. MA1-980, 1:1000) diluted in 5% Bovine Serum Albumin (Sigma-Aldrich, St. Louis, MO, USA) and POT1 (Novusbio No. NB500-176, 1:1000, Centennial, Colorado), GAPDH (Cell Signaling No. 5174S, 1:1000, Danvers, Massachusetts) diluted in 5% milk. The following day, the secondary antibody was applied after three consecutive washes (Cell Signaling Anti-Rabbit IgG No. 7075 1:1000, Cell Signaling Anti-Mouse IgG No. 7076) for 1 h in the dark, at room temperature, diluted in 5% milk. After three consecutive washes, the blot was imaged after the application of an HRP-linked solution (SuperSignal West Pico PLUS Chemiluminescent Substrate, Thermo Fisher Scientific, Waltham, MA, USA).

4.11. QRT-PCR of the Telomere Length (TL)

DNA was isolated from the fibroblast samples, the blood samples and the stably transfected HEK293T cells overexpressing POT1 using an AllPrep DNA/RNA/Protein Mini Kit (Qiagen) and stored at −20 °C. Afterwards, DNA was diluted to a concentration of approximately 25 ng/µL. The relative telomere length was measured by qPCR using a Relative Human Telomere Length Quantification qPCR Assay kit (Science Cell No. 8908) following the manufacturer's instructions. For the HL-60 cells, the annealing time was extended to 30 s. Three biological replicates of each sample were analysed. For quantification of the TL, the comparative $\Delta\Delta C_T$ method was applied.

4.12. Telomeric FISH on Metaphase Chromosomes

One million POT1 WT and POT1 p.Q199* HEK293T cells were seeded in T75 flasks and incubated for 48 h at 37 °C and 5% CO_2. Colchicine (Sigma-Aldrich) was added at a concentration of 5 µg/mL for 2 h. After trypsinisation, the cells were gently exposed to a hypotonic solution containing 0.075 M KCl and incubated for 10 min at 37 °C and 5% CO_2. The cells were first fixed in methanol/acetic acid (3:1), placed on slides and stored at −20 °C. For performing fluorescence in situ hybridisation, Telomere PNA FISH Kit/Cy3 (Dako/Agilent Technologies No. K5326, Santa Clara, California) was used following the manufacturer's instructions. The denaturation step was optimised to 90 °C for 10 min, and the hybridisation time was extended to 2 h at RT.

The slides were mounted in the ProLong Diamond Antifade mounting medium containing DAPI (Thermo Fisher Scientific) and stored at −20 °C. Metaphase spreads were captured with a Zeiss Axio Observer microscope using a Plan Apochromat objective with 63× magnificationaccompanying to the Core Facility Cellular Imaging (CFCI), Dresden, Germany). The DAPI images were used for featuring the metaphase chromosomes. At least 10 metaphase spreads per sample were captured and analysed. Chromosomal aberrations are presented as the frequency per metaphase.

4.13. Immunofluorescence Staining

Fibroblasts and stably transfected HEK293T cells overexpressing POT1 were plated onto Poly-L-Lysine pre-coated coverslips in 24-well-plates at a density of 2×10^4 cells (fibroblasts) or 6×10^4 cells (HEK293T) and cultured for 24 h at 37 °C and 5% CO_2. The cells were exposed to 3 Gy (HEK293T cells) or 6 Gy (fibroblasts) ionising radiation and cultured again for 24 h as described above. The following day, the cells were fixed for 15 min in 3% formaldehyde/PBS, blocked with 0.25% Triton X-100/PBS for 10 min and blocked again in 1% bovine serum albumin/PBS for 30 min. The samples were incubated with the primary antibodies for 1 h at RT. Using a mouse polyclonal anti-phospho-histone H2AX (Ser139) antibody (Merck Millipore No. 05-636, Burlington, Massachusetts) at a dilution of 1:100 and a rabbit 53BP1 antibody (Novusbio No. NB100-304) at a dilution of 1:1000, yH2AX and 53BP1 foci were detected. Coverslips were further stained with the secondary antibodies for 1 h at room temperature in the dark. The goat anti-mouse Alexa Fluor 488 IgG antibody (Invitrogen No. A-11029) and the goat anti-rabbit Alexa Fluor 594 IgG antibody (Invitrogen No. A-11037) were used as the secondary antibodies, each at a dilution of 1:200. The slides were mounted in the ProLong Diamond Antifade medium containing DAPI. Wide-field microscopy was performed with a Zeiss Axio Observer microscope using a Plan Apochromat objective with 20×/40× magnification accompanying to the Core Facility Cellular Imaging (CFCI), Dresden, Germany). The DAPI images were used to detect signals inside the nuclei.

4.14. Single-Cell RNA Sequencing Analysis

Healthy human bone marrow scRNA-seq data from eight donors were downloaded from the Human Cell Atlas (https://data.humancellatlas.org/explore/projects/cc95ff89-2e68-4a08-a234-480eca21ce79) (accessed on 12 August 2021) [36] and aligned to hg19 using Cell Ranger v3.0.0. Scanpy (https://doi.org/10.1186/s13059-017-1382-0) (accessed on 12

August 2021) was used to characterise the cell types in the data, correcting for possible batch effects with the mutual nearest neighbors (https://doi.org/10.1038/nbt.4091) (accessed on 12 August 2021) and filtering for outliers using the median absolute deviation. The cell clusters found with Louvain clustering (https://zenodo.org/record/1054103) (accessed on 12 August 2021) were mapped to cell types using the known marker genes. Cell cycle phases were annotated by scoring cell cycle marker gene sets from https://doi.org/10.1101/gr.192237.115 (accessed on 12 August 2021). Two-dimensional visualisation was performed with UMAP (https://doi.org/10.1038/nbt.4314) (accessed on 12 August 2021).

4.15. Statistical Analyses

For the statistical analysis, two-tailed Student's unpaired *t*-test was performed. Differences with a p-value < 0.05 were considered to be significant; ns = $p > 0.05$, * $p \leq 0.05$, ** $p \leq 0.01$, *** $p \leq 0.001$, **** $p \leq 0.0001$.

Supplementary Materials: The following are available online at https://www.mdpi.com/article/10.3390/ijms222111572/s1.

Author Contributions: Conceptualisation, J.H. and F.A.; methodology, P.M., A.S., M.W., U.A.F., R.W., J.M., D.N., G.P., M.E. and F.A.; software, U.A.F., R.W., J.M., C.W., M.D. and M.H.; validation, P.M., A.S. and M.W.; formal analysis, P.M.; investigation, P.M., A.S. and F.A.; resources, R.W., T.B., M.M., G.G., A.B., F.S. and J.H.; data curation, U.A.F., R.W., T.B., M.M., C.W. and F.A.; writing—original draft preparation, F.A.; writing—review and editing, P.M., A.S., U.A.F., J.H. and F.A.; visualisation, P.M., U.A.F. and J.M.; supervision, J.H. and F.A.; project administration, F.A.; funding acquisition, J.H. All authors have read and agreed to the published version of the manuscript.

Funding: J.H. is supported by ERC Stg 85222 "PreventALL", ERA Per Med.JTC 2018 "GEPARD" and Sonnenstrahl e.V. P.M. is supported by Mildred-Scheel-Nachwuchszentrum (MSNZ)—TU Dresden together with Deutsche Krebsstiftung. A.S. is supported by Deutsche José Carreras Leukämie-Stiftung e.V. together with Gesellschaft für Paediatrische Onkologie und Haematologie e.V. M.H. (funding number: PS (01/2020) and J.M. are supported by ERA Per Med.JTC 2018 "GEPARD". R.W. acknowledges the financial support of Forschungskommission (2020-28) of Heinrich-Heine-Universität Düsseldorf.

Institutional Review Board Statement: The study was conducted according to the guidelines of the Declaration of Helsinki and approved by the Ethics Committee of Carl Gustav Carus University Clinic, Dresden (EK 181042019).

Informed Consent Statement: Informed consent was obtained from all the subjects involved in the study.

Data Availability Statement: The novel *POT1* variant was submitted to ClinVar (https://www.ncbi.nlm.nih.gov/clinvar/) (accessed on 26 October 2021).

Acknowledgments: The authors would like to thank all the members of our groups for useful discussions and for their critical reading of the manuscript. Particular acknowledgement goes to Silke Furlan, Carolin Melzig and Rajanya Ghosh.

Conflicts of Interest: The authors declare no conflict of interest.

References

1. Lim, C.J.; Cech, T.R. Shaping human telomeres: From shelterin and CST complexes to telomeric chromatin organization. *Nat. Rev. Mol. Cell Biol.* **2021**, *22*, 283–298. [CrossRef]
2. de Lange, T. Shelterin: The protein complex that shapes and safeguards human telomeres. *Genes Dev.* **2005**, *19*, 2100–2110. [CrossRef]
3. Barbaro, P.M.; Ziegler, D.S.; Reddel, R.R. The wide-ranging clinical implications of the short telomere syndromes. *Intern. Med. J.* **2016**, *46*, 393–403. [CrossRef] [PubMed]
4. Baumann, P.; Cech, T.R. Pot1, the putative telomere end-binding protein in fission yeast and humans. *Science* **2001**, *292*, 1171–1175. [CrossRef]
5. Loayza, D.; De Lange, T. POT1 as a terminal transducer of TRF1 telomere length control. *Nature* **2003**, *423*, 1013–1018. [CrossRef] [PubMed]

6. Lei, M.; Podell, E.R.; Cech, T.R. Structure of human POT1 bound to telomeric single-stranded DNA provides a model for chromosome end-protection. *Nat. Struct. Mol. Biol.* **2004**, *11*, 1223–1229. [CrossRef]
7. Houghtaling, B.R.; Cuttonaro, L.; Chang, W.; Smith, S. A dynamic molecular link between the telomere length regulator TRF1 and the chromosome end protector TRF2. *Curr. Biol.* **2004**, *14*, 1621–1631. [CrossRef] [PubMed]
8. Liu, D.; Safari, A.; O'Connor, M.S.; Chan, D.W.; Laegeler, A.; Qin, J.; Songyang, Z. PTOP interacts with POT1 and regulates its localization to telomeres. *Nat. Cell Biol.* **2004**, *6*, 673–680. [CrossRef]
9. Ye, J.Z.; Hockemeyer, D.; Krutchinsky, A.N.; Loayza, D.; Hooper, S.M.; Chait, B.T.; de Lange, T. POT1-interacting protein PIP1: A telomere length regulator that recruits POT1 to the TIN2/TRF1 complex. *Genes Dev.* **2004**, *18*, 1649–1654. [CrossRef]
10. Denchi, E.L.; de Lange, T. Protection of telomeres through independent control of ATM and ATR by TRF2 and POT1. *Nature* **2007**, *448*, 1068–1071. [CrossRef]
11. Kratz, K.; de Lange, T. Protection of telomeres 1 proteins POT1a and POT1b can repress ATR signaling by RPA exclusion, but binding to CST limits ATR repression by POT1b. *J. Biol. Chem.* **2018**, *293*, 14384–14392. [CrossRef]
12. Wu, Y.; Poulos, R.C.; Reddel, R.R. Role of POT1 in Human Cancer. *Cancers* **2020**, *12*, 2739. [CrossRef]
13. Shen, E.; Xiu, J.; Lopez, G.Y.; Bentley, R.; Jalali, A.; Heimberger, A.B.; Bainbridge, M.N.; Bondy, M.L.; Walsh, K.M. POT1 mutation spectrum in tumour types commonly diagnosed among POT1-associated hereditary cancer syndrome families. *J. Med. Genet.* **2020**, *57*, 664–670. [CrossRef] [PubMed]
14. Henry, M.L.; Osborne, J.; Else, T. POT1 Tumor Predisposition. In *GeneReviews((R))*; Adam, M.P., Ardinger, H.H., Pagon, R.A., Wallace, S.E., Bean, L.J.H., Mirzaa, G., Amemiya, A., Eds.; Seattle, WA, USA, 1993; Available online: https://europepmc.org/article/NBK/nbk563529 (accessed on 30 September 2021).
15. Lim, T.L.; Lieberman, D.B.; Davis, A.R.; Loren, A.W.; Hausler, R.; Bigdeli, A.; Li, Y.; Powers, J.; Raper, A.; Regeneron Genetics, C.; et al. Germline POT1 variants can predispose to myeloid and lymphoid neoplasms. *Leukemia* **2021**. [CrossRef]
16. Wagener, R.; Taeubner, J.; Walter, C.; Yasin, L.; Alzoubi, D.; Bartenhagen, C.; Attarbaschi, A.; Classen, C.F.; Kontny, U.; Hauer, J.; et al. Comprehensive germline-genomic and clinical profiling in 160 unselected children and adolescents with cancer. *Eur. J. Hum. Genet.* **2021**, *29*, 1301–1311. [CrossRef]
17. Chen, C.; Gu, P.L.; Wu, J.; Chen, X.Y.; Niu, S.S.; Sun, H.; Wu, L.J.; Li, N.; Peng, J.H.; Shi, S.H.; et al. Structural insights into POT1-TPP1 interaction and POT1 C-terminal mutations in human cancer. *Nat. Commun.* **2017**, *8*. [CrossRef] [PubMed]
18. Aramburu, T.; Plucinsky, S.; Skordalakes, E. POT1-TPP1 telomere length regulation and disease. *Comput. Struct. Biotechnol. J.* **2020**, *18*, 1939–1946. [CrossRef]
19. Hockemeyer, D.; Daniels, J.P.; Takai, H.; de Lange, T. Recent expansion of the telomeric complex in rodents: Two distinct POT1 proteins protect mouse telomeres. *Cell* **2006**, *126*, 63–77. [CrossRef]
20. Pinzaru, A.M.; Hom, R.A.; Beal, A.; Phillips, A.F.; Ni, E.; Cardozo, T.; Nair, N.; Choi, J.; Wuttke, D.S.; Sfeir, A.; et al. Telomere Replication Stress Induced by POT1 Inactivation Accelerates Tumorigenesis. *Cell Rep.* **2016**, *15*, 2170–2184. [CrossRef]
21. Kim, W.T.; Hennick, K.; Johnson, J.; Finnerty, B.; Choo, S.; Short, S.B.; Drubin, C.; Forster, R.; McMaster, M.L.; Hockemeyer, D. Cancer-associated POT1 mutations lead to telomere elongation without induction of a DNA damage response. *EMBO J.* **2021**, *40*, e107346. [CrossRef] [PubMed]
22. Glousker, G.; Briod, A.S.; Quadroni, M.; Lingner, J. Human shelterin protein POT1 prevents severe telomere instability induced by homology-directed DNA repair. *EMBO J.* **2020**, *39*, e104500. [CrossRef] [PubMed]
23. Lei, M.; Zaug, A.J.; Podell, E.R.; Cech, T.R. Switching human telomerase on and off with hPOT1 protein in vitro. *J. Biol. Chem.* **2005**, *280*, 20449–20456. [CrossRef]
24. Telomeres Mendelian Randomization, C.; Haycock, P.C.; Burgess, S.; Nounu, A.; Zheng, J.; Okoli, G.N.; Bowden, J.; Wade, K.H.; Timpson, N.J.; Evans, D.M.; et al. Association Between Telomere Length and Risk of Cancer and Non-Neoplastic Diseases: A Mendelian Randomization Study. *JAMA Oncol.* **2017**, *3*, 636–651. [CrossRef]
25. Schratz, K.E.; Haley, L.; Danoff, S.K.; Blackford, A.L.; DeZern, A.E.; Gocke, C.D.; Duffield, A.S.; Armanios, M. Cancer spectrum and outcomes in the Mendelian short telomere syndromes. *Blood* **2020**, *135*, 1946–1956. [CrossRef]
26. Ramsay, A.J.; Quesada, V.; Foronda, M.; Conde, L.; Martinez-Trillos, A.; Villamor, N.; Rodriguez, D.; Kwarciak, A.; Garabaya, C.; Gallardo, M.; et al. POT1 mutations cause telomere dysfunction in chronic lymphocytic leukemia. *Nat. Genet.* **2013**, *45*, 526–530. [CrossRef] [PubMed]
27. Robles-Espinoza, C.D.; Harland, M.; Ramsay, A.J.; Aoude, L.G.; Quesada, V.; Ding, Z.; Pooley, K.A.; Pritchard, A.L.; Tiffen, J.C.; Petljak, M.; et al. POT1 loss-of-function variants predispose to familial melanoma. *Nat. Genet.* **2014**, *46*, 478–481. [CrossRef]
28. Chakravarti, D.; LaBella, K.A.; DePinho, R.A. Telomeres: History, health, and hallmarks of aging. *Cell* **2021**, *184*, 306–322. [CrossRef] [PubMed]
29. Bolger, A.M.; Lohse, M.; Usadel, B. Trimmomatic: A flexible trimmer for Illumina sequence data. *Bioinformatics* **2014**, *30*, 2114–2120. [CrossRef]
30. Li, H.; Durbin, R. Fast and accurate short read alignment with Burrows-Wheeler transform. *Bioinformatics* **2009**, *25*, 1754–1760. [CrossRef]
31. Li, H.; Handsaker, B.; Wysoker, A.; Fennell, T.; Ruan, J.; Homer, N.; Marth, G.; Abecasis, G.; Durbin, R.; Genome Project Data Processing, S. The Sequence Alignment/Map format and SAMtools. *Bioinformatics* **2009**, *25*, 2078–2079. [CrossRef] [PubMed]
32. Pedersen, B.S.; Quinlan, A.R. Who's Who? Detecting and Resolving Sample Anomalies in Human DNA Sequencing Studies with Peddy. *Am. J. Hum. Genet.* **2017**, *100*, 406–413. [CrossRef] [PubMed]

33. Koboldt, D.C.; Zhang, Q.; Larson, D.E.; Shen, D.; McLellan, M.D.; Lin, L.; Miller, C.A.; Mardis, E.R.; Ding, L.; Wilson, R.K. VarScan 2: Somatic mutation and copy number alteration discovery in cancer by exome sequencing. *Genome Res.* **2012**, *22*, 568–576. [CrossRef] [PubMed]
34. Nakken, S.; Saveliev, V.; Hofmann, O.; Møller, P.; Myklebost, O.; Hovig, E. Cancer Predisposition Sequencing Reporter (CPSR): A Flexible Variant Report Engine for Germline Screening in Cancer. *Bioinformatics* **2019**. [CrossRef]
35. Kircher, M.; Witten, D.M.; Jain, P.; O'Roak, B.J.; Cooper, G.M.; Shendure, J. A general framework for estimating the relative pathogenicity of human genetic variants. *Nat. Genet.* **2014**, *46*, 310–315. [CrossRef] [PubMed]
36. Hay, S.B.; Ferchen, K.; Chetal, K.; Grimes, H.L.; Salomonis, N. The Human Cell Atlas bone marrow single-cell interactive web portal. *Exp. Hematol.* **2018**, *68*, 51–61. [CrossRef] [PubMed]

International Journal of
Molecular Sciences

MDPI

Article

Altered DNA Methylation Profiles in *SF3B1* Mutated CLL Patients

Alicja Pacholewska [1,2,†], Christina Grimm [1,2,†], Carmen D. Herling [3], Matthias Lienhard [4], Anja Königs [1,2], Bernd Timmermann [5], Janine Altmüller [6], Oliver Mücke [7], Hans Christian Reinhardt [3,8,9], Christoph Plass [7,8], Ralf Herwig [4], Michael Hallek [3] and Michal R. Schweiger [1,2,*]

1 Institute for Translational Epigenetics, Faculty of Medicine, University Hospital Cologne, 50931 Cologne, Germany; alicja.pacholewska@uk-koeln.de (A.P.); cgrimm6@uni-koeln.de (C.G.); akoeni11@smail.uni-koeln.de (A.K.)
2 Center for Molecular Medicine Cologne (CMMC), University of Cologne, 50931 Cologne, Germany
3 Center for Integrated Oncology Aachen Bonn Cologne Duesseldorf, German CLL Study Group, Department I of Internal Medicine, Faculty of Medicine, University Hospital Cologne, 50931 Cologne, Germany; carmen.herling@uk-koeln.de (C.D.H.); Christian.Reinhardt@uk-essen.de (H.C.R.); michael.hallek@uni-koeln.de (M.H.)
4 Department of Computational Molecular Biology, Max Planck Institute for Molecular Genetics, 14195 Berlin, Germany; lienhard@molgen.mpg.de (M.L.); herwig@molgen.mpg.de (R.H.)
5 Sequencing Core Facility, Max Planck Institute for Molecular Genetics, 14195 Berlin, Germany; timmerma@molgen.mpg.de
6 Cologne Center for Genomics, University of Cologne, 50931 Cologne, Germany; janine.altmueller@uk-koeln.de
7 German Cancer Research Center, Cancer Epigenomics, 69120 Heidelberg, Germany; o.muecke@dkfz.de (O.M.); c.plass@dkfz.de (C.P.)
8 German Cancer Consortium (DKTK), 69120 Heidelberg, Germany
9 West German Cancer Center Essen, Department of Hematology and Stem Cell Transplantation, University Hospital Essen, 45147 Essen, Germany
* Correspondence: mschweig@uni-koeln.de
† These authors contributed equally.

Citation: Pacholewska, A.; Grimm, C.; Herling, C.D.; Lienhard, M.; Königs, A.; Timmermann, B.; Altmüller, J.; Mücke, O.; Reinhardt, H.C.; Plass, C.; et al. Altered DNA Methylation Profiles in *SF3B1* Mutated CLL Patients. *Int. J. Mol. Sci.* **2021**, *22*, 9337. https://doi.org/10.3390/ijms22179337

Academic Editor: Elixabet Lopez-Lopez

Received: 9 July 2021
Accepted: 25 August 2021
Published: 28 August 2021

Publisher's Note: MDPI stays neutral with regard to jurisdictional claims in published maps and institutional affiliations.

Abstract: Mutations in splicing factor genes have a severe impact on the survival of cancer patients. Splicing factor 3b subunit 1 (*SF3B1*) is one of the most frequently mutated genes in chronic lymphocytic leukemia (CLL); patients carrying these mutations have a poor prognosis. Since the splicing machinery and the epigenome are closely interconnected, we investigated whether these alterations may affect the epigenomes of CLL patients. While an overall hypomethylation during CLL carcinogenesis has been observed, the interplay between the epigenetic stage of the originating B cells and *SF3B1* mutations, and the subsequent effect of the mutations on methylation alterations in CLL, have not been investigated. We profiled the genome-wide DNA methylation patterns of 27 CLL patients with and without *SF3B1* mutations and identified local decreases in methylation levels in $SF3B1^{mut}$ CLL patients at 67 genomic regions, mostly in proximity to telomeric regions. These differentially methylated regions (DMRs) were enriched in gene bodies of cancer-related signaling genes, e.g., *NOTCH1*, *HTRA3*, and *BCL9L*. In our study, *SF3B1* mutations exclusively emerged in two out of three epigenetic stages of the originating B cells. However, not all the DMRs could be associated with the methylation programming of B cells during development, suggesting that mutations in *SF3B1* cause additional epigenetic aberrations during carcinogenesis.

Keywords: chronic lymphocytic leukemia; CLL; DNA methylation; *SF3B1* mutation; NOTCH; IKAROS

1. Introduction

Chronic lymphocytic leukemia (CLL) is the most common leukemia in the Western world and mainly affects elderly people [1]. Although the CLL phenotype is quite specific and homogenous, the clinical outcome is extremely heterogeneous [1]. The clinical outcome is partly associated with the mutational status of the immunoglobulin heavy chain variable region (*IGHV*) as patients with a high level of somatic mutations in IGHV (M-CLL) have a better prognosis than patients with no or a low level of somatic mutations in this region (U-CLL) [1–3].

Through the development of new high-throughput sequencing technologies, additional genomic alterations have been identified, which are associated with poor prognosis or insufficient therapy response. The strongest impact has been found for *del(17p)* and mutations in tumor protein P53 (*TP53*). Additionally, shorter progression-free survival is conferred by mutations in the splicing factor 3b subunit 1 (*SF3B1*), ATM serine/threonine kinase (*ATM*), ribosomal protein 15 (*RPS15*), and Notch receptor 1 (*NOTCH1*) [4–10]. Mechanistic insight into how these genomic alterations lead to poor prognosis or therapy resistance in CLL is largely missing. One of the most frequently mutated genes in CLL is *SF3B1*, encoding a component of the splicing machinery. Patients with *SF3B1* mutations (*SF3B1*^{mut}) have a poor prognosis [5] and *SF3B1* alterations are associated with chemotherapy refractory disease [7]. *SF3B1* mutations cluster in the HEAT repeat domain (Huntingtin, Elongation factor 3—EF3), protein phosphatase 2A (PP2A), and the yeast kinase TOR1) with the most common mutation in CLL being *p.K700E* [5,9]. Depletion of *SF3B1* transcripts by small interfering RNA (siRNA) in cell culture leads to an alteration of exon usage, in most cases causing a decrease in cassette exon inclusion [11]. Long read sequencing also revealed differential 3′ splice site changes and a strong downregulation of intron retention events associated with *SF3B1* mutations [12]. Recently, it has been shown that blood malignancies, in particular, such as leukemias, have a strong link between mutations in splicing factors and epigenetic dysregulation [13]. As such, SF3B1 interacts with the chromatin remodeling complex WICH (WSTF-SNF2H) [14] and with the polycomb group proteins: polycomb group ring finger 2 (PCGF2) and ring finger protein 2 (RNF2) [15]. A direct interaction between SF3B1 and polycomb repressive complex 2 (PRC2) was shown in mice. As such *Sf3b1*$^{+/-}$ mice exhibited a similar phenotype as *PcG*^{mut} mice, i.e., various skeletal alterations along the anterior-posterior axis [15]. Moreover, hypermethylation of polycomb-repressed regions was observed in the proliferating fraction of circulating CLL [16]. Furthermore, SF3B1 interacts with nucleosomes in an RNA-independent manner and is preferentially associated with GC-rich exons [11]. However, a detailed characterization of epigenomic changes associated with *SF3B1*^{mut} CLL is still largely missing.

Epigenomic alterations are emerging as powerful prognostic indicators in CLL [17,18]. It has been found that CLL genomes, when compared to normal B cells, are globally hypomethylated [19–21] and that M- and U-CLL classes show distinct methylomes [22]. Hypomethylations are found at gene bodies [23], especially at repetitive sequences, such as Alu, long interspersed nuclear elements-1 (LINE-1), and satellite-α (SAT-α) repeats [24]. Epigenetic profiling using DNA methylation arrays identified three subgroups of CLL, reflecting the developmental stage of the B cells from which the CLL cells originated [23,25]. Normal B cell maturation from naive to high-maturity memory B cells is accompanied by unidirectional DNA methylation changes, of which most show a decrease in DNA methylation. Such epigenetic changes during differentiation processes are also referred to as epigenetic programming. Epigenetic-defined CLL subgroups were therefore named low-programmed CLL (LP-CLL), intermediate-programmed CLL (IP-CLL), and high-programmed CLL (HP-CLL) [25]. LP-CLL is enriched for unmutated *IGHV* (U-CLL) and is associated with a poor prognosis, whereas HP-CLL is enriched for mutated *IGHV* (M-CLL) [23].

Recent studies have revealed wide-spread intratumor heterogeneity of the methylation in CLL [18,26,27]. Upon maintenance of DNA methylation, the methylation of one CpG is influenced by neighboring CpGs, yielding concordant methylation states [28]. In contrast, discordant methylation states are associated with active reprogramming. In cancer, a higher degree of discordant neighboring CpG methylation at promoter sites has been associated with worse prognosis [26]. According to this study, the median time for 'failure free survival' (FFS), meaning the time between the first and the second treatment or death, decreases from 44 months to 16.5 months for patients with a high proportion of discordant methylation in the promoter. This suggests that DNA methylation is a predictor for prognosis, but whether it is causative or just a confounding factor to the malignant process is so far unknown. Along this line, it is unclear whether DNA methylation and oncogenic mutations are independent prognostic factors or if are functionally related. Landau et al. observed that the presence of sub-clonal drivers overruled the increased risk associated with the elevated PDR (proportion of discordant reads), suggesting that either the heterogeneous methylation facilitates mutational processes, or that the mutations exert their functions through epigenetic mechanisms [26].

Moreover, all three DNA methyltransferases (DNMTs) are subjected to alternative splicing [29–31] and, therefore, mutations in splicing factors can potentially lead to changes in the methylation profiles, either directly by affecting the splicing of the DNA methyltransferases and DNA demethylases genes, or by alternatively spliced isoforms of the chromatin and methylation regulators, such as non-coding RNAs [32]. Thus, to achieve a better understanding on the methylome changes in CLL patients and how they might be connected to mutations in *SF3B1*, we compared genome-wide methylation profiles of CLL patients with (*SF3B1*^{mut}) and without (*SF3B1*^{WT}) *SF3B1* mutations.

2. Results

2.1. CLL Patients with SF3B1 Mutations Feature Local Hypomethylations

Although overall hypomethylation in CLL patients compared to normal B cells is known [23,24], the impact of the *SF3B1* mutations on the epigenome is unclear. We used DNA of 27 patients with (*SF3B1*^{mut}, $n = 13$) and without (*SF3B1*^{WT}, $n = 14$) *SF3B1* mutations (Table S1), and investigated genome-wide methylation profiles using the methylated DNA immunoprecipitation sequencing (MeDIP-seq) technology. Although principal component analysis with the 1,000 most variable regions within CG islands (CGIs) revealed that samples clustered according to the *IGHV* mutational status, and some separation could also be observed based on the *SF3B1* mutational status, no clear clusters grouped by the *SF3B1* genotype were visible (Figure 1A). This suggested that *SF3B1*^{mut} has no widespread effect on the methylation patterns of the CLL patients. The distribution of sex or age between the *SF3B1*^{WT} and *SF3B1*^{mut} patients was not significantly different (Mann–Whitney test p-value = 0.94) (Table S1). Using the QSEA package [33], we identified 67 significantly hypomethylated, but no hypermethylated regions (adjusted p-value (false discovery rate, FDR) < 0.05, and $|\log2(\text{fold change})| \geq 1$) (Table S2). We validated 16 of the significantly differentially methylated regions (DMRs) with the EpiTYPER (Tables S3 and S4) and observed a significant correlation ($R = 0.71$, p-value $< 2.2 \times 10^{-16}$) between the methylation levels of the regions estimated with MeDIP-seq/QSEA and EpiTYPER (Figures 1B,C and S1). Although the EpiTYPER data cannot provide methylation information for individual CpGs localized on the same DNA strand, the bulk methylation level of neighboring CpG sites at 14 DMRs with more than one CpG tested seems mostly concordant (Table S4). A hierarchical clustering using the beta normalized methylation values at the 67 DMRs clearly separated *SF3B1*^{mut} from *SF3B1*^{WT} (Figure 2A).

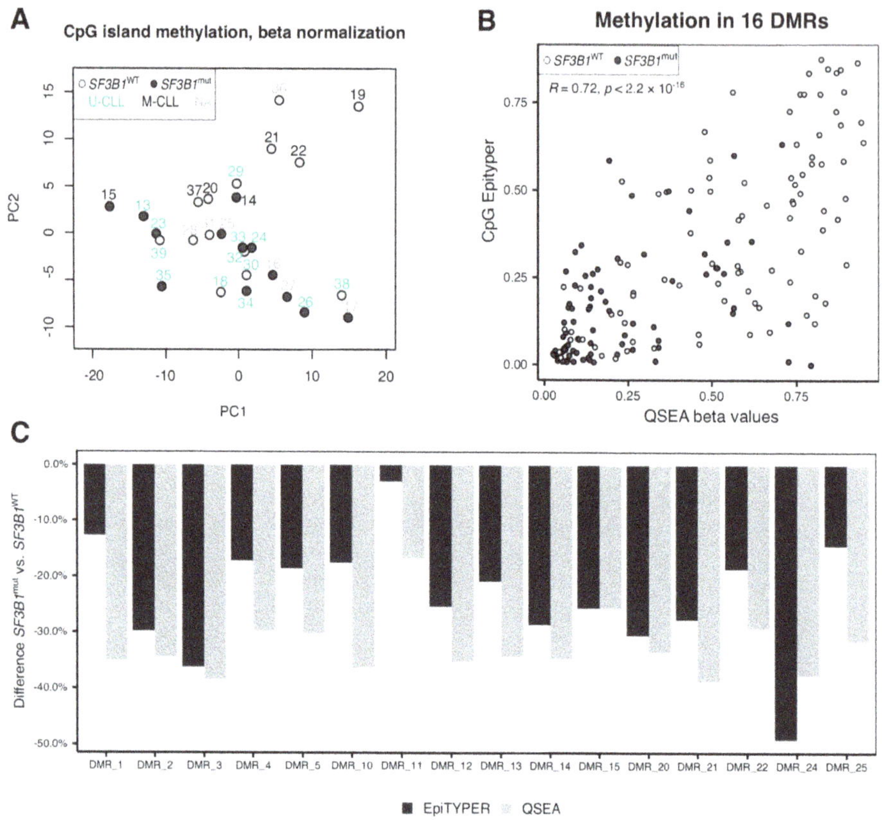

Figure 1. Hypomethylation in chronic lymphocytic leukemia (CLL) patients with *SF3B1* mutation. (**A**) Principal component analysis (PCA) of the samples revealed clustering of the *SF3B1*^{mut} samples with darker points corresponding to the samples with a mutation. PCA was performed using beta normalized values for the 1,000 most variable 250 bp windows within CpG islands. Points colors correspond to the *SF3B1* mutational status. Sample IDs were colored according to *IGHV* mutational status. (**B**,**C**) Validation of 16 from 67 differentially methylated regions (DMRs) identified with the subset of samples (6 *SF3B1*^{WT} and 6 *SF3B1*^{mut}) using the EpiTYPER to estimate CpGs methylation levels within the DMRs: (**B**)—correlation between EpiTYPER and beta methylation for each sample and every DMR/CpG; (**C**) comparison of the methylation difference between *SF3B1*^{mut} and *SF3B1*^{WT} CLL mean methylation levels.

We next compared our results to published methylation data created with methylation arrays that, even though they did not focus on the *SF3B1*^{mut} effect, did contain methylation data of *SF3B1*^{mut} patients. Kulis et al. used 139 CLL patients, 8 of which with *SF3B1* mutation [23]. The authors identified 64 hypo- and 30 hypermethylated differentially methylated CpGs between *SF3B1*^{mut} and *SF3B1*^{WT}. Although 74 of these 94 CpGs (79%) were within genomic regions sufficiently covered by sequencing reads in our data and tested for differential methylation, none of the DMRs identified overlapped the exact differentially methylated CpGs identified by Kulis et al. [23]. However, we identified DMRs within three genes that contained a differentially methylated CpG reported by Kulis et al. [23]: *BCL9L*, *MYB*, and *NCOR2*. The CpGs and DMRs within these genes had the same direction of the methylation change; they were hypomethylated in CLL with *SF3B1*^{mut} compared to CLL with *SF3B1*^{WT}. We next used an even larger dataset, one that encompasses the dataset from Kulis et al. [23], available from BloodCancerMultiOmics2017 R package [34]. In total, 174 CLL patients with known *SF3B1* mutational status (148 *SF3B1*^{WT} and 26 *SF3B1*^{mut}) were analyzed for 435,102 CpGs (all CpG sites with a single nucleotide polymorphism,

SNP, were removed by the authors). We used the limma R package [35] to analyze the array data sets and found that only 18 from our 67 DMRs (27%) overlapped at least one CpG (in total, 27 CpGs) included in the methylation array dataset, suggesting that the differential methylation might be located outside of regions captured with the array technology. The methylation values of the array data within all of these 18 regions were altered in the same direction as we have identified it within our dataset. Apart from one CpG with no change in methylation, the CpGs overlapping these DMRs showed a slight hypomethylation in CLL *SF3B1*^{mut} patients with a maximum $|\log2(\text{fold change})| = 0.3$, within the *ACOX3* gene (Table S2, column AI). This again puts our results in line with previously published data. Among the 27 CpGs overlapping the 18 DMRs, there were CpGs already reported by Wierzbińska as CLL-specific ($n = 8$), or B cell-specific ($n = 8$) [36] (Table S2). The mean methylation values for CLL *SF3B1*^{WT} and CLL *SF3B1*^{mut} of the 18 CpGs were significantly correlated between the array and the MeDIP-seq based data (Pearson R = 0.47, p-value = 0.0038) (Figure S2).

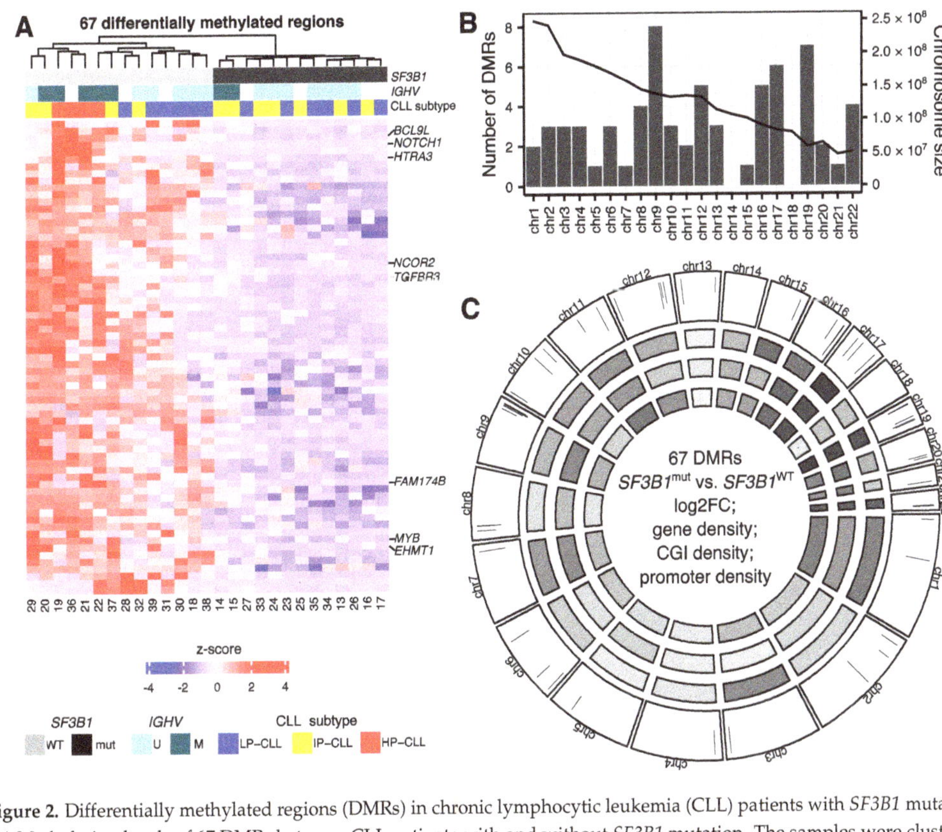

Figure 2. Differentially methylated regions (DMRs) in chronic lymphocytic leukemia (CLL) patients with *SF3B1* mutation. (**A**) Methylation levels of 67 DMRs between CLL patients with and without *SF3B1* mutation. The samples were clustered based on Euclidean distances with a complete linkage agglomeration method. White cells in the *IGHV* annotation denote missing information. CLL-subtype was defined by Oakes et al. based on the 18 loci selected by the authors [25]: low-programmed (LP-CLL), intermediate-programmed (IP-CLL), and high-programmed (HP-CLL). See Figures 4 and S7. (**B**) Numbers of DMRs between CLL patients with and without *SF3B1* mutation identified per chromosome. (**C**) Circos plot with the outer panel showing the log2 fold change in the 67 DMRs between *SF3B1*^{mut} and *SF3B1*^{WT} CLL patients in a −3.7 to 0.2 range. The next three panels show, from outside to inside, the average density of 250 bp regions tested for differential methylation per chromosome within (i) genes; (ii) CpG islands (CGI); (iii) gene promoters. The darker the grey color, the higher the density.

2.2. Hypomethylations in CLL Patients with SF3B1 Mutation Are Enriched in Gene Bodies and Subtelomeric Regions

We next asked if the changes in DNA methylation in *SF3B1*^{mut} samples are enriched in certain genomic regions or if they are evenly distributed across the genome. While we identified DMRs on almost all chromosomes, except chromosomes 14 and 18 (Figure 2B,C), the density of DMRs for the chromosomes varied (chi-squared test p-value = 0.00015). We observed the highest number of DMRs on chromosomes 9 and 19 (Figure 2B,C) (Figure S3A,B). Chromosome 19 is known for its high density of genes and CG content [37]. We therefore corrected the analysis for the CGI content and identified the largest enrichment of DMRs on chromosome 9 (Figure S3C). We did not observe any major deletions or insertions on chromosomes 9 or 19 based on the QSEA estimation (Figure S4). All eight DMRs identified on chromosome 9 (as well as many DMRs on other chromosomes) were located within 15 Mbp from the chromosome end in proximity of telomeric regions (Figure 2C) [38]. In fact, 43 of our 67 DMRs (64%) were located within 10 Mbp from the chromosomal start/end, and 28 of these (42% total) were located even more peripheral, within 5Mbp from the chromosomal ends. This result suggested that the hypomethylation may involve spatially and therefore possibly functionally related chromosomal regions. The subtelomeric DMRs on chr9 overlapped gene bodies of *EHMT1, NOTCH1, VAV2, PRRC2B, NEK6*, and the promoter region of *LCN10*.

Furthermore, to examine the distribution of DMRs in the genomic context, we annotated the DMRs and counted how many DMRs spanned different genomic features (Figure 3A). DMRs were significantly enriched in gene bodies (Figure 3B) with more than half (n = 48, 72%) of the DMRs located within gene bodies, mostly intronic regions (n = 40). Furthermore, we observed that DMRs were enriched in transcription factor binding sites (TFBS, n = 29, 43%). These TFBS were mostly downstream to gene promoters, except six, which were located in the promoters of *TGFBR3, RGPD8, HTRA3, LCN10, FAM174B*, and *IL17C* (Table S2). Of those, four had a CpG island annotated within the promoter region (*RGPD8, HTRA3, FAM174B, IL17C*). We next performed TFBS enrichment analysis of the 122 transcription factors (TFs) included in the ENCODE Uniform TFBS track [39] derived from ChIP-seq experiments [40–42]. We identified IKZF1 (IKAROS) and BHLHE40 as the most significantly enriched transcription factors with 12 and 8 hypomethylated DMRs at their binding sites, respectively (Figure 3C). Both TFs are critical for B cell development [43,44]. This is particularly interesting in regard to the question if the detected differential methylations are due to different B cell developmental stages where the *SF3B1*^{mut} cases develop from. However, the highest odds ratio was identified for the histone deacetylase HDAC6 with two DMRs at the promoter regions of *HTRA3* and *FAM174B*. HDACs' role in the initiation and progression of cancer has been extensively studied, as reviewed in [45]. Among chromatin states, we found weak and strong enhancers as predominant sites for DNA hypomethylations (Figure 3D). Among the 29 DMRs within enhancer regions, 21 (72%) were within 1 Mb from a promotor of a gene with significantly different expression levels (FDR < 0.05, no threshold on log2(fold change, Table S2), e.g., *XXYLT1, HTRA3*, and *ARID3A* genes. Eleven DMRs (16%) were located at ten unique promoter regions (*BCL9L, HTRA3, HPCAL1, IL17C, RGPD8, NCOR2, LCN10, AC107959.1, FAM1748, TGFBR3*) and two genes (*HTRA3* and *FAM1748*) had higher expression and hypomethylated promoters in CLL patients with *SF3B1*^{mut} compared to *SF3B1*^{WT} (Table S2).

Subsequently, to gain more biological insight into the differential methylations in CLL patients with *SF3B1* mutation, we used all 40 genes containing at least one DMR within their gene body or promoter region and subjected the list to a gProfiler functional enrichment analysis. This analysis identified the NOTCH signaling pathway (KEGG pathway, g:SCS adjusted [46] p-value = 1.49×10^{-2}) containing three of the genes (*DTX1, NCOR2, NOTCH1*) as significantly affected by differential methylations (Figure S5).

Motif enrichment analysis of the 67 regions did not reveal any significantly enriched transcription factor motif after multiple testing correction (Table S5).

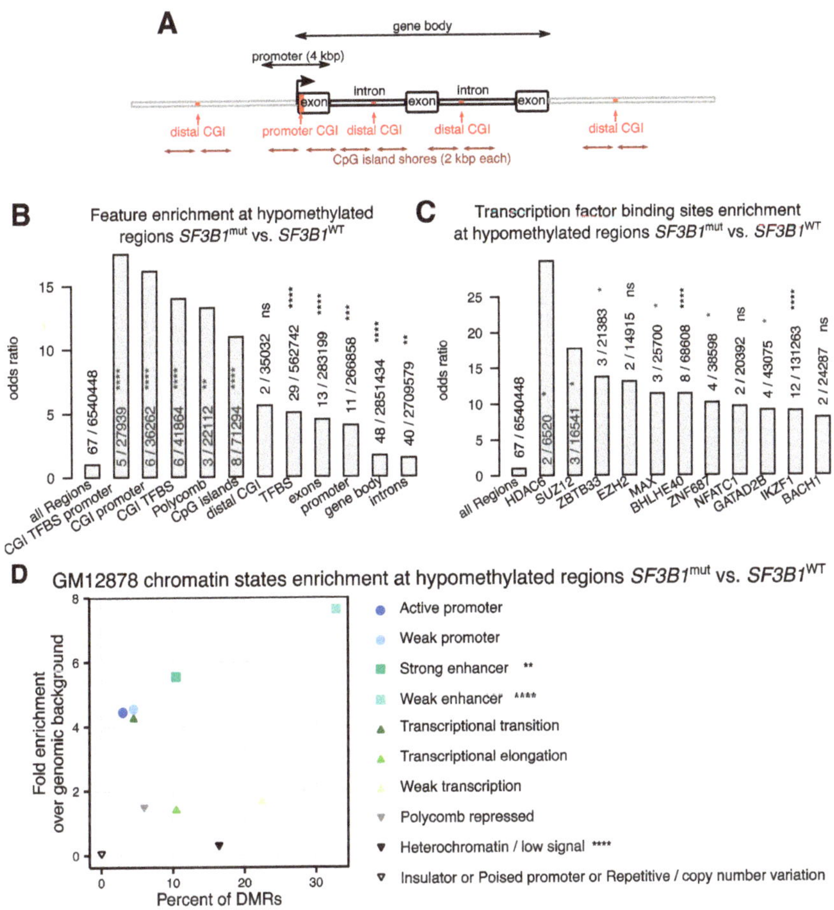

Figure 3. Genomic features of *SF3B1*^{mut} vs. *SF3B1*^{WT} differentially methylated regions (DMRs) at binding sites. (**A**) Schematic visualization of features used in (**B**) A promoter was defined as a region +/− 2000 base pairs (kbp) from a transcription start site. CpG islands (CGI) were obtained from [47]. A CGI shore was defined as a 2-kbp region flanking a CGI up—and downstream. A distal CGI is any CGI outside of promoter regions. (**B**) Enrichment analysis of genomic features with differentially methylated regions (DMRs) shown by odds ratio. Enhancer regions are taken from [48]. Transcription factor binding sites (TFBS, n = 122) were derived from ENCODE Encyclopedia v.3. PRC2 binding sites were defined as binding sites for EZH2 or SUZ12. (**C**) Enrichment analysis of the TFBS listed in (**B**)—only 11 TFBSs with the highest odds ratio are shown. (**D**) Enrichment analysis of the chromatin states as derived from GM12878 cell line [49,50]. Significance is denoted by stars with adjusted p-value (false discovery rate) < 0.05 = "*"; < 0.01 = "**"; < 0.001 = "***"; < 0.0001 = "****"; ≥ 0.05 = "ns".

2.3. SF3B1^{mut} *Is Associated with Aberrant Methylation and Is Partially Related to the Developmental B Cell Epigenetic State*

It has been previously shown that the methylation profile in CLL reflects, besides tumor specific alterations, the developmental state of the B cells from which the tumor has derived [25]. Accordingly, CLL cases have been classified into three stages of B cell development: low-programmed (LP-CLL), intermediate-programmed (IP-CLL), and high-programmed (HP-CLL) [25]. This classification is based on clustering of methylation levels from regions containing binding sites for transcription factors involved in B cell development (AP-1, EBF1, RUNX3) and transcriptional elongation [25]. The authors identified 18 regions, which sufficiently separate the three B cell developmental stages [25]. Using beta-transformed methylation levels calculated by the QSEA software [33] for these 18 regions, we clustered our 27 samples together with the 329 CLL Research Consortium samples analyzed by Oakes et al. [25] (Figures 4A and S6).

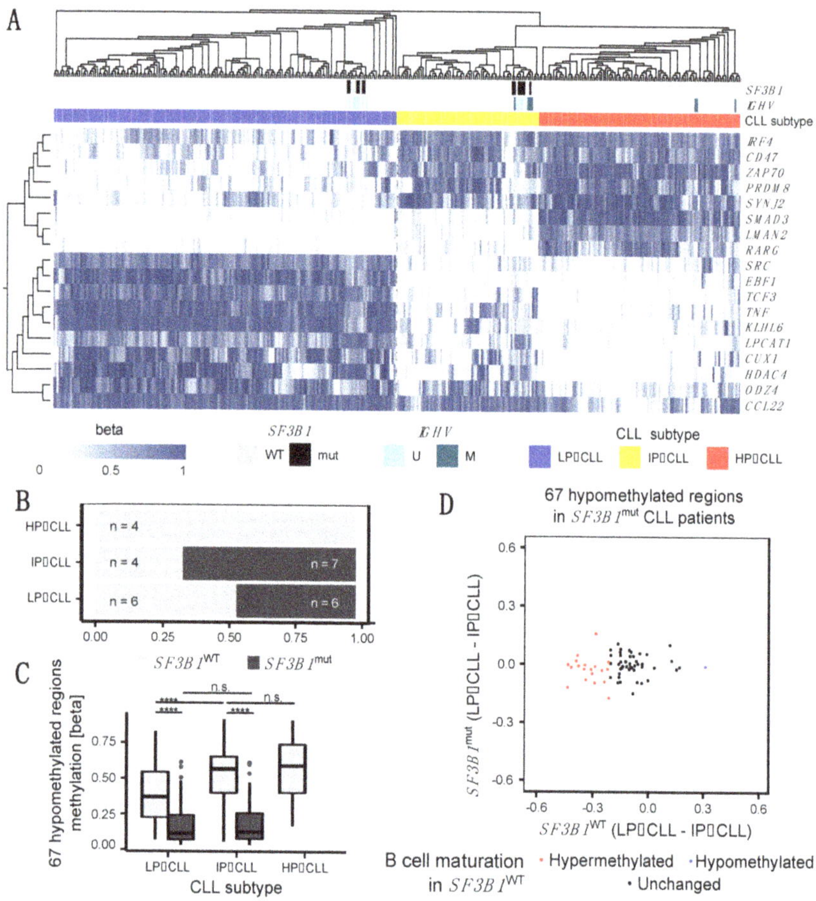

Figure 4. Changes in the DNA methylome in *SF3B1*^{mut} chronic lymphocytic leukemia (CLL) samples occur partly independent of B cell maturation. (**A**) Heatmap showing unsupervised clustering of 27 samples from this study and 329 CLL Research Consortium (CRC) samples from Oakes et al. [25] based on beta methylation values of the 18 most variable regions among the three CLL subtypes. White cells in the *SF3B1* and *IGHV* annotations denote missing information. The inclusion of our 27 samples led to re-classification of 5 out of the 329 CRC samples (<2%) from the LP-CLL to the IP-CLL subtype. (**B**) Proportion of *SF3B1*^{WT} and *SF3B1*^{mut} samples among the three CLL subtypes. (**C**) Methylation values of the 67 hypomethylated regions *SF3B1*^{mut} samples according to the *SF3B1* mutational status and CLL subtype. Wilcoxon test significance is denoted by stars with p-value < 0.0001 = "****"; ≥ 0.05 = "ns". (**D**) Comparison of the methylation programming between CLL patients with and without *SF3B1* mutation within 67 differentially methylated regions (DMRs) identified between *SF3B1*^{mut} CLL and *SF3B1*^{WT} CLL samples. Due to the lack of high-programmed CLL samples with *SF3B1*^{mut} in our dataset (**B**), the methylation programming is shown as the difference between the average methylation of low-programmed (LP-) and the average methylation of intermediate-programmed (IP-) CLL samples in each of the *SF3B1* mutational groups. (**B**) cell development-related DMRs hypo- or hyper-methylated in LP-CLL vs. IP-CLL in wild type samples (difference ≥ 20% in *SF3B1*^{WT}) are shown in blue and red, respectively.

The occurrence of *SF3B1* mutations have been associated with less differentiated states of B cells with the *SF3B1*^{mut} CLL resembling more naïve B cells whereas *SF3B1*^{WT} CLL resembling more memory B cells based on their methylation profiles [51–53]. We therefore tested to what extent the *SF3B1*^{mut}-associated hypomethylated regions can be explained by the developmental stage, and to what extend they can be attributed to the effects of the *SF3B1*^{mut} during carcinogenesis. While four of the *SF3B1*^{WT} samples clustered within the HP-CLL cluster, we did not observe any sample with a *SF3B1* mutation in this cluster (Figure 4A,B). This was confirmed by independent clustering of our 27 samples, as all HP-CLL samples (19, 21, 22, 36), none of which carried *SF3B1*^{mut}, created a separate cluster (Figures 1A and S6).

In order to evaluate if the identified 67 DMRs between *SF3B1*^{mut} and *SF3B1*^{WT} were due to different developmental stages of the originating B cells, we looked at the methylation differences based on beta-normalized methylation values at the 67 regions among the developmental stages represented by LP-, IP-, and HP-CLL subtypes (Figure 4C). We observed that, for *SF3B1*^{WT} CLL, the methylation levels at these regions significantly increased between the LP- and IP-CLL stages, whereas *SF3B1*^{mut} CLL samples showed stable methylation levels, which were significantly lower compared to the corresponding stage in the *SF3B1*^{WT} CLL (Figure 4C). At 22 from the 67 DMRs (33%) identified between *SF3B1*^{mut} CLL and *SF3B1*^{WT} CLL samples the methylation level changed (≥20%) between LP- and IP-CLL among the *SF3B1*^{WT} CLL samples (Figure 4D, Table S2), suggesting that these DMRs are involved in the normal B cell developmental process.

These 22 B cell development-related DMRs are located in 16 genes, including genes involved in B cell-specific functionality, epigenetic remodeling, and carcinogenesis, such as *TGFBR3*, *EHMT1*, *ACOX3*, *SEPTIN9*, *MYB*, *FAM174B*, *ARID3A*, *XXYLT1*, and *RHOBTB2*. Although we observed ≥ 20% change in the methylation at the 22 of 67 DMRs between LP-CLL and IP CLL among the *SF3B1*^{WT} CLL samples, the difference was lost in *SF3B1*^{mut} (Figure 4C). An additional 12 of our DMRs showed methylation differences (≥20%) when we compared HP-CLL to IP-CLL among the *SF3B1*^{WT} CLL samples, and 33 of the 67 DMRs (49%) showed stable methylation profiles among the *SF3B1*^{WT} CLL samples (LP-CLL vs. IP-CLL and IP-CLL vs. HP-CLL, Table S2), indicating that about half of the methylation changes observed between *SF3B1*^{WT} and *SF3B1*^{mut} are independent of B cell developmental stages.

Furthermore, we compared our list of DMRs with the 10,000 regions from Oakes et al. and concluded that most significantly change their methylation levels during physiological B cell maturation [25]. Only 4 of our 67 DMRs (6%) overlapped the 10,000 regions associated with B cell development reported by the authors [25] (Table S2). It is worth noting that all four overlapping DMRs, including DMRs within *BCL9L* and *NOTCH1*, were hypermethylated in high-maturity memory B cells and hypomethylated when comparing *SF3B1*^{mut} vs. *SF3B1*^{WT} CLL, suggesting an association of *SF3B1* mutations with less mature B cell developmental stages in CLL.

In addition, to further investigate if the observed DMRs between *SF3B1*^{mut} and *SF3B1*^{WT} were related to the B cell maturation or not, we compared our list of DMRs with epigenetic B cell programming sites identified using a methylome-based cell-of-origin modelling framework [36]. The authors identified linear dynamics of the methylation changes at 59,329 CpGs occurring during normal B cell development across six B cell differentiation stages, from naive to memory B cells (B cell-specific sites). CpGs with deviations from the expected methylation levels in CLL (CLL-specific) were classified into four classes: B cell-specific developmental sites hypomethylated (class A, n = 5757) and hypermethylated (class B, n = 183); and non B cell-specific sites hypomethylated in CLL (class C, n = 4238) and hypermethylation in CLL (class D, n = 157). The CLL-specific CpGs are expected to be associated with the tumorigenic transformation to CLL [36]. Eight (12%) of our DMRs overlapped the B cell-specific developmental sites reported by the authors [36], and seven (10%) overlapped CLL-specific CpGs not related to the B cell differentiation program (Table S2). This indicates that a part of the differential methylations detected in *SF3B1*^{mut} compared to *SF3B1*^{WT} patients is related to the normal B cell differentiation process, and that the other part is specific to *SF3B1*^{mut} CLL.

3. Discussion

Methylation is known to regulate splicing [54–57], but alternative isoforms of DNA methyltransferases genes or genes regulating chromatin conformation or methylation can also modulate methylation profiles [32]. Although mutations in genes required for splicing and methylation commonly occur in leukemia, and a mutation in a splicing factor (*SRSF2*) has been shown to impact methylation in acute myeloid leukaemia [58], the interaction of these two processes has not been described in CLL patients carrying *SF3B1*^{mut}. To acquire additional insight into the methylome differences in CLL patients with and without *SF3B1* mutations, we analyzed genome-wide methylation profiles using MeDIP-seq. We identified 67 regions with significantly lower methylation levels in *SF3B1*^{mut} CLL (Figure 2A) which we partly validated with the EpiTYPER assay.

The question remains what the cause for the altered methylation pattern might be. So far, there are no reports on the *SF3B1* mutation causing differentially spliced isoforms of DNMTs in CLL patients, which is also in line with our data. However, altered splice variants in euchromatic histone lysine methyltransferase 1 (*EHMT1*) have been significantly associated with *SF3B1*^{mut} CLL patients [59], and were also detected by us to contain a hypomethylated region (Table S2). In fact, the *EHMT1* and *UCKL1* genes were the only genes that overlapped the DMRs reported here, and were listed as genes with altered splicing associated with *SF3B1* mutation in CLL patients [59]. Although most histone methyltransferases are independent from DNA methylation, they are involved in gene repression [60] and DNA damage [61]. Furthermore, *EHMT1*, in particular, has also been associated with DNA methylation [62,63] and, therefore, this enzyme requires further investigation in *SF3B1*^{mut} CLL patients.

The methylation changes observed seemed to be only partially affected by the *IGHV* mutational status and allowed to clearly separate the samples by the *SF3B1* mutational status (Figure 2A). The hypomethylated regions were distributed across the genome; however, chromosomes 9 and 19 showed higher numbers of DMRs with many DMRs located close to telomeric regions (Figure 2C). The hypomethylation of regions in CLL compared to normal B cells [19,24] and further hypomethylation of CLL patients with *SF3B1*^{mut} may have a role in the worse prognosis of these patients. Interestingly, all eight DMRs on chr9, which contained, for example, a DMR within the *NOTCH1* gene, as well as many DMRs on other chromosomes, were located in a close proximity to telomeric regions. Short telomers have been already associated with anomalies in *SF3B1* showing worse prognosis in CLL patients [59,64,65]. However, the potential link between differential methylation close to the subtelomeric region and shorter telomeres in CLL patients with *SF3B1*^{mut} requires further studies. We also observed hypomethylation in two other genes involved in NOTCH signaling: *DTX1* and *NCOR2*. Mutations in the *NOTCH1* gene are frequent in CLL [4–10] and NOTCH signaling was associated with CLL progression [66–68]. In addition, NOTCH1 pathway was shown to be activated in CLL patients with *SF3B1* mutation [59,69].

In agreement with previous studies, most of our DMRs were located in genic regions (Figure 3B). The genes with DMRs were significantly enriched within the NOTCH signaling pathway (Figure S5). Interestingly, 12 DMRs included a binding site for the IKAROS (IKZF1) transcription factor (TF). These 12 DMRs were associated with eight genes: *ACOX3, ADAT3, ARID3A, FAM174B, NOTCH1, RGPD8, SCAMP4*, and *XXYLT1*. IKAROS is involved in B cell development [59] and has previously been implicated with CLL [70]. IKAROS expression increases during B cell differentiation and half of all genes upregulated during B cell development are IKAROS targets [71]. Moreover, IKAROS proteins are destructed by lenalidomide [72,73], a drug shown to act on CLL cells in vitro [74] and tested in the CLLM1 trial, where it improved progression-free survival. However, since a subset of treated patients developed B cell acute lymphoblastic leukemia, lenalidomide treatment was terminated in the CLLM1 study [75]. Of note, none of the patients analyzed in this study were treated with lenalidomide. Not surprisingly, 9 of the 12 DMRs within IKAROS binding sites were in regions identified as associated with B cell development and reported here or in previous studies [25,36]. The differential methylation between *SF3B1*^{mut} and *SF3B1*^{WT} CLL patients may, at least in part, be influenced by the differentiation state of the originating B cells [51–53]. In agreement with this hypothesis, all of our 14 *SF3B1*^{mut} samples were classified as early- (LP-CLL) and intermediate- (IP-CLL) programmed CLLs, and the role of IKAROS in B cell development has been highlighted in the early stages [43,76–78]. However, about half of the DMRs reported here seem to be independent from the B cell differentiation process.

Induction of IKAROS in CLL cells is associated with poor disease outcome [79] and promotes BCR signaling [80]. Taking into account that IKAROS tumor suppressive capacity includes an induction of enhancers in T cells [81], the hypomethylation in *SF3B1*^{mut} patients at IKAROS binding sites and the enrichment of the hypomethylation sites in weak enhancers is noticeable and requires further investigation.

IKAROS family proteins interact with nucleosome remodeling and deacetylase (NuRD) and PRC2 [82]. In T cells, IKAROS interacts with PRC2, thereby mediating epigenetic repression at stem cell-associated genes [83]. Such an interaction of IKAROS with PRC2 links IKAROS to DNA methylation regulation. Although binding of IKAROS to PRC2 was not observed in B cells, a loss of IKAROS function results in ectopic enhancer activation accompanied by a loss of the PRC2-mediated repressive histone modification H3K27me3 in the corresponding promoter regions [84]. In addition, the IKAROS family member, IKZF3 (AIOLOS), is recurrently mutated in CLL, with an incidence of 2% carrying the hotspot mutation *p*.L162R [5], emphasizing the importance of this transcription factor family for lymphoid malignancies.

Although significant, but with only two DMRs in TFBSs, is the histone deacetylase, HDAC6. The hypomethylated two sites for this region were located within promoter regions *HTRA3* and *FAM174B*. Although *FAM174B* has binding sites for more TFs, *HTRA3* has binding sites in this region only for HDAC6 and the PRC2 component, SUZ12. Histone deacetylase inhibitors have been already in phase II clinical trials to treat patients with breast cancer [85]. Furthermore, 30% of the DMRs had a TCF7L2 binding motif (Table S5), a key player in Wnt signaling [86–88]. A DMR was also identified in the *BCL9L* gene—an activator of Wnt signaling associated with B cell malignancies that have been implicated in cancer development and epithelial-mesenchymal transition through a down-regulation of c-Myc, cyclin D1, CD44, and vascular growth factor in tumor cells [89]. Hypomethylation within this gene in CLL patients with *SF3B1* mutation has been previously reported [23] and it has been recently shown that BCL9 and BCL9L promote tumorigenicity in a triple negative breast cancer mouse model through immune-dependent (TGF-β) and immune-independent (Wnt) pathways [90]. Interestingly, 25 of the 67 DMRs (34%) contained a TCF7L2 TF motif, and transcription factor 7-like 2 plays a key role in the Wnt signaling pathway [86–88]. A potential role of the *BCL9L* hypomethylation in poor prognosis of the CLL patients with *SF3B1*^{mut} should be further investigated.

In comparison with previous studies, our MeDIP-seq approach covers a broader part of the genome. For example, 450-K methylome arrays analyze 482,486 CpGs, most of which are located in genic regions and CpG islands [91]. Since the human genome contains roughly 28 million CpGs, around 1.6% of all CpGs are amenable by 450-K arrays. In contrast, MeDIP-seq accesses the mappable genome with a CpG content of at least 3% [92] at a resolution of approximately 250 bp. Only 435,102 CpGs from the 450-K methylation array [34] were amenable for the differential methylation testing, whereas MeDIP-seq analyzed by QSEA allowed for testing of 6,540,448 250 bp windows, covering 335,033 CpGs (77%) from the 450-K methylation array. This difference in coverage may also explain the small overlap of our results with previous studies. The larger fraction of the epigenome analyzed our study added new insight into the understanding of the disease.

It has been previously suggested that methylation differences identified among CLL samples may derive from a different maturation status of the B cells at the time of tumorigenesis [25,36]. We therefore classified our patients to the three CLL subtypes as defined previously [25]. Similarly to other studies, *SF3B1*^{mut} CLL samples were classified as LP- and IP-CLL subtypes [51–53]. Even though this dependency is not significant in our data due to a low number of samples (chi-squared test *p*-value = 0.09), it is in agreement with other studies, which reported the highest occurrence of CLL with *SF3B1*^{mut} in naïve B cell-like CLL (LP-CLL), and lowest in memory B cell-like CLL (HP-CLL) [51–53]. In line with this, we observed an intersection between DMRs reported here and regions differentially methylated during physiological B cell maturation [25,36], indicating that *SF3B1* mutations are, at least in part, associated with B cell developmental stages (Table S2). Some of them have been already associated with CLL-specific methylation changes [36]. However, there remains a large fraction of DMRs not related to B cell development, indicating that these regions are associated with *SF3B1* specific functions. How *SF3B1*^{mut} and changes in the methylome are exactly related needs further investigation. So far, we have shown that CLLs carrying *SF3B1*^{mut} contain differences in their DNA methylation patterns and that these changes affect genes involved in BCL9 and NOTCH signaling, among other processes. Thus, our findings provide a rich insight for further studies of the causes and consequences of *SF3B1*^{mut} induced changes in gene expression. This might, in the long term, provide the basis for the development of new therapeutic options.

4. Materials and Methods

4.1. Sample Preparation

Clinical information of the patients is summarized in Table S1. Staging was performed, according to Binet et al. [93], using blood cell counts. Patients were classified into Binet stage C when patients were anemic (hemoglobin < 10 g/mL) and/or displayed thrombocytopenia (thrombocytes < 100,000/µL), and into stage A/B when patients had more hemoglobin or thrombocytes. One patient had exactly 100,000 thrombocytes/µL and was therefore staged B/C (Table S1). *TP53*, *SF3B1*, *ATM*, *XPO1*, and *NOTCH1* mutational status were analyzed by a PCR panel followed by next-generation sequencing, as described in [94]. In particular, the complete coding region for *SF3B1*, *TP53*, and *ATM* was analyzed, and *XPO1* exons 12, 13, and 15 were evaluated. *IGHV* mutational status was determined, as previously described [95]. Peripheral blood B cells were isolated via negative selection using RosetteSep immunodensity-based cell separation (Stemcell Technologies, Vancouver, BC, Canada). The cell purity of CLL B cells was analyzed by flow cytometry, and cells co-expressing CD5/CD19 were $\geq$ 90%. DNA was isolated from frozen B cells using the QIAamp DNA Mini Kit (Qiagen, Hilden, Germany) or obtained from the local biobank. Informed consent was obtained from all patients and the study was approved by the local ethics committee. RNA was isolated using TRIzol (Thermo Fisher Scientific, Waltham, MA, USA) according to the manufacturer's instructions. A DNAse digest was performed using 2 µg of RNA, and the quality of the RNA was assessed using a Bioanalyzer (Agilent Technologies, Waldbronn, Germany).

4.2. Methylated DNA Immunoprecipitation Sequencing (MeDIP-Seq)

The integrity of DNA was evaluated on a 1% agarose gel and 1.3 µg was subjected to MeDIP. Methylated DNA immunoprecipitation (MeDIP) was based on a method developed by [96] with modifications. In brief, 1.3 µg of DNA in 65 µL of TE was sheared to a size of 100–300 bp using a Covaris M220 (Covaris Ltd., Brighton, UK). The size was controlled on a 1% agarose gel. Library preparation was performed using the TruSeq DNA sample preparation kit (Illumina, San Diego, CA, USA) and unmethylated TruSeq indexed adaptors. Library preparation reactions were purified using AMPure XP beads (Beckman Coulter GmbH, Krefeld, Germany) and the adapter-ligated DNA was denatured at 95 °C for 10 min and subjected to MeDIP. For the MeDIP reaction, 5 µg of the monoclonal antibody clone 33D3 directed against 5-methylcytidine (Eurogentec GmbH, Cologne, Germany) was coupled over night at 4 °C to magnetic Dynabeads M-280 sheep anti-mouse IgG (Thermo Fisher Scientific, Waltham, MA, USA). Subsequently, denatured DNA and antibody coupled Dynabeads were incubated at 4 °C for 4 h in immunoprecipitation buffer (IP: 10 mM sodium phosphate buffer (pH 7.0), 140 mM NaCl, 0.25% Triton X100) followed by three washes with IP buffer. DNA was eluted from the beads in 50 mM of Tris-HCl (pH 7.5), 10 mM of EDTA, and 1% SDS at 65 °C for 15 min. The eluted DNA was diluted 1 to 1 with 10 mM of Tris at pH 8.0 and 1 mM of EDTA, and treated with proteinase K (0.2 µg/µL) for 2 h at 55 °C. The methylated DNA was purified using the QIAquick PCR Purification Kit (Qiagen, Hilden, Germany). Following MeDIP enrichment, libraries were amplified using 10 PCR-cycles, size-selected using an agarose gel, and purified DNA was quantified using the Quant-iT dsDNA HS Assay Kit and a Qubit 1.0 Fluorometer (Thermo Fisher Scientific, Waltham, MA, USA). Next, 50 bp single-end reads were generated on a HiSeq4000 (Illumina, San Diego, CA, USA).

4.3. Sequencing Reads Processing

MeDIP sequencing reads were aligned to the hg38 reference genome (Genome Reference Consortium GRCh38) with BWA v0.7.15 aln followed by samse modules [97]. Patients' sequencing data are available from the corresponding author upon reasonable request.

4.4. Differential Methylation Analysis

MeDIP-seq reads were processed in R with QSEA v.1.14.0 [33], according to the package recommendations. In brief, reads where counted per genomic 250 base window, and CpG enrichment profiles were calibrated based on highly methylated genomic regions in 196 primary CLL samples of the PACE project, as retrieved from the Bioconductor package CLLmethylation [34] (methylation > 80% in at least 95% of the samples). Differentially methylated regions (DMRs) were called with the implemented likelihood ratio test, based on generalized linear models, and p-values were corrected for multiple testing by false discovery rate (FDR) [98]. We considered a region to be differentially methylated if the FDR was smaller than 0.05 and $|\log 2$ fold change$| \geq 1$. Moreover, we excluded regions with expected CpG density below 4 CpGs per sequencing fragment and all fragments from X and Y chromosomes. DMRs were annotated with BSgenome v.1.56.0. and RefSeq release 71, ENCODE Encyclopedia v.3 as of 24th April 2014, enhancer data from [48], and CGI as described in [47], transcription factor binding sites (TFBS) from ENCODE Uniform TFBS track [39–42]. Chromatin states coordinates for GM12878 cell line [49,50] were converted from hg19 to hg38 reference genome with liftOver UCSC tool [99]. Promoters were defined as 2 kb upstream and downstream from transcription start sites. Copy number variation (CNV) was calculated from MeDIP-seq data by considering only fragments without any CpG, based on 2-Mb windows and a fragment size of 250 bp.

Methylation microarray data from 196 CLL patients' samples (CLLmethylation) covering 435,155 CpGs (we were able to determine hg38 positions for 435,102 CpGs) were obtained via BloodCancerMultiOmics2017 R packages [34] with ExperimentHub and filtered out 53 CpGs without hg38 genomic information. We excluded data from 22 patients without *SF3B1* mutational status information. Differentially methylated probes were called using lmFit and eBayes functions of limma [35] and filtered according to adjusted p-value and fold change thresholds applied for the MeDIP-seq data, as described above.

4.5. RNA-Seq and Differential Expression Analysis

RNA libraries were generated using the TruSeq Stranded Total RNA sample preparation kit (Illumina, San Diego, CA, USA) which includes a Ribo-Zero depletion of ribosomal RNAs prior to library preparation. Sequencing of 50 bp paired-end reads was performed on a HiSeq2000 (Illumina, San Diego, CA, USA). RNA-seq reads were mapped to the same reference with STAR v2.6.0c [100] and GENCODE gene annotation v36 [101]. Differential expression analysis was performed with edgeR v. 3.30.3 [102]. A gene was considered significantly differentially expressed if the FDR-adjusted p-value was < 0.05. To identify possible genes affected by the differential methylation of weak enhancers, no threshold on log2(fold change) was set.

4.6. Bisulfite Mass Spectrometry (BS-MS) with Agena Bioscience EpiTYPER-Assay

We selected 16 DMRs with |log2 fold change| > 2 for validation by the EpiTYPER. Primers were designed for CpGs within 16 DMRs. We used a subset of six CLL samples with and six CLL samples without *SF3B1*^{mut}.

For the validation of the differentially methylated regions identified by MeDIP-seq, we used the EpiTYPER assay, which is based on bisulfite conversion followed by PCR amplification using one primer containing a T7 promoter sequence, followed by in vitro transcription and Uracil-specific cleavage of the RNA. Fragments were then analyzed by matrix-assisted time-of-flight mass spectrometry (MALDI-TOF) mass array analysis [103,104].

Primers for the EpiTYPER assay were designed using the online tool EpiDesigner with default settings (www.epidesigner.com accessed on 24 March 2021) and purchased from Integrated DNA Technologies (Leuven, Belgium). Oligo sequences, genomic coordinates, and annealing temperatures are given in Table S3. For the assay, 1 µg of genomic DNA was bisulfite-converted using the EZ DNA Methylation kit (Zymo Research Europe GmbH, Freiburg, Germany) according to the manufacturer's recommendations. Bisulfite converted DNA was eluted in 60 µL and 1 µL of the dilution was used for amplification in 384-well plates in a 5-µL reaction volume using 0.2 U of HotStarTaq (Qiagen, Hilden, Germany), 1 pmol of each oligo, and 1 nmol of dNTPs. The reaction was run in a thermocycler at 95 °C for 15 min and 35 to 45 cycles at 94 °C for 30 s, annealing (52, 56, or 60 °C) for 30 s, 72 °C for 1 min, and a final extension of 5 min at 72 °C. Subsequently, the PCR product was in vitro transcribed and enzymatically cleaved using the MASSCleave T7 Kit (Agena Bioscience GmbH, Hamburg, Germany) and run on a MassArrayDX (Agena Bioscience GmbH, Hamburg, Germany). A DNA methylation standard was generated by whole genome amplification, WGA using Repli-G (Qiagen, Hilden, Germany) and in vitro methylation using *M.SssI* (CpG) methyltransferase (New England Biolabs, Frankfurt am Main, Germany). Standards of 0%, 20%, 40%, 60%, 80%, and 100% were generated by mixing WGA DNA with WGA and in vitro-methylated DNA. For each assay, a methylation standard was run in parallel. Methylation for individual CpG units was calculated with the EpiTYPER 1.3 software. Subsequently, methylation values (0 to 1) for a given region were calculated as the mean of the analyzed CpG units that passed the quality criteria. CpG units were excluded from the analysis when: (i) less than 50% of all samples had values; (ii) CpG units within one amplicon had an identical mass; and (iii) > 3 CpGs within one CpG unit. DNA methylation values for the amplicons are given in Table S4.

4.7. Gene Set Enrichment Analysis

We collected unique genes that contained at least one significant DMR (FDR < 0.05, |log2 fold change| ≥ 1) within their gene bodies or promoter regions and subjected to functional enrichment analysis with gProfiler2 with default options (version e103_eg50_p15_eadf141) [105,106].

4.8. Motif Enrichment Analysis

Differentially methylated 250-bp regions in CLL patients with *SF3B1* mutation compared to CLL without *SF3B1* mutation were used as input for findMotifsGenome.pl module of HOMER v4.11.1 [107] with the following additional options: -nomotif -known -cpg -size 250. The DMRs were analyzed against 440 known motifs identified in vertebrate genomes and available for hg38 annotation provided by HOMER.

4.9. Assessing the CLL Subtype

For this part of analysis, we remapped samples to the hg19 reference genome (Genome Reference Consortium—GRCh37) and ran the QSEA R package v.1.14.0 [33], as before. We used mean beta normalized values of the 250 bp bins that overlapped at least 145 bp with the loci defined to classify CLL samples into methylation programming subtypes from Oakes et al. [25]. The merged table with beta values of 27 samples from this study and 329 from the CLL Research Consortium (CRC) in Oakes et al., available in Supplementary Table S10, was used to cluster the samples and draw a heat map with the same software (Qlucore, Lund, Sweden, trial version) and settings used by the authors [25].

Supplementary Materials: The following are available online at https://www.mdpi.com/article/10.3390/ijms22179337/s1.

Author Contributions: Acquisition of Data A.K., C.G., O.M., B.T. and J.A.; Analysis of Data A.P., M.L., C.G., R.H. and M.R.S.; Collection of Specimens and Clinical Data C.D.H., H.C.R. and M.H.; Bioinformatics Analyses A.P., M.L. and R.H., Drafting of the Manuscript A.P., C.G., M.L. and M.R.S.; Reviewing of the Manuscript A.P., C.G., C.D.H. M.L., A.K., B.T., J.A., O.M., C.P., R.H., H.C.R., M.H. and M.R.S.; Supervision H.C.R., C.P., M.H. and M.R.S.; Study Design and Coordination A.P., C.G. and M.R.S. All authors have read and agreed to the published version of the manuscript.

Funding: This work was funded through the German Research Foundation (KFO-286-RP2 and RE 2246/13-1 to H.C.R.; KFO-286-RP8 to M.R.S, KFO-286-RP6 to M.H., KFO-286-CP to C.D.H., SFB1399 to M.R.S.), the GO-LONG project (SCHW 1605/4-1 to M.R.S. and HE 4607/7-1 to R.H), the Centre for Molecular Medicine CMMC (A12 to M.R.S.), the German-Israeli Foundation for Research and Development (I-65-412.20-2016 to H.C.R.), the Deutsche Jose Carreras Leukämie Stiftung (R12/08 to H.C.R.), the Deutsche Krebshilfe (1117240 and 70113041 to H.C.R.), the German Ministry of Education and Research (BMBF e:Med 01ZX1303A to H.C.R.), the Deutsche Krebshilfe (70113869 to C.P.).

Institutional Review Board Statement: The study was approved by the Ethics Committee of the University of Cologne (Ethikvotum 11-319 from 11 December 2011, with an amendment from 7 June 2016). For investigations involving human subjects, informed consent has been obtained from the participants involved. Patient samples were obtained from the Biobank of the University Clinic Cologne, Germany.

Informed Consent Statement: Informed consent was obtained from all subjects involved in the study.

Data Availability Statement: The authors will provide the data upon reasonable request.

Acknowledgments: We are grateful to the patients and their families for their contribution in the study. We would like to also acknowledge the Cologne High Efficient Operating Platform for Science for their computational resources provided by the Regional Computing Centre of Cologne (RRZK) together with the HPC expert Bull. We thank Elena Wasserburger for technical support.

Conflicts of Interest: H.C.R. received consulting and lecture fees from Abbvie, Astra-Zeneca, Vertex and Merck. H.C.R. received research funding from Gilead Pharmaceuticals. H.C.R. is a co-founder of CDL Therapeutics GmbH. M.H. received research funding, consulting and lecture fees from Roche, Abbvie, Gilead, Janssen, Celgene and Astra Zeneca. The remaining authors declare no competing financial interest.

References

1. Damle, R.N.; Wasil, T.; Fais, F.; Ghiotto, F.; Valetto, A.; Allen, S.L.; Buchbinder, A.; Budman, D.; Dittmar, K.; Kolitz, J.; et al. Ig V gene mutation status and CD38 expression as novel prognostic indicators in chronic lymphocytic leukemia. *Blood* **1999**, *94*, 1840–1847. [CrossRef] [PubMed]
2. Hamblin, T.J.; Davis, Z.; Gardiner, A.; Oscier, D.G.; Stevenson, F.K. Unmutated Ig V(H) genes are associated with a more aggressive form of chronic lymphocytic leukemia. *Blood* **1999**, *94*, 1848–1854. [CrossRef]
3. Fabbri, G.; Dalla-Favera, R. The molecular pathogenesis of chronic lymphocytic leukaemia. *Nat. Rev. Cancer* **2016**, *16*, 145–162. [CrossRef]
4. Guièze, R.; Robbe, P.; Clifford, R.; de Guibert, S.; Pereira, B.; Timbs, A.; Dilhuydy, M.S.; Cabes, M.; Ysebaert, L.; Burns, A.; et al. Presence of multiple recurrent mutations confers poor trial outcome of relapsed/refractory CLL. *Blood* **2015**, *126*, 2110–2117. [CrossRef] [PubMed]
5. Landau, D.A.; Tausch, E.; Taylor-Weiner, A.N.; Stewart, C.; Reiter, J.G.; Bahlo, J.; Kluth, S.; Bozic, I.; Lawrence, M.; Böttcher, S.; et al. Mutations driving CLL and their evolution in progression and relapse. *Nature* **2015**, *526*, 525–530. [CrossRef]
6. Ljungström, V.; Cortese, D.; Young, E.; Pandzic, T.; Mansouri, L.; Plevova, K.; Ntoufa, S.; Baliakas, P.; Clifford, R.; Sutton, L.A.; et al. Whole-exome sequencing in relapsing chronic lymphocytic leukemia: Clinical impact of recurrent RPS15 mutations. *Blood* **2016**, *127*, 1007–1016. [CrossRef]

7. Rossi, D.; Bruscaggin, A.; Spina, V.; Rasi, S.; Khiabanian, H.; Messina, M.; Fangazio, M.; Vaisitti, T.; Monti, S.; Chiaretti, S.; et al. Mutations of the SF3B1 splicing factor in chronic lymphocytic leukemia: Association with progression and fludarabine-refractoriness. *Blood* **2011**, *118*, 6904–6908. [CrossRef] [PubMed]
8. Stilgenbauer, S.; Schnaiter, A.; Paschka, P.; Zenz, T.; Rossi, M.; Döhner, K.; Bühler, A.; Böttcher, S.; Ritgen, M.; Kneba, M.; et al. Gene Mutations and Treatment Outcome in Chronic Lymphocytic Leukemia. *Blood* **2014**, *123*, 3247–3254. [CrossRef] [PubMed]
9. Quesada, V.; Conde, L.; Villamor, N.; Ordóñez, G.R.; Jares, P.; Bassaganyas, L.; Ramsay, A.J.; Beà, S.; Pinyol, M.; Martínez-Trillos, A.; et al. Exome sequencing identifies recurrent mutations of the splicing factor SF3B1 gene in chronic lymphocytic leukemia. *Nat. Genet.* **2012**, *44*, 47–52. [CrossRef] [PubMed]
10. Wang, L.; Lawrence, M.S.; Wan, Y.; Stojanov, P.; Sougnez, C.; Stevenson, K.; Werner, L.; Sivachenko, A.; DeLuca, D.S.; Zhang, L.; et al. SF3B1 and other novel cancer genes in chronic lymphocytic leukemia. *N. Engl. J. Med.* **2011**, *365*, 2497–2506. [CrossRef] [PubMed]
11. Kfir, N.; Lev-Maor, G.; Glaich, O.; Alajem, A.; Datta, A.; Sze, S.K.; Meshorer, E.; Ast, G. SF3B1 Association with Chromatin Determines Splicing Outcomes. *Cell Rep.* **2015**, *11*, 618–629. [CrossRef] [PubMed]
12. Tang, A.D.; Soulette, C.M.; van Baren, M.J.; Hart, K.; Hrabeta-Robinson, E.; Wu, C.J.; Brooks, A.N. Full-length transcript characterization of SF3B1 mutation in chronic lymphocytic leukemia reveals downregulation of retained introns. *Nat. Commun.* **2020**, *11*, 1438. [CrossRef]
13. Ntziachristos, P.; Abdel-Wahab, O.; Aifantis, I. Emerging concepts of epigenetic dysregulation in hematological malignancies. *Nat. Immunol.* **2016**, *17*, 1016–1024. [CrossRef] [PubMed]
14. Cavellán, E.; Asp, P.; Percipalle, P.; Farrants, A.K.Ö. The WSTF-SNF2h chromatin remodeling complex interacts with several nuclear proteins in transcription. *J. Biol. Chem.* **2006**, *281*, 16264–16271. [CrossRef]
15. Isono, K.; Mizutani-Koseki, Y.; Komori, T.; Schmidt-Zachmann, M.S.; Koseki, H. Mammalian Polycomb-mediated repression of Hox genes requires the essential spliceosomal protein Sf3b1. *Genes Dev.* **2005**, *19*, 536–541. [CrossRef]
16. Bartholdy, B.A.; Wang, X.; Yan, X.J.; Pascual, M.; Fan, M.; Barrientos, J.; Allen, S.L.; Martinez-Climent, J.A.; Rai, K.R.; Chiorazzi, N.; et al. CLL intraclonal fractions exhibit established and recently acquired patterns of DNA methylation. *Blood Adv.* **2020**, *4*, 893–905. [CrossRef] [PubMed]
17. Cahill, N.; Rosenquist, R. Uncovering the DNA methylome in chronic lymphocytic leukemia. *Epigenetics* **2013**, *8*, 138–148. [CrossRef]
18. Oakes, C.C.; Claus, R.; Gu, L.; Assenov, Y.; Hüllein, J.; Zucknick, M.; Bieg, M.; Brocks, D.; Bogatyrova, O.; Schmidt, C.R.; et al. Evolution of DNA methylation is linked to genetic aberrations in chronic lymphocytic leukemia. *Cancer Discov.* **2014**, *4*, 348–361. [CrossRef]
19. Pastore, A.; Gaiti, F.; Lu, S.X.; Brand, R.M.; Kulm, S.; Chaligne, R.; Gu, H.; Huang, K.Y.; Stamenova, E.K.; Béguelin, W.; et al. Corrupted coordination of epigenetic modifications leads to diverging chromatin states and transcriptional heterogeneity in CLL. *Nat. Commun.* **2019**, *10*, 1874. [CrossRef]
20. Wernig-Zorc, S.; Yadav, M.P.; Kopparapu, P.K.; Bemark, M.; Kristjansdottir, H.L.; Andersson, P.-O.; Kanduri, C.; Kanduri, M. Global distribution of DNA hydroxymethylation and DNA methylation in chronic lymphocytic leukemia. *Epigenetics Chromatin* **2019**, *12*, 4. [CrossRef] [PubMed]
21. Subhash, S.; Andersson, P.-O.; Kosalai, S.T.; Kanduri, C.; Kanduri, M. Global DNA methylation profiling reveals new insights into epigenetically deregulated protein coding and long noncoding RNAs in CLL. *Clin. Epigenetics* **2016**, *8*, 106. [CrossRef]
22. Cahill, N.; Bergh, A.C.; Kanduri, M.; Göransson-Kultima, H.; Mansouri, L.; Isaksson, A.; Ryan, F.; Smedby, K.E.; Juliusson, G.; Sundström, C.; et al. 450K-array analysis of chronic lymphocytic leukemia cells reveals global DNA methylation to be relatively stable over time and similar in resting and proliferative compartments. *Leukemia* **2013**, *27*, 150–158. [CrossRef]
23. Kulis, M.; Heath, S.; Bibikova, M.; Queirós, A.C.; Navarro, A.; Clot, G.; Martínez-Trillos, A.; Castellano, G.; Brun-Heath, I.; Pinyol, M.; et al. Epigenomic analysis detects widespread gene-body DNA hypomethylation in chronic lymphocytic leukemia. *Nat. Genet.* **2012**, *44*, 1236–1242. [CrossRef]
24. Fabris, S.; Bollati, V.; Agnelli, L.; Morabito, F.; Motta, V.; Cutrona, G.; Matis, S.; Recchia, A.G.; Gigliotti, V.; Gentile, M.; et al. Biological and clinical relevance of quantitative global methylation of repetitive DNA sequences in chronic lymphocytic leukemia. *Epigenetics* **2011**, *6*, 188–194. [CrossRef]
25. Oakes, C.C.; Seifert, M.; Assenov, Y.; Gu, L.; Przekopowitz, M.; Ruppert, A.S.; Wang, Q.; Imbusch, C.D.; Serva, A.; Koser, S.D.; et al. DNA methylation dynamics during B cell maturation underlie a continuum of disease phenotypes in chronic lymphocytic leukemia. *Nat. Genet.* **2016**, *48*, 253–264. [CrossRef]
26. Landau, D.A.; Clement, K.; Ziller, M.J.; Boyle, P.; Fan, J.; Gu, H.; Stevenson, K.; Sougnez, C.; Wang, L.; Li, S.; et al. Locally Disordered Methylation Forms the Basis of Intratumor Methylome Variation in Chronic Lymphocytic Leukemia. *Cancer Cell* **2014**, *26*, 813–825. [CrossRef]
27. Gaiti, F.; Chaligne, R.; Gu, H.; Brand, R.M.; Kothen-Hill, S.; Schulman, R.C.; Grigorev, K.; Risso, D.; Kim, K.T.; Pastore, A.; et al. Epigenetic evolution and lineage histories of chronic lymphocytic leukaemia. *Nature* **2019**, *569*, 576–580. [CrossRef] [PubMed]
28. Jones, P.A. Functions of DNA methylation: Islands, start sites, gene bodies and beyond. *Nat. Rev. Genet.* **2012**, *13*, 484–492. [CrossRef] [PubMed]
29. Hsu, D.W.; Lin, M.J.; Lee, T.L.; Wen, S.C.; Chen, X.; Shen, C.K. Two major forms of DNA (cytosine-5) methyltransferase in human somatic tissues. *Proc. Natl. Acad. Sci. USA* **1999**, *96*, 9751–9756. [CrossRef] [PubMed]

30. Franchina, M.; Hooper, J.; Kay, P.H. Five novel alternatively spliced transcripts of DNA (cytosine-5) methyltransferase 2 in human peripheral blood leukocytes. *Int. J. Biochem. Cell Biol.* **2001**, *33*, 1104–1115. [CrossRef]
31. Weisenberger, D.J.; Velicescu, M.; Preciado-Lopez, M.A.; Gonzales, F.A.; Tsai, Y.C.; Liang, G.; Jones, P.A. Identification and characterization of alternatively spliced variants of DNA methyltransferase 3a in mammalian cells. *Gene* **2002**, *298*, 91–99. [CrossRef]
32. Zhang, J.; Zhang, Y.-Z.; Jiang, J.; Duan, C.-G. The Crosstalk Between Epigenetic Mechanisms and Alternative RNA Processing Regulation. *Front. Genet.* **2020**, *11*, 998. [CrossRef]
33. Lienhard, M.; Grasse, S.; Rolff, J.; Frese, S.; Schirmer, U.; Becker, M.; Börno, S.; Timmermann, B.; Chavez, L.; Sültmann, H.; et al. QSEA-modelling of genome-wide DNA methylation from sequencing enrichment experiments. *Nucleic Acids Res.* **2017**, *45*, e44. [CrossRef] [PubMed]
34. Dietrich, S.; Oleś, M.; Lu, J.; Sellner, L.; Anders, S.; Velten, B.; Wu, B.; Hüllein, J.; da Silva Liberio, M.; Walther, T.; et al. Drug-perturbation-based stratification of blood cancer. *J. Clin. Investig.* **2018**, *128*, 427–445. [CrossRef]
35. Ritchie, M.E.; Phipson, B.; Wu, D.; Hu, Y.; Law, C.W.; Shi, W.; Smyth, G.K. Limma powers differential expression analyses for RNA-sequencing and microarray studies. *Nucleic Acids Res.* **2015**, *43*, e47. [CrossRef]
36. Wierzbinska, J.A.; Toth, R.; Ishaque, N.; Rippe, K.; Mallm, J.P.; Klett, L.C.; Mertens, D.; Zenz, T.; Hielscher, T.; Seifert, M.; et al. Methylome-based cell-of-origin modeling (Methyl-COOM) identifies aberrant expression of immune regulatory molecules in CLL. *Genome Med.* **2020**, *12*, 29. [CrossRef] [PubMed]
37. Grimwood, J.; Gordon, L.A.; Olsen, A.; Terry, A.; Schmutz, J.; Lamerdin, J.; Hellsten, U.; Goodstein, D.; Couronne, O.; Tran-Gyamil, M.; et al. The DNA sequence and biology of human chromosome 19. *Nature* **2004**, *428*, 529–535. [CrossRef]
38. Humphray, S.J.; Oliver, K.; Hunt, A.R.; Plumb, R.W.; Loveland, J.E.; Howe, K.L.; Andrews, T.D.; Searle, S.; Hunt, S.E.; Scott, C.E.; et al. DNA sequence and analysis of human chromosome 9. *Nature* **2004**, *429*, 369–374. [CrossRef]
39. ENCODE Uniform TFBS Track. Available online: http://hgdownload.cse.ucsc.edu/goldenpath/hg19/encodeDCC/wgEncodeAwgTfbsUniform/ (accessed on 9 December 2020).
40. Gerstein, M.B.; Kundaje, A.; Hariharan, M.; Landt, S.G.; Yan, K.K.; Cheng, C.; Mu, X.J.; Khurana, E.; Rozowsky, J.; Alexander, R.; et al. Architecture of the human regulatory network derived from ENCODE data. *Nature* **2012**, *489*, 91–100. [CrossRef]
41. Wang, J.; Zhuang, J.; Iyer, S.; Lin, X.Y.; Whitfield, T.W.; Greven, M.C.; Pierce, B.G.; Dong, X.; Kundaje, A.; Cheng, Y.; et al. Sequence features and chromatin structure around the genomic regions bound by 119 human transcription factors. *Genome Res.* **2012**, *22*, 1798–1812. [CrossRef]
42. Wang, J.; Zhuang, J.; Iyer, S.; Lin, X.Y.; Greven, M.C.; Kim, B.H.; Moore, J.; Pierce, B.G.; Dong, X.; Virgil, D.; et al. Factorbook.org: A Wiki-based database for transcription factor-binding data generated by the ENCODE consortium. *Nucleic Acids Res.* **2013**, *41*, D171–D176. [CrossRef] [PubMed]
43. Wang, J.H.; Nichogiannopoulou, A.; Wu, L.; Sun, L.; Sharpe, A.H.; Bigby, M.; Georgopoulos, K. Selective defects in the development of the fetal and adult lymphoid system in mice with an Ikaros null mutation. *Immunity* **1996**, *5*, 537–549. [CrossRef]
44. Cook, M.E.; Jarjour, N.N.; Lin, C.-C.; Edelson, B.T. Transcription Factor Bhlhe40 in Immunity and Autoimmunity. *Trends Immunol.* **2020**, *41*, 1023–1036. [CrossRef] [PubMed]
45. Glozak, M.A.; Seto, E. Histone deacetylases and cancer. *Oncogene* **2007**, *26*, 5420–5432. [CrossRef]
46. Reimand, J.; Kull, M.; Peterson, H.; Hansen, J.; Vilo, J. g:Profiler-a web-based toolset for functional profiling of gene lists from large-scale experiments. *Nucleic Acids Res.* **2007**, *35*, W193–W200. [CrossRef]
47. Irizarry, R.A.; Wu, H.; Feinberg, A.P. A species-generalized probabilistic model-based definition of CpG islands. *Mamm. Genome* **2009**, *20*, 674–680. [CrossRef]
48. Visel, A.; Minovitsky, S.; Dubchak, I.; Pennacchio, L.A. VISTA Enhancer Browser—A database of tissue-specific human enhancers. *Nucleic Acids Res.* **2007**, *35*, D88–D92. [CrossRef]
49. Ernst, J.; Kellis, M. Discovery and characterization of chromatin states for systematic annotation of the human genome. *Nat. Biotechnol.* **2010**, *28*, 817–825. [CrossRef] [PubMed]
50. Ernst, J.; Kheradpour, P.; Mikkelsen, T.S.; Shoresh, N.; Ward, L.D.; Epstein, C.B.; Zhang, X.; Wang, L.; Issner, R.; Coyne, M.; et al. Mapping and analysis of chromatin state dynamics in nine human cell types. *Nature* **2011**, *473*, 43–49. [CrossRef]
51. Wojdacz, T.K.; Amarasinghe, H.E.; Kadalayil, L.; Beattie, A.; Forster, J.; Blakemore, S.J.; Parker, H.; Bryant, D.; Larrayoz, M.; Clifford, R.; et al. Clinical significance of DNA methylation in chronic lymphocytic leukemia patients: Results from 3 UK clinical trials. *Blood Adv.* **2019**, *3*, 2474–2481. [CrossRef]
52. Bhoi, S.; Ljungström, V.; Baliakas, P.; Mattsson, M.; Smedby, K.E.; Juliusson, G.; Rosenquist, R.; Mansouri, L. Prognostic impact of epigenetic classification in chronic lymphocytic leukemia: The case of subset #2. *Epigenetics* **2016**, *11*, 449–455. [CrossRef] [PubMed]
53. Queirós, A.C.; Villamor, N.; Clot, G.; Martinez-Trillos, A.; Kulis, M.; Navarro, A.; Penas, E.M.M.; Jayne, S.; Majid, A.; Richter, J.; et al. A B-cell epigenetic signature defines three biologic subgroups of chronic lymphocytic leukemia with clinical impact. *Leukemia* **2015**, *29*, 598–605. [CrossRef]
54. Shayevitch, R.; Askayo, D.; Keydar, I.; Ast, G. The importance of DNA methylation of exons on alternative splicing. *RNA* **2018**, *24*, 1351–1362. [CrossRef] [PubMed]
55. Lev Maor, G.; Yearim, A.; Ast, G. The alternative role of DNA methylation in splicing regulation. *Trends Genet.* **2015**, *31*, 274–280. [CrossRef] [PubMed]

56. Sun, X.; Tian, Y.; Wang, J.; Sun, Z.; Zhu, Y. Genome-wide analysis reveals the association between alternative splicing and DNA methylation across human solid tumors. *BMC Med. Genomics* **2020**, *13*, 4. [CrossRef] [PubMed]
57. Maunakea, A.K.; Chepelev, I.; Cui, K.; Zhao, K. Intragenic DNA methylation modulates alternative splicing by recruiting MeCP2 to promote exon recognition. *Cell Res.* **2013**, *23*, 1256–1269. [CrossRef]
58. Yoshimi, A.; Lin, K.-T.; Wiseman, D.H.; Rahman, M.A.; Pastore, A.; Wang, B.; Lee, S.C.-W.; Micol, J.-B.; Zhang, X.J.; de Botton, S.; et al. Coordinated alterations in RNA splicing and epigenetic regulation drive leukaemogenesis. *Nature* **2019**, *574*, 273–277. [CrossRef]
59. Wang, L.; Brooks, A.N.; Fan, J.; Wan, Y.; Gambe, R.; Li, S.; Hergert, S.; Yin, S.; Freeman, S.S.; Levin, J.Z.; et al. Transcriptomic Characterization of SF3B1 Mutation Reveals Its Pleiotropic Effects in Chronic Lymphocytic Leukemia. *Cancer Cell* **2016**, *30*, 750–763. [CrossRef]
60. Tachibana, M.; Ueda, J.; Fukuda, M.; Takeda, N.; Ohta, T.; Iwanari, H.; Sakihama, T.; Kodama, T.; Hamakubo, T.; Shinkai, Y. Histone methyltransferases G9a and GLP form heteromeric complexes and are both crucial for methylation of euchromatin at H3-K9. *Genes Dev.* **2005**, *19*, 815–826. [CrossRef]
61. Yang, Q.; Zhu, Q.; Lu, X.; Du, Y.; Cao, L.; Shen, C.; Hou, T.; Li, M.; Li, Z.; Liu, C.; et al. G9a coordinates with the RPA complex to promote DNA damage repair and cell survival. *Proc. Natl. Acad. Sci. USA* **2017**, *114*, E6054–E6063. [CrossRef]
62. Ferry, L.; Fournier, A.; Tsusaka, T.; Adelmant, G.; Shimazu, T.; Matano, S.; Kirsh, O.; Amouroux, R.; Dohmae, N.; Suzuki, T.; et al. Methylation of DNA Ligase 1 by G9a/GLP Recruits UHRF1 to Replicating DNA and Regulates DNA Methylation. *Mol. Cell* **2017**, *67*, 550–565.e5. [CrossRef]
63. Goodman, S.J.; Cytrynbaum, C.; Chung, B.H.-Y.; Chater-Diehl, E.; Aziz, C.; Turinsky, A.L.; Kellam, B.; Keller, M.; Ko, J.M.; Caluseriu, O.; et al. *EHMT1* pathogenic variants and 9q34.3 microdeletions share altered DNA methylation patterns in patients with Kleefstra syndrome. *J. Transl. Genet. Genomics* **2020**, *4*, 144–158. [CrossRef]
64. Mansouri, L.; Grabowski, P.; Degerman, S.; Svenson, U.; Gunnarsson, R.; Cahill, N.; Smedby, K.E.; Geisler, C.; Juliusson, G.; Roos, G.; et al. Short telomere length is associated with NOTCH1/SF3B1/TP53 aberrations and poor outcome in newly diagnosed chronic lymphocytic leukemia patients. *Am. J. Hematol.* **2013**, *88*, 647–651. [CrossRef]
65. Jebaraj, B.M.C.; Stilgenbauer, S. Telomere Dysfunction in Chronic Lymphocytic Leukemia. *Front. Oncol.* **2021**, *10*, 612665. [CrossRef]
66. Tardivon, D.; Antoszewski, M.; Zangger, N.; Nkosi, M.; Sordet-Dessimoz, J.; Hendriks, R.W.; Koch, U.; Radtke, F. Notch Signaling Promotes Disease Initiation and Progression in Murine Chronic Lymphocytic Leukemia. *Blood* **2020**, *137*, 3079–3092. [CrossRef]
67. Fabbri, G.; Holmes, A.B.; Viganotti, M.; Scuoppo, C.; Belver, L.; Herranz, D.; Yan, X.-J.; Kieso, Y.; Rossi, D.; Gaidano, G.; et al. Common nonmutational NOTCH1 activation in chronic lymphocytic leukemia. *Proc. Natl. Acad. Sci. USA* **2017**, *114*, E2911–E2919. [CrossRef] [PubMed]
68. Del Papa, B.; Baldoni, S.; Dorillo, E.; de Falco, F.; Rompietti, C.; Cecchini, D.; Cantelmi, M.G.; Sorcini, D.; Nogarotto, M.; Adamo, F.M.; et al. Decreased NOTCH1 Activation Correlates with Response to Ibrutinib in Chronic Lymphocytic Leukemia. *Clin. Cancer Res.* **2019**, *25*, 7540–7553. [CrossRef] [PubMed]
69. Pozzo, F.; Bittolo, T.; Tissino, E.; Vit, F.; Vendramini, E.; Laurenti, L.; D'Arena, G.; Olivieri, J.; Pozzato, G.; Zaja, F.; et al. *SF3B1*-mutated chronic lymphocytic leukemia shows evidence of NOTCH1 pathway activation including CD20 downregulation. *Haematologica* **2020**, Online ahead of print. [CrossRef]
70. De Oliveira, V.C.; de Lacerda, M.P.; Moraes, B.B.M.; Gomes, C.P.; Maricato, J.T.; Souza, O.F.; Schenkman, S.; Pesquero, J.B.; Moretti, N.S.; Rodrigues, C.A.; et al. Deregulation of Ikaros expression in B-1 cells: New insights in the malignant transformation to chronic lymphocytic leukemia. *J. Leukoc. Biol.* **2019**, *106*, 581–594. [CrossRef] [PubMed]
71. Ferreirós-Vidal, I.; Carroll, T.; Taylor, B.; Terry, A.; Liang, Z.; Bruno, L.; Dharmalingam, G.; Khadayate, S.; Cobb, B.S.; Smale, S.T.; et al. Genome-wide identification of Ikaros targets elucidates its contribution to mouse B-cell lineage specification and pre-B-cell differentiation. *Blood* **2013**, *121*, 1769–1782. [CrossRef]
72. Lu, G.; Middleton, R.E.; Sun, H.; Naniong, M.; Ott, C.J.; Mitsiades, C.S.; Wong, K.-K.; Bradner, J.E.; Kaelin, W.G.J. The myeloma drug lenalidomide promotes the cereblon-dependent destruction of Ikaros proteins. *Science* **2014**, *343*, 305–309. [CrossRef]
73. Krönke, J.; Udeshi, N.D.; Narla, A.; Grauman, P.; Hurst, S.N.; McConkey, M.; Svinkina, T.; Heckl, D.; Comer, E.; Li, X.; et al. Lenalidomide causes selective degradation of IKZF1 and IKZF3 in multiple myeloma cells. *Science* **2014**, *343*, 301–305. [CrossRef]
74. Fecteau, J.-F.; Corral, L.G.; Ghia, E.M.; Gaidarova, S.; Futalan, D.; Bharati, I.S.; Cathers, B.; Schwaederlé, M.; Cui, B.; Lopez-Girona, A.; et al. Lenalidomide inhibits the proliferation of CLL cells via a cereblon/p21(WAF1/Cip1)-dependent mechanism independent of functional p53. *Blood* **2014**, *124*, 1637–1644. [CrossRef] [PubMed]
75. Fürstenau, M.; Fink, A.M.; Schilhabel, A.; Weiss, J.; Robrecht, S.; Eckert, R.; de la Serna, J.; Crespo, M.; Coscia, M.; Vitale, C.; et al. B-cell acute lymphoblastic leukemia in patients with chronic lymphocytic leukemia treated with lenalidomide. *Blood* **2021**, *137*, 2267–2271. [CrossRef]
76. Yoshida, T.; Ng, S.Y.-M.; Zuniga-Pflucker, J.C.; Georgopoulos, K. Early hematopoietic lineage restrictions directed by Ikaros. *Nat. Immunol.* **2006**, *7*, 382–391. [CrossRef] [PubMed]
77. Kirstetter, P.; Thomas, M.; Dierich, A.; Kastner, P.; Chan, S. Ikaros is critical for B cell differentiation and function. *Eur. J. Immunol.* **2002**, *32*, 720–730. [CrossRef]
78. Schwickert, T.A.; Tagoh, H.; Gültekin, S.; Dakic, A.; Axelsson, E.; Minnich, M.; Ebert, A.; Werner, B.; Roth, M.; Cimmino, L.; et al. Stage-specific control of early B cell development by the transcription factor Ikaros. *Nat. Immunol.* **2014**, *15*, 283–293. [CrossRef]

79. Fiorcari, S.; Benatti, S.; Zucchetto, A.; Zucchini, P.; Gattei, V.; Luppi, M.; Marasca, R.; Maffei, R. Overexpression of CD49d in trisomy 12 chronic lymphocytic leukemia patients is mediated by IRF4 through induction of IKAROS. *Leukemia* **2019**, *33*, 1278–1302. [CrossRef] [PubMed]
80. Maffei, R.; Fiorcari, S.; Benatti, S.; Atene, C.G.; Martinelli, S.; Zucchini, P.; Potenza, L.; Luppi, M.; Marasca, R. IRF4 modulates the response to BCR activation in chronic lymphocytic leukemia regulating IKAROS and SYK. *Leukemia* **2021**, *35*, 1330–1343. [CrossRef] [PubMed]
81. Ding, Y.; Zhang, B.; Payne, J.L.; Song, C.; Ge, Z.; Gowda, C.; Iyer, S.; Dhanyamraju, P.K.; Dorsam, G.; Reeves, M.E.; et al. Ikaros tumor suppressor function includes induction of active enhancers and super-enhancers along with pioneering activity. *Leukemia* **2019**, *33*, 2720–2731. [CrossRef]
82. Heizmann, B.; Kastner, P.; Chan, S. The Ikaros family in lymphocyte development. *Curr. Opin. Immunol.* **2018**, *51*, 14–23. [CrossRef]
83. Oravecz, A.; Apostolov, A.; Polak, K.; Jost, B.; Le Gras, S.; Chan, S.; Kastner, P. Ikaros mediates gene silencing in T cells through Polycomb repressive complex 2. *Nat. Commun.* **2015**, *6*, 8823. [CrossRef]
84. Hu, Y.; Zhang, Z.; Kashiwagi, M.; Yoshida, T.; Joshi, I.; Jena, N.; Somasundaram, R.; Emmanuel, A.O.; Sigvardsson, M.; Fitamant, J.; et al. Superenhancer reprogramming drives a B-cell-epithelial transition and high-risk leukemia. *Genes Dev.* **2016**, *30*, 1971–1990. [CrossRef] [PubMed]
85. Jiang, Z.; Li, W.; Hu, X.; Zhang, Q.; Sun, T.; Cui, S.; Wang, S.; Ouyang, Q.; Yin, Y.; Geng, C.; et al. Tucidinostat plus exemestane for postmenopausal patients with advanced, hormone receptor-positive breast cancer (ACE): A randomised, double-blind, placebo-controlled, phase 3 trial. *Lancet. Oncol.* **2019**, *20*, 806–815. [CrossRef]
86. Molenaar, M.; van de Wetering, M.; Oosterwegel, M.; Peterson-Maduro, J.; Godsave, S.; Korinek, V.; Roose, J.; Destrée, O.; Clevers, H. XTcf-3 Transcription Factor Mediates β-Catenin-Induced Axis Formation in Xenopus Embryos. *Cell* **1996**, *86*, 391–399. [CrossRef]
87. Miller, J.R.; Hocking, A.M.; Brown, J.D.; Moon, R.T. Mechanism and function of signal transduction by the Wnt/β-catenin and Wnt/Ca2+ pathways. *Oncogene* **1999**, *18*, 7860–7872. [CrossRef] [PubMed]
88. Polakis, P. Wnt signaling and cancer. *Genes Dev.* **2000**, *14*, 1837–1851. [CrossRef]
89. Mani, M.; Carrasco, D.E.; Zhang, Y.; Takada, K.; Gatt, M.E.; Dutta-Simmons, J.; Ikeda, H.; Diaz-Griffero, F.; Pena-Cruz, V.; Bertagnolli, M.; et al. BCL9 promotes tumor progression by conferring enhanced proliferative, metastatic, and angiogenic properties to cancer cells. *Cancer Res.* **2009**, *69*, 7577–7586. [CrossRef]
90. Wang, X.; Feng, M.; Xiao, T.; Guo, B.; Liu, D.; Liu, C.; Pei, J.; Liu, Q.; Xiao, Y.; Rosin-Arbesfeld, R.; et al. BCL9/BCL9L promotes tumorigenicity through immune-dependent and independent mechanisms in triple negative breast cancer. *Oncogene* **2021**, *40*, 2982–2997. [CrossRef] [PubMed]
91. Bibikova, M.; Barnes, B.; Tsan, C.; Ho, V.; Klotzle, B.; Le, J.M.; Delano, D.; Zhang, L.; Schroth, G.P.; Gunderson, K.L.; et al. High density DNA methylation array with single CpG site resolution. *Genomics* **2011**, *98*, 288–295. [CrossRef]
92. Keshet, I.; Schlesinger, Y.; Farkash, S.; Rand, E.; Hecht, M.; Segal, E.; Pikarski, E.; Young, R.A.; Niveleau, A.; Cedar, H.; et al. Evidence for an instructive mechanism of de novo methylation in cancer cells. *Nat. Genet.* **2006**, *38*, 149–153. [CrossRef] [PubMed]
93. Binet, J.L.; Auquier, A.; Dighiero, G.; Chastang, C.; Piguet, H.; Goasguen, J.; Vaugier, G.; Potron, G.; Colona, P.; Oberling, F.; et al. A new prognostic classification of chronic lymphocytic leukemia derived from a multivariate survival analysis. *Cancer* **1981**, *48*, 198–206. [CrossRef]
94. Vollbrecht, C.; Mairinger, F.D.; Koitzsch, U.; Peifer, M.; Koenig, K.; Heukamp, L.C.; Crispatzu, G.; Wilden, L.; Kreuzer, K.A.; Hallek, M.; et al. Comprehensive analysis of disease-related genes in chronic lymphocytic leukemia by multiplex PCR-based next generation sequencing. *PLoS ONE* **2015**, *10*, e0129544. [CrossRef]
95. Rosenquist, R.; Ghia, P.; Hadzidimitriou, A.; Sutton, L.A.; Agathangelidis, A.; Baliakas, P.; Darzentas, N.; Giudicelli, V.; Lefranc, M.P.; Langerak, A.W.; et al. Immunoglobulin gene sequence analysis in chronic lymphocytic leukemia: Updated ERIC recommendations. *Leukemia* **2017**, *31*, 1477–1481. [CrossRef]
96. Weber, M.; Davies, J.J.; Wittig, D.; Oakeley, E.J.; Haase, M.; Lam, W.L.; Schübeler, D. Chromosome-wide and promoter-specific analyses identify sites of differential DNA methylation in normal and transformed human cells. *Nat. Genet.* **2005**, *37*, 853–862. [CrossRef] [PubMed]
97. Li, H.; Durbin, R. Fast and accurate short read alignment with Burrows-Wheeler transform. *Bioinformatics* **2009**, *25*, 1754–1760. [CrossRef] [PubMed]
98. Benjamini, Y.; Hochberg, Y. Controlling the False Discovery Rate: A Practical and Powerful Approach to Multiple Testing. *J. R. Stat. Soc. Ser. B* **1995**, *57*, 289–300. [CrossRef]
99. Hinrichs, A.S.; Karolchik, D.; Baertsch, R.; Barber, G.P.; Bejerano, G.; Clawson, H.; Diekhans, M.; Furey, T.S.; Harte, R.A.; Hsu, F.; et al. The UCSC Genome Browser Database: Update 2006. *Nucleic Acids Res.* **2006**, *34*, D590–D598. [CrossRef]
100. Dobin, A.; Davis, C.A.; Schlesinger, F.; Drenkow, J.; Zaleski, C.; Jha, S.; Batut, P.; Chaisson, M.; Gingeras, T.R. STAR: Ultrafast universal RNA-seq aligner. *Bioinformatics* **2013**, *29*, 15–21. [CrossRef] [PubMed]
101. Frankish, A.; Diekhans, M.; Jungreis, I.; Lagarde, J.; Loveland, J.E.; Mudge, J.M.; Sisu, C.; Wright, J.C.; Armstrong, J.; Barnes, I.; et al. GENCODE 2021. *Nucleic Acids Res.* **2021**, *49*, D916–D923. [CrossRef] [PubMed]
102. Robinson, M.D.; McCarthy, D.J.; Smyth, G.K. edgeR: A Bioconductor package for differential expression analysis of digital gene expression data. *Bioinformatics* **2010**, *26*, 139–140. [CrossRef] [PubMed]

103. Radpour, R.; Haghighi, M.M.; Fan, A.X.-C.; Torbati, P.M.; Hahn, S.; Holzgreve, W.; Zhong, X.Y. High-throughput hacking of the methylation patterns in breast cancer by in vitro transcription and thymidine-specific cleavage mass array on MALDI-TOF silico-chip. *Mol. Cancer Res.* **2008**, *6*, 1702–1709. [CrossRef]
104. Ehrich, M.; Nelson, M.R.; Stanssens, P.; Zabeau, M.; Liloglou, T.; Xinarianos, G.; Cantor, C.R.; Field, J.K.; van den Boom, D. Quantitative high-throughput analysis of DNA methylation patterns by base-specific cleavage and mass spectrometry. *Proc. Natl. Acad. Sci. USA* **2005**, *102*, 15785–15790. [CrossRef]
105. Raudvere, U.; Kolberg, L.; Kuzmin, I.; Arak, T.; Adler, P.; Peterson, H.; Vilo, J. g:Profiler: A web server for functional enrichment analysis and conversions of gene lists (2019 update). *Nucleic Acids Res.* **2019**, *47*, W191–W198. [CrossRef]
106. Peterson, H.; Kolberg, L.; Raudvere, U.; Kuzmin, I.; Vilo, J. gprofiler2—An R package for gene list functional enrichment analysis and namespace conversion toolset g: Profiler. *F1000Research* **2020**, *9*. [CrossRef]
107. Heinz, S.; Benner, C.; Spann, N.; Bertolino, E.; Lin, Y.C.; Laslo, P.; Cheng, J.X.; Murre, C.; Singh, H.; Glass, C.K. Simple Combinations of Lineage-Determining Transcription Factors Prime cis-Regulatory Elements Required for Macrophage and B Cell Identities. *Mol. Cell* **2010**, *38*, 576–589. [CrossRef] [PubMed]

International Journal of *Molecular Sciences*

MDPI

Article

A Comprehensive Assessment of Genetic and Epigenetic Alterations Identifies Frequent Variations Impacting Six Prototypic SCF Complex Members

Rubi Campos Gudiño [1,2,†], Ally C. Farrell [1,2,†], Nicole M. Neudorf [1,2,†] and Kirk J. McManus [2,*]

1 CancerCare Manitoba Research Institute, CancerCare Manitoba, Winnipeg, MB R3E 0V9, Canada; camposgr@myumanitoba.ca (R.C.G.); farrella@myumanitoba.ca (A.C.F.); neudorfn@myumanitoba.ca (N.M.N.)

2 Department of Biochemistry & Medical Genetics, University of Manitoba, Winnipeg, MB R3E 0J9, Canada

* Correspondence: Kirk.McManus@umanitoba.ca; Tel.: +1-204-787-2833

† These authors contributed equally to this work.

Abstract: The SKP1, CUL1, F-box protein (SCF) complex represents a family of 69 E3 ubiquitin ligases that poly-ubiquitinate protein substrates marking them for proteolytic degradation via the 26S proteasome. Established SCF complex targets include transcription factors, oncoproteins and tumor suppressors that modulate cell cycle activity and mitotic fidelity. Accordingly, genetic and epigenetic alterations involving SCF complex member genes are expected to adversely impact target regulation and contribute to disease etiology. To gain novel insight into cancer pathogenesis, we determined the prevalence of genetic and epigenetic alterations in six prototypic SCF complex member genes (*SKP1, CUL1, RBX1, SKP2, FBXW7* and *FBXO5*) from patient datasets extracted from The Cancer Genome Atlas (TCGA). Collectively, ~45% of observed SCF complex member mutations are predicted to impact complex structure and/or function in 10 solid tumor types. In addition, the distribution of encoded alterations suggest SCF complex members may exhibit either tumor suppressor or oncogenic mutational profiles in a cancer type dependent manner. Further bioinformatic analyses reveal the potential functional implications of encoded alterations arising from missense mutations by examining predicted deleterious mutations with available crystal structures. The SCF complex also exhibits frequent copy number alterations in a variety of cancer types that generally correspond with mRNA expression levels. Finally, we note that SCF complex member genes are differentially methylated across cancer types, which may effectively phenocopy gene copy number alterations. Collectively, these data show that SCF complex member genes are frequently altered at the genetic and epigenetic levels in many cancer types, which will adversely impact the normal targeting and timely destruction of protein substrates, which may contribute to the development and progression of an extensive array of cancer types.

Keywords: SKP1; CUL1; RBX1; SKP2; FBXW7; FBXO5; SCF complex; genome instability; chromosome instability; cancer

Citation: Campos Gudiño, R.; Farrell, A.C.; Neudorf, N.M.; McManus, K.J. A Comprehensive Assessment of Genetic and Epigenetic Alterations Identifies Frequent Variations Impacting Six Prototypic SCF Complex Members. *Int. J. Mol. Sci.* **2022**, *23*, 84. https://doi.org/10.3390/ijms23010084

Academic Editor: Elixabet Lopez-Lopez

Received: 19 November 2021
Accepted: 20 December 2021
Published: 22 December 2021

1. Introduction

In 2020, ~20 million individuals throughout the world were newly diagnosed with cancer, while ~10 million succumbed to the disease [1]. Despite these statistics, the molecular determinants (i.e., aberrant genes, proteins and pathways) underlying cancer development and progression remain poorly understood. Accordingly, new insight into the molecular events driving oncogenesis is needed to aid in the development of novel therapeutic strategies aimed at ultimately improving the lives and outcomes of those living with cancer. Decades of biochemical and genetic data have shown that aberrant expression/abundance of key proteins involved in critical biological processes (e.g., cell cycle regulation, DNA damage repair, and apoptosis [2–8]) are drivers of disease development and progression [9–12]. In this regard, the ubiquitin-proteasome system, an essential protein degradation system in eukaryotes, is a fundamental regulator of cell function at the protein level and its aberrant assembly and function are associated with malignant transformation and disease progression [13–17]. More specifically, the aberrant expression and function of the SCF (SKP1, S-phase Kinase associated Protein 1; CUL1, Cullin 1; F-box protein) complex occurs in an extensive array of cancer types [9,10,12,18–20]. However, additional studies are required to understand its role in disease pathogenesis as the molecular mechanisms underlying aberrant SCF complex function remain largely unknown.

The ubiquitin-proteasome system is best understood for its role in regulating protein abundance, where polyubiquitin chains are covalently attached to protein substrates through the activities of the E1 (ubiquitin activating), E2 (ubiquitin conjugating), and E3 (ubiquitin ligase) enzymes, targeting them for degradation by the 26S proteasome (reviewed in [13,21,22]). Ubiquitin ligation to a specific protein substrate requires recognition and binding by the E3 ligase enzyme [13,23], and so it is primarily responsible for conferring substrate specificity and ultimately, proteasomal degradation [24]. The SCF complex comprises the largest group of E3 ubiquitin ligases [25] that encompasses a diverse array substrate specificities that includes cell cycle regulators (e.g., Cyclin E1, P27) [2,18], transcription factors (e.g., c-MYC) [7,26] and regulators of DNA damage-response (e.g., RAD51) [3]. The SCF complex is comprised of three invariable core subunits, namely SKP1, CUL1 and RBX1 (Really Interesting New Gene [RING]-Box protein 1), and one of 69 variable F-box proteins conferring substrate specificity [4]. Conceptually, RBX1 interacts with the E2 enzyme and CUL1, a scaffolding protein that brings the F-box protein and its ligand within close spatial proximity of the E2 enzyme harboring the ubiquitin moiety to be transferred to the substrate protein upon binding with SKP1 [27,28]. In this regard, SKP1 serves as an adaptor between the SCF complex and the F-box proteins via its conserved F-box motif [29–31]. F-box proteins are classified into one of three families based on the presence (or absence) of specific protein motifs, including: 1) leucine-rich repeat (FBXL), 2) WD-40 (FBXW) and 3) other (FBXO) [27,30,32,33]. As there are 69 distinct F-box proteins there are 69 unique SCF complexes, each with distinct target specificities that are collectively proposed to regulate hundreds to thousands of protein substrates for ubiquitination [34]. Thus, aberrant expression of individual SCF complex members driven by genetic and epigenetic changes are expected to adversely impact an extensive array of proteins, which in turn is predicted to adversely impact biological pathways expected to contribute to the development and progression of cancer. Indeed, alterations in various SCF complex members occur in a vast array of cancer types and their altered expression has more recently been shown to promote various forms of genome instability and corresponds with early etiological events including cellular transformation [9–12] (reviewed in [20]).

In this study, we assessed both genetic and epigenetic changes in six key SCF complex member genes including the three invariable core members (i.e., *SKP1*, *CUL1* and *RBX1*) and a single representative member from each of the three F-box families, including *SKP2* (*FBXL1*), *FBXW7* (*CDC4*) and *FBXO5* (*EMI1*). Importantly, *SKP2* and *FBXW7* encode well studied F-box proteins and prototypical examples of an oncogene and tumor suppressor gene, respectively, while *FBXO5* encodes an F-box protein whose substrate specificities and role in disease development is poorly understood and requires further in-depth study. Using publicly available, patient-derived data from The Cancer Genome Atlas (TCGA) extracted from numerous cancer types [35–37], we assessed the prevalence and types of genetic and epigenetic alterations (gene mutations, copy number alterations [CNAs] and methylation status) within each gene to determine their potential implications in disease pathology. Overall, we determined that the six SCF complex member genes are frequently altered in solid tumor samples and predominantly harbor missense mutations predicted to be damaging. Overall, the mutations are generally distributed across the entirety of their coding sequences, with *FBXW7* exhibiting several alteration hot-spots. When combined with tertiary and quaternary structure (crystallographic) data, many of the encoded amino acid substitutions are predicted to adversely impact protein–protein interactions occurring between individual SCF complex members and/or their protein substrates, which is expected to prevent substrate protein degradation leading to their aberrant accumulation. Beyond mutations, we also discovered that CNAs occur frequently for all six genes and that while *CUL1* and *SKP2* predominantly exhibit copy number gains, the remaining genes (*SKP1, RBX1, FBXW7* and *FBXO5*) exhibit more copy number losses that correspond with reduced mRNA expression. Collectively, these data suggest that SCF complex members may act as oncogenes and/or tumor suppressor genes in a context-dependent or cancer-specific manner. Finally, we examined the methylation status of all six genes and determined that each is differentially methylated across cancer types, suggesting epigenetic misregulation may also contribute to cancer pathogenesis in a manner like that observed following gene copy number losses. Collectively, these findings provide novel insight into genetic and epigenetic alterations affecting the SCF complex and support the possibility that aberrant expression and/or function of key SCF complex members may be an early etiological event driving the development and progression of many cancer types.

2. Results

2.1. Genes Encoding SCF Complex Members Are Mutated Frequently in Cancer

To determine the mutational frequency and types of alterations occurring in the six prototypic and representative SCF complex member genes (*SKP1, CUL1, RBX1, SKP2, FBXW7* and *FBXO5*) across various cancers, TCGA data [35–37] were assessed as detailed in Materials and Methods. Briefly, mutational data, including frameshift deletions/insertions, fusions, in-frame deletions/insertions, missense mutations (benign and damaging), nonsense mutations and splice site/regions mutations were extracted from 10 solid tumor types, and analyzed (Figure 1A), with the respective mutational frequencies provided in Table S1. As shown in Figure 1A (left), each cancer type harbors mutations in the six SCF complex member genes that typically range from ~2% in ovarian cancer to ~38% in uterine cancer. Additionally, cursory analyses revealed that the core SCF complex member genes typically harbor fewer than 20 mutations within a given cancer type (Figure 1A; right), with specific genes, namely *SKP1* and *RBX1* having fewer than 10 mutations in each cancer type. Conversely, *CUL1* exhibits a higher mutational load in some cancers, as evidenced by the 23 and 35 mutations observed in stomach and uterine cancers, respectively. Similarly, the three F-box protein genes typically harbor few mutations, (1–20 total mutations) within most cancer types (Figure 1A; right); however, *FBXW7* is the most frequently mutated gene with 107 and 117 cases in colorectal and uterine cancers, respectively. Collectively, these data reveal that SCF complex member genes are mutated in a diverse array of cancer types with *FBXW7* being the most frequently mutated gene in many cases.

Next, to determine whether SCF complex member genes predominantly exhibit specific mutational subtypes (i.e., frameshift deletion/insertion, fusion, in-frame deletion/insertion, missense, nonsense or splice site/regions), TCGA sequencing data [35–37] were scrutinized for each of the six genes. The mutational frequencies and the predicted functional impacts of the single nucleotide polymorphisms (SNPs) ("benign", "possibly damaging" and "probably damaging") were determined using Sorting Intolerant From Tolerant (SIFT) and Polymorphism Phenotyping V2 (PolyPhen-2) (detailed in Materials and Methods). Unfortunately, PolyPhen-2 and SIFT score estimates are derived from protein alignments utilizing different protein databases, which sometimes results in inconsistent predictions between approaches. Accordingly, only PolyPhen-2 data were employed for mutation classification analyses, where missense mutations with "not available" (N/A) prediction scores were excluded from all downstream analyses. In general, the overall (Total) distributions of the various mutation categories (Figure 1B) are predominantly enriched for "probably damaging" mutations (42.4%), followed by "benign" (20.6%) and nonsense (14.8%) mutations, while individual genes are more variable. For example, *SKP1* predominantly harbors missense "benign" mutations (47.8%), while other mutations (i.e., missense "probably" damaging; 8.7%) are less common. *CUL1* exhibits numerous missense mutations (36.6% "probably damaging"; 32.41% "benign"), while a majority of *SKP2* mutations are missense "benign" mutations (47.4%). In contrast, *FBXW7* SNPs are primarily missense "probably damaging" mutations (54.5%), whereas *FBXO5* alterations are largely comprised of missense "benign" mutations (40%). In summary, while there is variation in the frequency of each mutational class between the six genes, there is an overall bias towards "probably damaging" missense mutations that supports the possibility that mutations within SCF complex member genes may have adverse implications for complex function and disease pathogenesis.

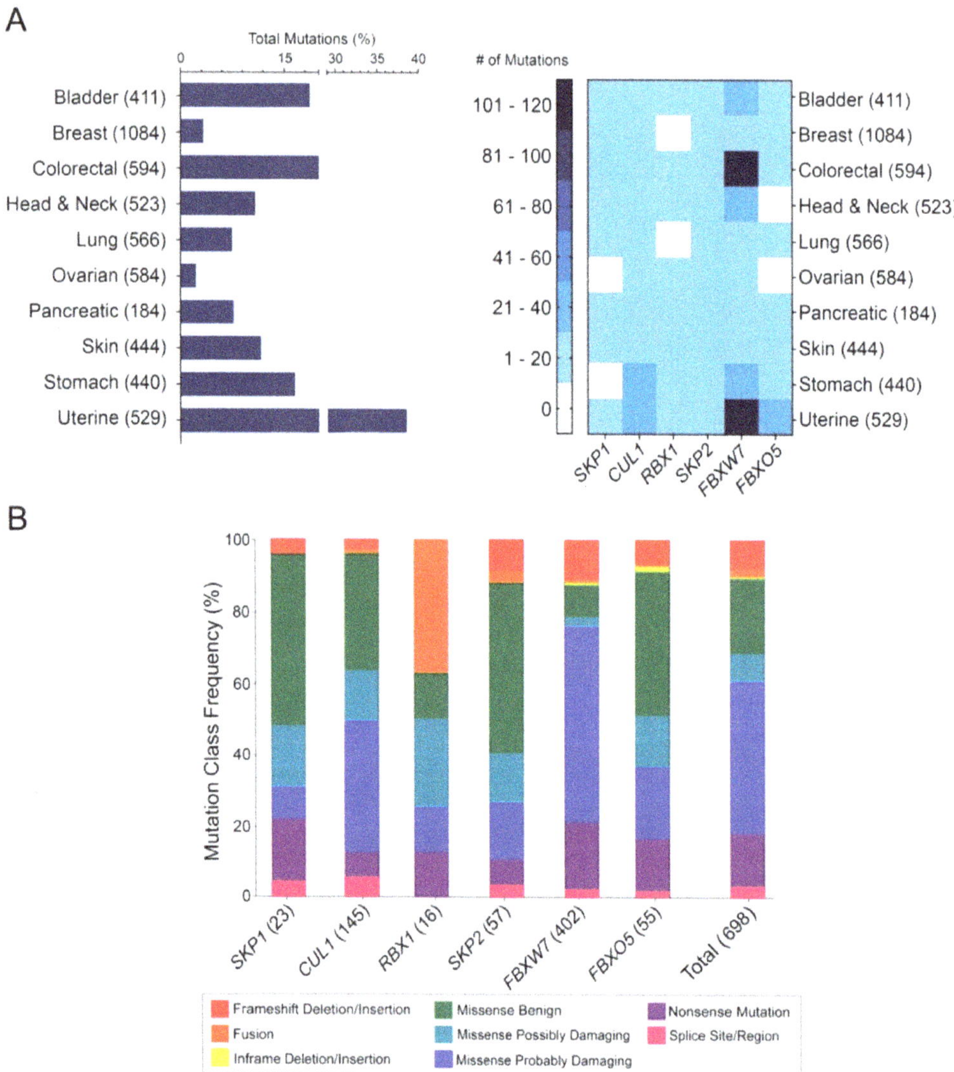

Figure 1. Frequency and types of mutations of six SCF complex member genes in 10 cancer types. (**A**) Bar graph (left) and heatmap (right) displaying the frequency and prevalence, respectively, of SCF complex member gene mutations in 10 cancer types with the total number of cancer cases indicated within brackets. Alteration categories include frameshifts, fusions, in-frame deletions/insertions, missense, nonsense and splice site mutations. (**B**) Bar graph presenting the frequency of the alteration categories from 10 cancer types, with the aggregate total frequencies provided in the last column (Total). The total number of alterations identified for each gene are indicated within brackets.

2.2. The Distributions of Encoded SCF Complex Alterations Are Consistent with a Tumor Suppressor Mutational Load

To gain a greater insight into the overall distribution of encoded missense, truncating and splice site alterations within the SCF complex members, the frequency and location (i.e., amino acid position) of encoded alterations were assessed in 10 cancer types (Figure 2; Table S2). In general, all SCF complex members exhibit alterations that span the length of the encoded protein, an overall distribution that is consistent with a tumor suppressor mutational load [38]. Interestingly, however, several alteration hot-spots were observed in FBXW7, a feature that is more commonly associated with oncogenes [38] (Figure 2). For example, encoded alterations involving R465C/H/G occur in 58 cancer patient samples (bladder, breast, colorectal, head and neck, stomach and uterine). Similarly, the encoded alterations R479Q/G/L/P occur in 22 patient samples (bladder, breast, colorectal, head and neck, pancreatic, stomach and uterine cancers). R505G/C/H/L substitutions account for 39 of the encoded alterations, in colorectal, head and neck, ovarian, stomach and uterine cancers, whereas D520N/Y/H/E occurs in 22 bladder, colorectal, head and neck and ovarian patient cases. Notably, the alteration hotspots detailed above reside within the WD-40 domain of FBXW7, which in some cases may adversely impact protein structure (e.g., D520; discussed further below). Taken together, these data suggest that FBXW7 (and perhaps other SCF complex members) may exhibit tumor suppressor and/or oncogene-like functions in a context/tissue-dependent manner.

2.3. Encoded Alterations in SCF Complex Members May Adversely Impact Protein Structure

Having described the frequency and location of encoded alterations in the preceding sections, we next sought to explore the potential impact these changes have on the structure and function by analyzing crystallographic data extracted from The Protein Data Bank (PDB) [31,39]. Mutations deemed "deleterious" or "possibly/probably damaging" by SIFT and Polyphen-2 (Table S2), respectively, are presented in Figures 3 and 4. SIFT deems an alteration "deleterious" based on conserved amino acid positions, where alterations at strongly conserved positions are expected to be intolerant to most substitutions [40]. By contrast, PolyPhen-2 predictions are based on the stability and function of human proteins using functional annotation of SNPs, maps coding SNPs to gene transcripts, extracts protein sequence annotations and structural attributes [41]. Importantly, these specific amino acid substitutions could critically impact SCF complex formation and function by disrupting secondary structure, tertiary structure and protein–protein interactions.

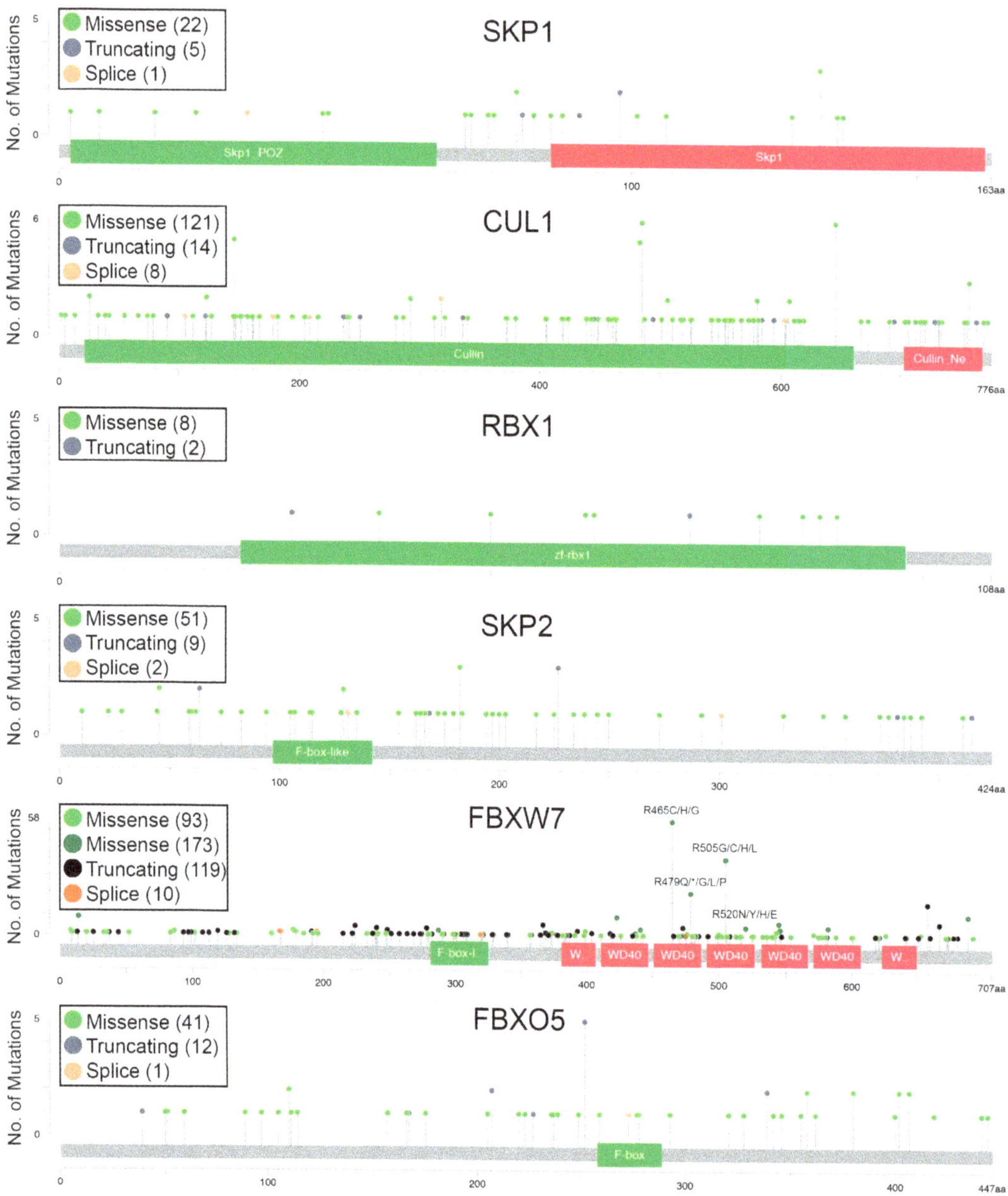

Figure 2. Distribution of the encoded alterations of six SCF complex members. Schematics depicting the amino acid position (x-axis) and frequency (y-axis) of encoded alterations from 10 cancer types, with the total number of each mutational subtype presented within brackets in the associated bounding box. Key protein motifs of SCF complex members are denoted in green and red. Note that light colors correspond to mutations of unknown significance and dark colors correspond to putative driver mutations (* indicates a premature stop codon).

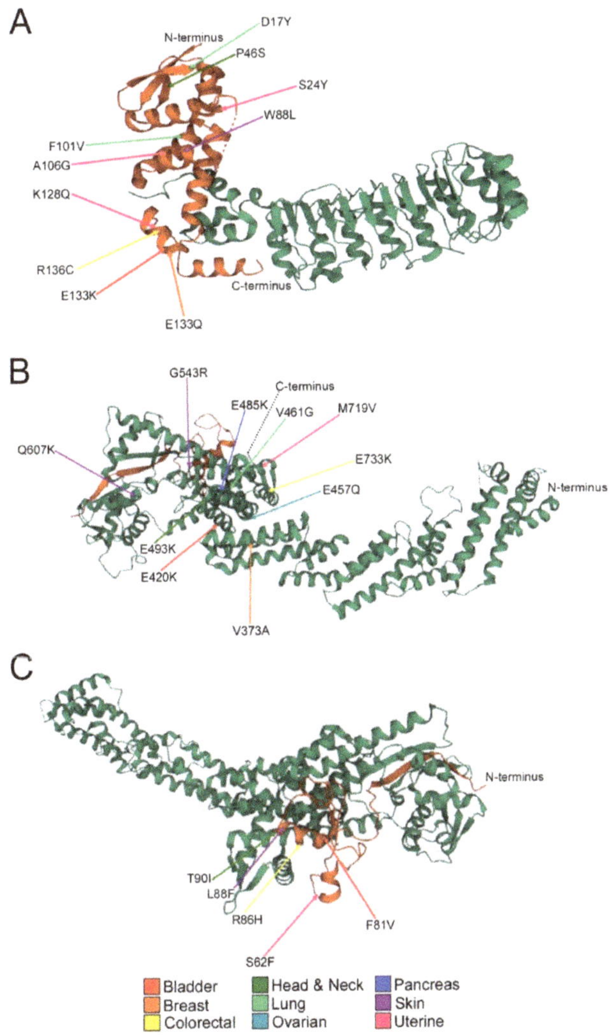

Figure 3. Missense mutations underlie potentially damaging alterations in core SCF complex members. (**A**) Partial crystal structure (missing amino acids 1-14; ribbon model) of the SKP1 (brown)-SKP2 (green) complex shows predicted detrimental alterations. Indicated amino acid substitutions are deemed deleterious, possibly damaging or probably damaging by SIFT and PolyPhen-2 databases (Table S2) and are predicted to impact protein–protein interactions between SKP1 and CUL1 or SKP1 and SKP2. Amino acids are denoted by their single letter code, numbers indicate amino acid position in the SKP1 protein. Colored arrows identify cancer type in which the underlying mutation was identified (see key). (**B**) Partial crystal structure (missing amino acids 1-16; ribbon model) of the CUL1 (green)-RBX1 (brown) complex presents predicted detrimental alterations. Indicated alterations are deemed deleterious, possibly/probably damaging by SIFT and PolyPhen-2 and are predicted to impact protein–protein interactions between CUL1-RBX1 or CUL1-SKP1. (**C**) Partial crystal structure (missing amino acids 1-18; ribbon model) of the CUL1 (green)-RBX1 (brown) complex presents predicted detrimental alterations. Indicated alterations are deemed deleterious, possibly/probably deleterious by SIFT and PolyPhen-2 and are predicted to impact protein–protein interactions between RBX1-CUL1 or RBX1-E2 conjugase.

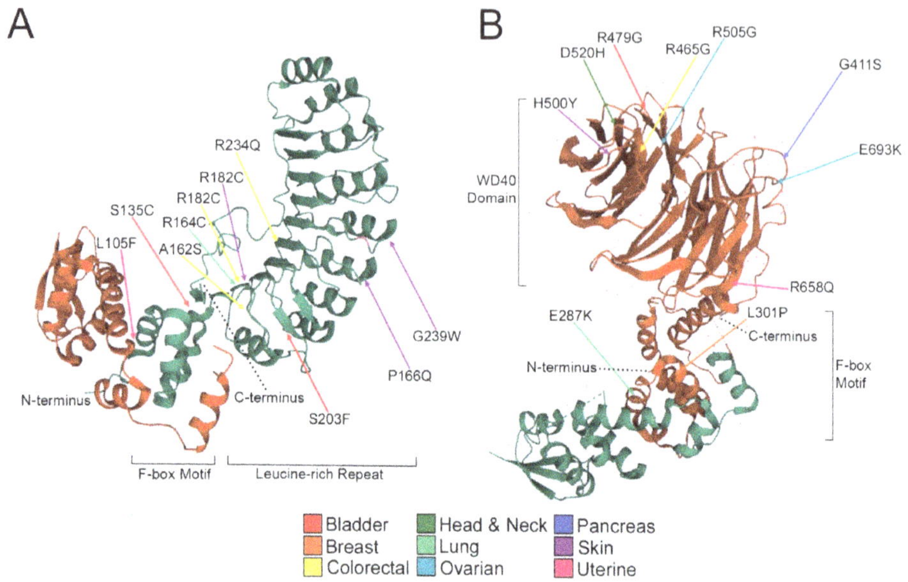

Figure 4. Missense mutations encode potentially detrimental amino acid substitutions in SKP2 and FBXW7. (**A**) Partial crystal structure (missing amino acids 1-88; ribbon model) of the SKP1 (brown)-SKP2 (green) complex presenting predicted damaging alterations. Indicated amino acid substitutions are deemed deleterious, possibly/probably damaging by SIFT and PolyPhen-2 (Table S2) and are predicted to impact protein–protein interactions between SKP1 and SKP2 or SKP2 and its target proteins. Amino acids are denoted by their single letter code, numbers indicate amino acid position in the SKP2 protein. Colored arrows identify cancer type in which the underlying mutation was identified (see key). (**B**) Partial crystal structure (missing amino acids 1-262; ribbon model) of the SKP1 (green)-FBXW7 (brown) complex presents predicted damaging alterations. Indicated alterations are deemed deleterious, possibly/probably damaging by SIFT and PolyPhen-2 and are predicted to impact protein–protein interactions between SKP1 and FBXW7 or FBXW7 and its substrate proteins.

Recall that SKP1 is an invariable member of the SCF complex that recruits the variable F-box proteins and their protein ligands [31]. Although SKP1 alterations occur most frequently in uterine and lung cancers, most substitutions only occur in a single cancer type, except for E133K/Q (bladder, breast; Figure 3A). Notably, ~50% of the encoded alterations deemed "deleterious" and "probably/possibly damaging" occur within the F-box protein recognition domain and affect net charge or alpha helicity including E133K/Q (bladder; breast), K128Q (uterine), R126C (colorectal) and A106G (uterine). CUL1 is the scaffolding member of the SCF complex [29] and alterations that potentially affect its ability to bind with either SKP1 or RBX1 could critically impact SCF complex formation and function. As with SKP1, most CUL1 alterations only occur within a given cancer type with the exceptions of E485K (bladder; head and neck; pancreatic; uterine) and Q607H/K (bladder; skin) (Figure 3B). Strikingly, ~50% of alterations involve glutamate residues that are frequently converted to lysine residues, resulting in a change in net charge and include E420K (bladder), E485K (uterine), E493K (head and neck), E733K (colorectal) and E457Q (lung). Only five RBX1 alterations were identified as "deleterious" or "possibly/probably damaging" across all patient samples including S62F (uterine), F81V (bladder), R86H (colorectal), L88F (skin) and T90I (head and neck) (Figure 3C). These alterations occur within the RBX1 zinc finger binding domain, which may impact interactions between RBX1 and the E2 enzyme or CUL1. Finally, FBXO5 is an F-box protein whose substrates remain largely unknown and has many "deleterious" and "possibly/probably damaging" alterations identified within colorectal and uterine cancers. Approximately half of the substitutions in colorectal cancer convert a leucine residue to a larger residue such as L205F, L220R and L259F. Conversely, there is a wide range of FBXO5 alterations in uterine cancer that includes a serine to glycine conversion within the F-box motif (S278G), which could potentially impact protein–protein interactions with SKP1 and the rest of the SCF complex. Unfortunately, there is no available crystal structure of SKP1-FBXO5 to date, therefore it is difficult to speculate how these alterations may impact either substrate interactions or SCF complex formation and function.

SKP2 is an F-box protein that has been primarily classified as an oncoprotein and targets the products of proto-oncogenes such as Cyclin E1 and c-MYC for proteolytic degradation [7,26,42,43]. *SKP2* mutations occur most frequently in bladder, colorectal, skin and uterine cancers and result in key amino acid substitutions, including L105F (uterine) and S135C (bladder; Figure 4A), that are located within the F-box motif, which is critical for interactions between SKP2 and the SCF complex [44]. Another important SKP2 protein domain is the leucine-rich repeat, consisting of a hydrophilic cap and a hydrophobic concave surface that are essential for ligand interaction [45]. Therefore, non-conservative alterations within this region may impact the ability of SKP2 to recognize and ubiquitinate protein targets. Substitutions in the leucine-rich repeat include the elimination of arginine residues (positively charged aliphatic chain) and consist of R164C (lung), R182C (colorectal; skin; uterine) and R234Q (colorectal). Other alterations in this domain include S203F (bladder), P166Q and G239W (uterine). Importantly, amino acid substitutions within SKP2 have the potential to impede normal function which can result in the aberrant accumulation of its substrate proteins.

FBXW7 is an F-box protein with a WD-40 protein domain that is critical for substrate specificity and whose target proteins includes cell cycle regulators and oncoproteins such as Cyclin E1 and c-MYC [26,46,47]. Unlike *SKP2*, there are "deleterious" and "possibly/probably damaging" substitutions in all 10 cancer types assessed, with the highest incidences occurring in breast, head and neck and uterine cancers. Substitutions within the FBXW7 F-box motif include E287V/K (colorectal; lung; stomach), which induces a change in net charge from negative to positive and L301P (breast), which is predicted to disrupt alpha helicity (Figure 4B). As with SKP2, there is a trend for arginine substitutions to occur across a diverse array of cancer types (bladder; colorectal; head and neck; stomach; uterine). These positively charged hydrophilic residues often protrude from the protein surface to participate in hydrogen bonding Van der Waals interactions in addition to aiding in salt bridge formation [48,49]. Therefore, arginine substitutions could critically impact the ability of the SCF complex to recognize substrates and regulate the abundance of target proteins. Notably, three of the four arginine substitutions are conversions to glycine residues, all occurring in close spatial proximity of one another (40 residues), at the exterior of the WD-40 ligand binding domain, which effectively eliminates a net positive charge by replacing it with a small hydrophobic residue. Additional substitutions in the WD-40 domain include changes in net charge (E693K; ovarian), gain of a hydrogen bonding group (G411S; pancreatic), or both (D520N/Y/H/E; bladder; colorectal; head and neck; ovarian). Collectively, these encoded amino acid substitutions are predicted to adversely impact the F-box proteins from interacting with the SCF complex and/or their protein substrates, which is expected to promote the aberrant accumulation of substrates and contribute to early disease development in certain instances.

2.4. SCF Complex Members Exhibit Frequent Copy Number Alterations in Cancer

While the preceding sections focused on specific mutations and their encoded amino acid substitutions within the six SCF complex members, we sought to investigate the prevalence of gene CNAs, which are expected to impact their overall expression levels. Like the previous sections, TCGA data [35–37] from 10 cancer types were scrutinized for CNAs including deep deletions, shallow deletions, gains and amplifications as described within Materials and Methods, with the full list of CNAs provided in Table S3. In general,

CNAs occur frequently in all 10 cancer types and collectively range from ~19% (pancreatic) to ~85% (ovarian) (Figure 5A). More specifically, *SKP1* and *RBX1* predominantly exhibit shallow deletions that range from ~12% (pancreatic) to ~46% (bladder) and ~7% (skin) to ~79% (ovarian), respectively. In contrast, copy number gains occur most frequently with *CUL1* and range from ~13% (uterine) to ~50% (colorectal), whereas amplifications are rare, ranging from ~1.5% (lung) to ~7% (ovarian). *SKP2* also exhibits amplifications ranging from ~0.5% (colorectal) to ~8% (lung) and a prevalence of gains ranging from ~12% (uterine) to ~46% (lung). Conversely, *FBXW7* and *FBXO5* typically harbor shallow deletions ranging from ~15% (pancreatic) to ~66% (ovarian) and ~7% (uterine) to ~59% (ovarian), respectively. Collectively, these data show that *CUL1* and *SKP2* predominantly exhibit copy number gains (gains and amplifications), whereas *SKP1*, *RBX1*, *FBXW7* and *FBXO5* tend to exhibit more copy number losses (shallow and deep deletions). Importantly, these data lend further support to the possibility that aberrant expression of SCF complex member genes adversely impacts complex function and may contribute to the development and/or progression of many cancer types.

It should also be noted that there is a compounding effect when the individual frequencies of shallow deletions and gains are concurrently assessed for the six genes. That is, the combined frequencies are >50% for losses and gains in breast, colorectal and lung cancers, the three most prevalent cancer types (Figure 5B). More specifically, the combined frequencies of shallow deletions are 54% (colorectal), 74% (breast) and 84% (lung), while combined frequencies of gains are 55% (breast), 61% (colorectal) and 72% (lung), which suggests that aberrant SCF complex expression (i.e., decreases or increases) may be a significant, yet underappreciated contributor to disease pathology in these cancer types. In support of this possibility, mRNA expression levels are significantly altered (reduced or increased) in a manner consistent with the specific CNA (i.e., loss or gain). For example, reduced mRNA expression is typically associated with copy number losses (deep and shallow deletions) for *SKP2*, *FBXW7* and *FBXO5* relative to the diploid state, while increased expression typically occurs with copy number gains (gains and amplifications) (Figure 6). Consistent with these findings, CNAs involving *SKP1*, *CUL1* and *RBX1* also correspond with overall changes in mRNA expression levels. Collectively, these data show that CNAs and corresponding changes in mRNA expression occur frequently for SCF complex member genes, highlighting how CNAs likely adversely affect SCF complex expression and/or function.

2.5. SCF Complex Members Are Differentially Methylated in Cancer

To assess the methylation status of SCF complex member genes, β-values and corresponding CNA data were exported from the TCGA Firehose Legacy dataset (https://gdac.broadinstitute.org/, accessed on 20 October 2021) using cBioPortal [35,36] (Figure 7; Tables S4–S9). Samples were characterized as hypomethylated ($\beta < 0.2$), partially methylated ($0.2 \leq \beta \leq 0.7$), or hypermethylated ($\beta > 0.7$) as described within Materials and Methods. Overall, SCF complex members exhibit diverse methylation profiles across the 10 cancer types. As expected, the methylation status of *SKP1* varies by cancer type, with *SKP1* being hypomethylated (i.e., expressed) in the majority of colorectal, head and neck, ovarian, stomach and uterine cases, whereas it tends to be partially or hypermethylated (i.e., repressed) in bladder, breast, lung and pancreatic cancers. Note that for all cancer types investigated, samples exhibiting *SKP1* copy number losses also tend to correspond with hypermethylated states (i.e., greater β-values) relative to diploid samples or those harboring copy number gains (Figure 7, Table S4). Overall, the methylation status of *CUL1* is very similar to that of *SKP1*, as tumor samples with copy number losses also tend to be hypermethylated; however, *CUL1* is generally partially methylated or hypermethylated in all uterine cancer cases (Figure 7, Table S5). Interestingly, *RBX1* (Figure 7, Table S6) and *SKP2* (Figure 7, Table S7) tend to be hypomethylated across all cancer types, irrespective of copy number status, whereas *FBXW7* is differentially methylated across cancer types (Figure 7, Table S8). Briefly, *FBXW7* is hypomethylated in colorectal, head and neck, ovarian, stomach and uterine patient samples and partially or hypermethylated in bladder, breast, lung, pancreatic and skin cancers. Finally, *FBXO5* is partially methylated in bladder and lung cancers, but is predominantly hypomethylated in all remaining cancer types (Figure 7, Table S9). As methylation is a negative regulator of gene expression [50], increases in methylation status may phenotypically mimic copy number losses and/or result in loss of heterozygosity, whereas aberrant decreases in methylation status are expected to correspond with increases in gene expression, which may phenocopy, to a limited extent, copy number gains. Thus, aberrant increases or decreases in methylation status may adversely impact SCF complex members harboring tumor suppressor [9,10,51,52] or oncogene-like functions [53,54], respectively. Collectively, these data show that SCF complex member genes are differentially methylated across a variety of cancer types and supports the possibility that changes in methylation status may induce aberrant SCF complex function and contribute to cancer pathogenesis.

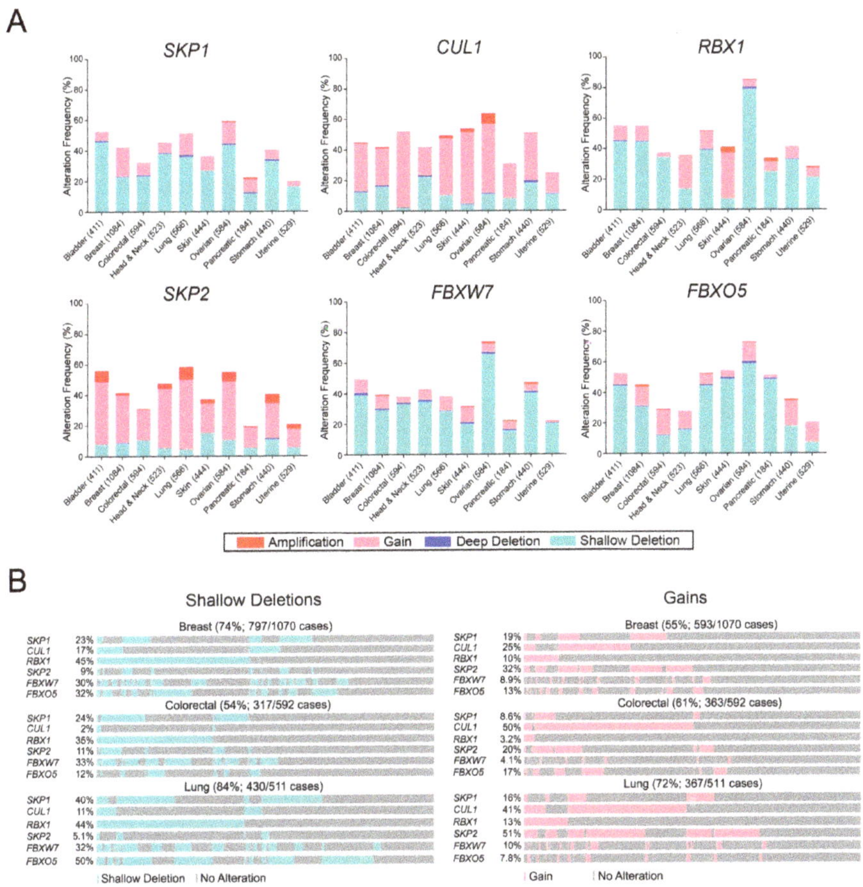

Figure 5. The frequency of gene copy number alterations of SCF complex members in 10 cancer types. (**A**) Bar graphs depicting the frequency of CNAs (deep deletions; shallow deletions; gains; amplifications) for six SCF complex member genes in 10 cancer types (Table S3). Cancer types are listed along the x-axis with the number of cases of each indicated in brackets. (**B**) OncoPrint data for breast, colorectal and lung cancer depicting the individual and cumulative frequencies for only shallow deletions (left) and gains (right) of six SCF complex member genes. Vertical alignments within a given cancer type identify samples from the same patient; patient-specific comparisons cannot be made between categories (i.e., shallow deletions versus gains).

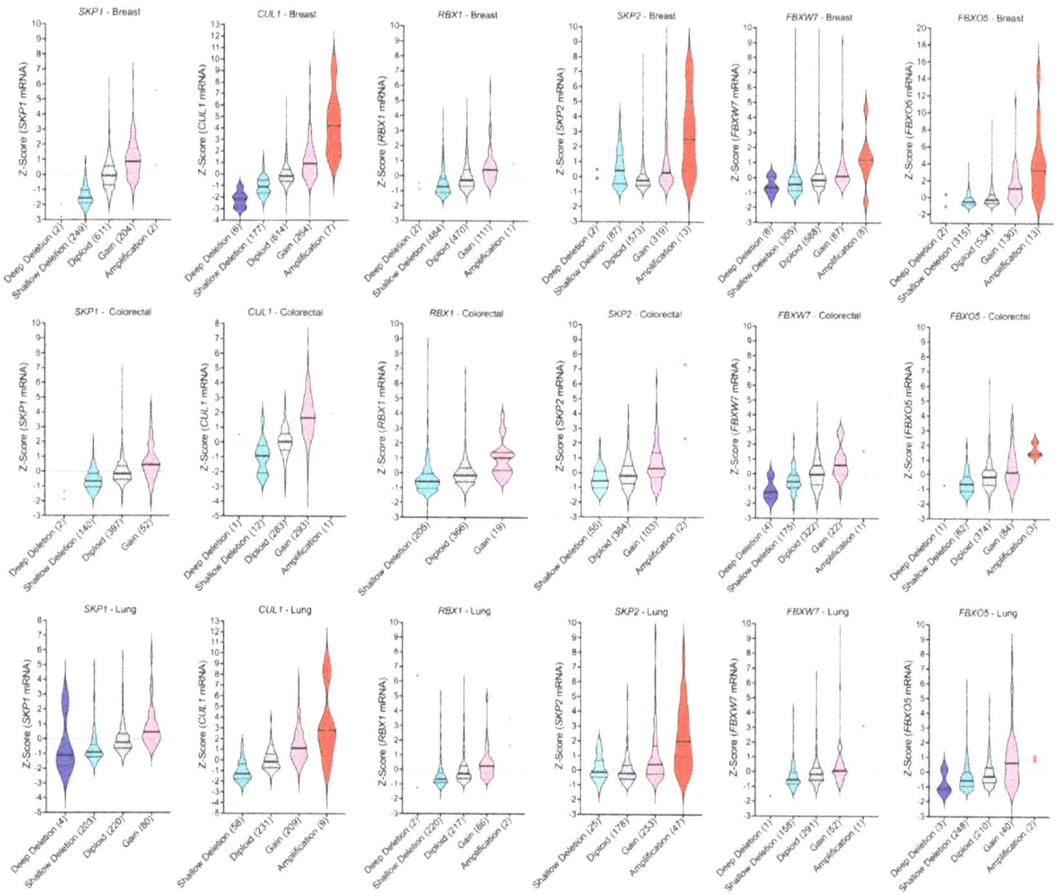

Figure 6. SCF complex copy number alterations correspond with changes in mRNA expression levels. Violin plots of mRNA expression data [35–37] from the three commonly diagnosed cancers (i.e., breast [top row], colorectal [middle] and lung [bottom]). *SKP1*, *CUL1*, *RBX1*, *SKP2*, *FBXW7*, and *FBXO5* CNAs (deep deletions; shallow deletions; gains, amplifications) and diploid cases are presented along the *x*-axis with total case numbers indicated within brackets. Note that categories with $\leq$ 2 cases are identified by dots and that in general, deep deletions, and amplifications are rare. Note that some of the *y*-axes are differentially scaled to better present the data.

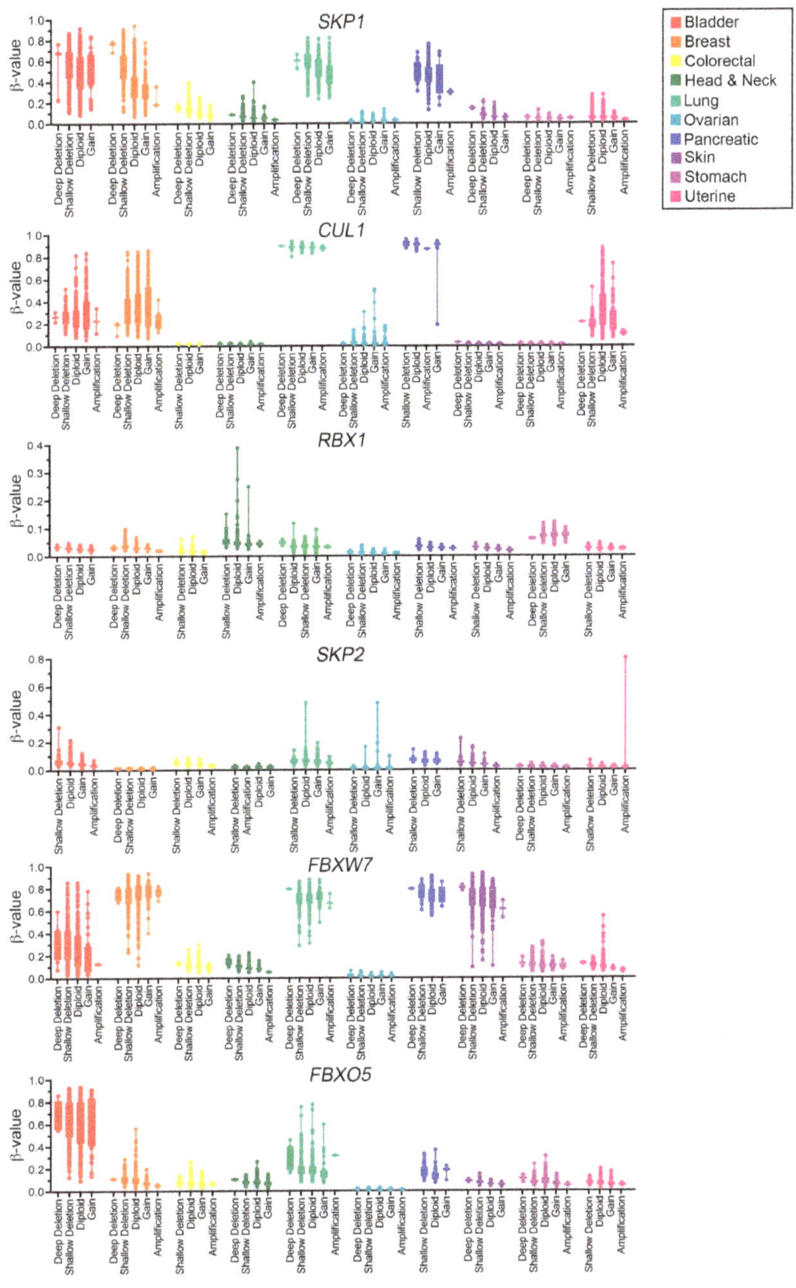

Figure 7. The SCF complex is differentially methylated in human cancers. Box-and-whisker plots presenting the methylation status of cancer samples derived from TCGA Firehose Legacy data (https://gdac.broadinstitute.org/, accessed on 18 November 2021) (see Tables S4–S9). Associated β-values of SCF complex members are presented along the *y*-axis. Note that the *y*-axis scale for *RBX1* and *SKP2* (0.4 and 0.8, respectively) differ relative to the remaining genes (1.0) for better data visualization. Methylation data are grouped according to gene CNAs (deep deletion; shallow deletion; gain; amplification) or diploid status (*x*-axis).

3. Discussion

In this study, we examined the genetic and epigenetic changes associated with six prototypic SCF complex member genes across 10 common cancer types. First, we determined that each member is somatically mutated across multiple cancer types, with *FBXW7* exhibiting the most frequent number of mutations. Additionally, we determined that the mutations are typically distributed across the entirety of their respective coding regions, which is consistent with the mutational distribution of a tumor suppressor gene [38]; however, *FBXW7* also harbored focal hot-spots that are consistent with the mutational profile of an oncogene [38]. Interestingly, in silico analyses determined that many of the encoded amino acid substitutions are predicted to adversely impact protein structure and/or function of the SCF complex, supporting a potential role in cancer pathogenesis. Next, CNA analyses revealed that each gene is frequently altered in 10 common cancer types, and that *SKP1, RBX1, FBXW7* and *FBXO5* tend to exhibit more losses, while *CUL1* and *SKP2* exhibit more gains. It is also important to note that the specific categories of CNAs corresponded with similar changes in gene expression at the mRNA levels. That is, cancers with deep or shallow deletions exhibited reduced mRNA expression levels relative to diploid cases, while those harboring gains or amplifications exhibited increases. Finally, examination of their DNA methylation profiles revealed a large range of patterns for each gene. In general, samples harboring copy number losses (particularly shallow deletions) were frequently hypermethylated relative to the diploid cases or those with copy number gains (gains or amplifications). In summary, these findings determined that the six SCF complex member genes exhibit frequent mutations, CNAs and/or aberrant methylation profiles that collectively are predicted to adversely impact complex member expression and/or function that is consistent with an etiological role in cancer development and progression.

When assessing the amino acid substitutions in conjunction with available crystal structure data, our results highlight the potential functional impact that encoded alterations may have on the SCF complex and identifies a possible mechanism by which aberrant SCF complex structure and function may contribute to cancer pathogenesis. Importantly, additional functional studies are highly warranted as not every mutation (and encoded alteration) will have pathogenic implications in cancer; however, it is well established that specific amino acid substitutions can adversely impact protein structure and function, and thus have the potential to contribute to disease pathogenesis. For example, the mutations we identified that alter key amino acid characteristics, including net charge, hydrogen bonding ability and steric interactions (i.e., protein–protein interactions) within the leucine-rich repeat domain of SKP2 or the WD-40 domain of FBXW7 can decrease the affinity and specificity of these binding sites for their protein substrates, including the oncoproteins Cyclin E1 and c-MYC [26,47,55]. SKP2 also targets P27 (*CDKN1B*) for proteolytic degradation, a cell cycle regulating protein that inhibits Cyclin Dependent Kinase activity [56]. Importantly, aberrant P27 accumulation is associated with chromosome instability (CIN; ongoing changes in chromosome numbers) and mitotic defects [2,53,57], whereas aberrant Cyclin E1 and c-MYC turnover lead to cell cycle and apoptotic defects that promote cancer development and progression [9,10,12,58–61]. In fact, increased *c-Myc* expression in mouse embryonic fibroblasts increases growth rate, enhances clonogenic growth, decreases in contact inhibition and spontaneously forms tumors in mice within 10 days [62], providing strong evidence that deregulated *c-MYC* expression exhibits a significant role in tumorigenesis. Importantly, while the c-MYC protein is overexpressed in ~70% of human cancers, typically only ~20% of tumors harbor *c-MYC* gene amplifications or translocations and thus it remains possible that aberrant c-MYC turnover/degradation may account for this discrepancy [63]. Thus, these findings highlight the importance of c-MYC regulation at the protein level and the possible malignant consequences of abnormal SCF complex function. Moreover, the detrimental impact encoded alterations have on protein structure and function is perhaps best exemplified by Wang and colleagues [55], who determined that D331 within the leucine-rich repeat domain of SKP2 is essential for its interaction with CKS1 (Cyclin-dependent Kinases regulatory protein 1) and the subsequent ubiquitination of P27. More specifically, they noted that a D331A substitution converted the normally negative surface potential to a positive potential, which ablated the SKP2-CKS1 interaction, while a conserved substitution, D331E, maintained the interaction. Similarly, amino acid substitutions involving R417 and R457 within the WD40 domain of FBXW7 abolish Cyclin E binding, while R495 substitutions reduce binding affinity [47,64]. While none of these alterations were detected within the current study, their results highlight the need and value of downstream genetic and functional studies aimed at assessing the plethora of individual substitutions to accurately determine their impacts on protein–protein interactions and overall SCF complex function. This is particularly important as the SCF complex regulates numerous biological processes required for genome stability, including cell cycle progression, centrosome biology and DNA repair [9,10,12,19,20,65]. Thus, it is conceivable that specific amino acid substitutions occurring in key protein motifs may adversely impact SCF complex function and underlie the accumulation of protein substrates, especially (proto-) oncogenes whose increase in abundance promotes genome instability and contributes to cancer pathogenesis [9,10,12,19,20]. For example, *CCNE1* (Cyclin E1) is an established oncogene [58,66,67] that is genomically amplified in many cancer types [35–37] and whose overexpression induces CIN, which is associated with cell cycle misregulation, genome instability, cellular transformation and tumor formation in mice [59]. More specifically, Karst and colleagues [67] determined that increased abundance of Cyclin E1 leads to inappropriate cell growth including accelerated growth, loss of contact-inhibition, clonogenic growth and anchorage independent growth. Moreover, we recently demonstrated that reduced expression of *SKP1*, *CUL1* and *RBX1* prevents Cyclin E1 degradation leading to an increase in abundance that induces CIN and promotes cellular transformation in colorectal and ovarian cancer contexts [9,10,12]. As the three core SCF complex members interact in an epistatic manner, it is reasonable to assume that specific alterations to any of these proteins will adversely impact complex formation and lead to abnormal increases in Cyclin E1 abundance. Additionally, it was also shown through the use of genetic rescue experiments that the aberrant increases in Cyclin E1 observed following *FBXW7* inactivation in colorectal cancer cells were an underlying mechanism driving increases in CIN [68]. Importantly, SKP2 and FBXW7 regulate Cyclin E1 turnover [2,18,25,46] and thus specific *SKP2* and *FBXW7* mutations, CNAs or altered methylation (i.e., hypermethylation) status are predicted to adversely impact Cyclin E1 abundance and contribute to disease pathogenesis. Thus, it is the collective lack of Cyclin E1 turnover arising from defects in the expression and/or function of SCF complex members that is expected to phenocopy genomic amplification of *CCNE1* and contribute to cancer pathogenesis. Given estimates that the SCF complex targets hundreds to thousands of proteins, it is possible that the aberrant regulation (i.e., turnover) of additional protein substrates, such as P27, RAD51, c-MYC may also contribute to cancer pathogenesis [3–5,7,43,69].

While TCGA data reveal that CNAs of the six SCF complex member genes are common in cancer and are either predominantly gained or lost, it is important to highlight the phenotypic consequences that these CNAs may have in cancer irrespective of their traditionally established oncogenic or tumor suppressive roles. Previous genetic studies have shown that heterozygous knockout of *SKP1*, *CUL1*, *RBX1* and *FBXO7* induces CIN [9–11], an enabling hallmark of cancer [70] frequently associated with cellular transformation, drug resistance, metastasis and poor patient prognosis in many cancer types [71–75]. Indeed, CIN is a dynamic form of genome instability that drives ongoing changes in genetic and cell-to-cell heterogeneity that is proposed to contribute to cancer development by increasing the rate at which oncogenes, tumor suppressor genes and cell cycle regulators are gained, lost, or altered [76–78]. Importantly, the phenotypic consequences corresponding with cancer development (e.g., CIN) following aberrant expression of SCF complex members may vary and are likely to arise via the misregulation of protein substrates in a context-dependent manner. For example, overexpression of SCF complex members may result in excess degradation of protein substrates with tumor suppressor functions, and has been associated with cancer cell stemness, tumor progression and worse patient survival [79–86].

Similarly, amplification and overexpression of *SKP2* promotes the increased degradation of P27 and disease progression [56] and is associated with worse survival outcomes and poor response to therapy in numerous cancer types [87–89], contributing to *SKP2* being traditionally classified as an oncogene. Conversely, other studies have shown that reduced *SKP2* expression leads to aberrant increases in P27 abundance that is suggested to prevent mitotic entry and result in increased nuclear areas [2], a CIN-associated phenotype [70,90–92] suggestive of large increases in DNA/chromosome content (i.e., polyploidy) [69]. In contrast, *FBXW7* has been primarily classified as a tumor suppressor gene [51,52] and loss of SCF^{FBXW7} activity corresponds with aneuploidy, micronucleus formation (extranuclear bodies found outside the primary nucleus and a hallmark of CIN), sister chromatid cohesion defects and chromosome segregation defects (i.e., CIN) [68,93]. However, our data support the possibility that *FBXW7* may also exhibit oncogene-like roles as gains, amplifications and mutational hot-spots were observed [38]. Indeed, Galindo-Moreno and colleagues [94] determined that *FBXW7* amplification in tumors harboring wild-type *TP53* expression reduced breast cancer patient survival but did not impact survival of patients with skin or bladder cancers. Accordingly, these collective data highlight that losses or gains of *FBXW7* (and perhaps other SCF complex member genes), may adversely contribute to cancer pathogenesis in a context-dependent manner [95]. With respect to *FBXO5*, Marzio et al. [3] recently determined that reduced *FBXO5* expression corresponds with a DNA re-replication phenotype (i.e., endoreduplication) in a breast cancer model and that loss of its F-box dependent function leads to increases in RAD51 abundance (a homologous recombination repair protein). In contrast, Vaidyanathan and co-workers [96] determined that *FBXO5* overexpression corresponds with increases in mitotic defects (e.g., lagging and incorrect segregation of chromosomes) and aneuploidy that promotes CIN in transgenic mouse models. Additionally, it is important to note that while each F-box protein targets a distinct array of protein substrates, there is some functional redundancy between F-box proteins, in that multiple F-box proteins can share specific substrates. For example, while SKP2 and FBXW7 exhibit distinct substrate profiles, both target Cyclin E1 and c-MYC for proteolytic degradation [2,7,26,43]. Thus, this inherent functional redundancy may serve as a compensatory mechanism to limit the functional outcomes associated with the aberrant expression and/or function of a given F-box protein. Therefore, with the knowledge that 69 distinct SCF complexes are predicted to target hundreds to thousands of protein substrates, aberrant expression of the core SCF complex members are expected to evoke greater phenotypic consequences than alterations involving the individual F-box proteins. In this regard, future functional studies are required to determine the impacts aberrant expression and function of each individual SCF complex member have on protein substrates, genome instability and disease pathogenesis are now highly warranted.

In summary, this study revealed that the six SCF complex member genes predominantly exhibit missense damaging mutations, a subset of which likely impact SCF complex function by disrupting protein–protein interactions between SCF complex members and their protein substrates. Further, the distribution of the encoded amino acid substitutions and CNA data support the possibility that SCF complex members may exhibit both tumor suppressive or oncogenic roles that will likely be disease and context dependent. Finally, we determined that SCF complex members tend to be differentially methylated across multiple cancer types, which may phenotypically mimic certain CNAs (e.g., shallow deletions) and contribute to cancer pathogenesis but in an epigenetic manner. Ultimately, these alterations may underlie the aberrant accumulation of protein substrates that are implicated in CIN [2,9,10,12,59] and cancer pathogenesis [3,7,26,67,97,98]. Unfortunately however, the complete spectrum of target specificities of all 69 F-box proteins and their implications for disease development remain largely unknown or poorly characterized. Accordingly, functional studies are now required to determine the specific outcomes driven by aberrant expression and/or function of each SCF complex member gene and its impact on substrate accumulation to better understand the molecular determinants giving rise to cancer development, which will be critical to develop innovative precision medicine strategies to better combat the disease. For instance, as aberrant SCF complex function is associated with CIN [9,10,12], developing biomarkers to detect abnormal SCF complex expression could aid early diagnosis in cancers that exhibit high levels of CIN, including colorectal [99] and ovarian cancers [100,101], where early disease detection will be critical to ultimately improve patient outcomes [102,103]. Thus, future studies need to be designed that compile and analyze patient sample datasets for mutations, CNAs and methylation changes with matched normal samples and patient outcomes to gain a comprehensive understanding of how genetic alterations are associated with disease and their impact on patient outcomes. In addition, studies aimed at investigating the pathogenic implications of SCF complex alterations are needed to gain a greater understanding of the etiological events contributing to oncogenesis.

4. Materials and Methods

4.1. Genomic Datasets and Data Collection

Two publicly available TCGA patient-based datasets were utilized for this study, including TCGA Pan-Cancer Atlas [37] (mutation data and gene CNAs) and TCGA Firehose Legacy (methylation data; https://gdac.broadinstitute.org/, accessed on 20 October 2021). Genetic analyses were performed in cBioPortal (https://www.cbioportal.org, accessed on 20 October 2021) [35,36] for SCF complex members *SKP1*, *CUL1*, *RBX1*, *SKP2*, *FBXW7* and *FBXO5*. Patient datasets included bladder urothelial carcinoma, breast invasive carcinoma, colorectal adenocarcinoma, head and neck squamous cell carcinoma, lung adenocarcinoma, ovarian serous cystadenocarcinoma, pancreatic adenocarcinoma, skin cutaneous melanoma, stomach adenocarcinoma and uterine corpus endometrial carcinoma patient samples. All TCGA datasets were accessed no later than 10 November 2021.

4.2. Assessing the Frequency, Distribution and Predicted Functional Impact of Encoded SCF Complex Mutations

SCF complex member SNPs were evaluated to assess the frequency and the predicted functional impact of encoded alterations. Mutational frequencies were calculated as follows: total mutations within a given mutation class (n)/total number of mutations (N) for each gene × 100%. SNPs were assessed using SIFT (https://sift.bii.a-star.edu.sg/, accessed on 9 November 2021) [104] and PolyPhen-2 (http://genetics.bwh.harvard.edu/pph2/) [105]; Note that only single amino acid substitutions can be assessed using PolyPhen-2 and SIFT, where the remaining alterations (e.g., fusions, splice sites) are denoted with not applicable (N/A). For reference purposes, SIFT classifies SNPs as tolerated or deleterious [104], whereas PolyPhen-2 classifies SNPs as benign, possibly damaging or probably damaging [105]. Conceptually, SIFT assesses the conservation of an amino acid position (i.e., sequence identity/similarity) across species, where substitutions occurring at highly conserved positions that are deemed "deleterious" [40]. By contrast, PolyPhen-2 predicts the impact of substitutions on the stability and function of proteins using functional annotation of SNPs, maps coding SNPs to gene transcripts, extracts protein sequence annotations and structural attributes to create a score ranging from 0.0 (benign) to 1.0 (damaging) to ultimately assign it as "benign", "possibly damaging" or "probably damaging". A prediction of "possibly damaging" means that the substitution is predicted to be damaging, but with low confidence [41]. To display the spatial distribution and category of the encoded alterations (e.g., missense, truncating, splice or fusion), lollipop diagrams were extracted from cBioPortal [35–37] and alterations were classified as: (1) missense (unknown significance); (2) missense (putative driver); (3) truncating (unknown significance); (4) truncating (putative driver); (5) splice (unknown significance); or (6) or splice (putative diver). Figures presenting mutational data were generated with Prism v9 (GraphPad, San Diego, CA, USA) and assembled in Photoshop 2022 (Adobe, Toronto, ON, Canada).

4.3. Assessing the Predicted Structural Impact of Encoded SCF Complex Alterations

Crystal structures for SCF complex members were retrieved and visualized using the PDB (https://www.rcsb.org/, accessed on 9 November 2021) [39]: (1) SKP1 (PDB ID: 1FQV [31,39]); (2) CUL1 (PDB ID: ILDJ [39,106]); (3) RBX1 (PDB ID: ILDJ [39,106]); (4) SKP2 (PDB ID: 1FQV [31,39]); and (5) FBXW7 (PDB ID: 2OVP [39,107]). No crystal structure is available for FBXO5 (EMI1). All datasets were accessed no later than 10 November 2021. Figures were assembled in Photoshop.

4.4. Gene Copy Number Alteration and mRNA Expression Analyses

Gene CNA data were extracted from TCGA Pan-Cancer Atlas [37] with CNAs identified using the following Onco Query Language commands in cBioPortal [35,36]: (1) HOMDEL, deep deletion (i.e., loss of 2 alleles); (2) HETLOSS, shallow deletion (i.e., loss of 1 allele); (3) GAIN, small gain (i.e., gain of 1 allele); and (4) AMP, large amplification (i.e., gain of $\geq$ 2 alleles). OncoPrint and mRNA expression data were retrieved using cBioPortal (no later than 10 November 2021) to identify the individual and cumulative frequencies of CNAs (shallow deletions and gains) which were compared with their respective mRNA expression z-scores (z = expression in tumor sample-mean expression in reference sample[diploid])/standard deviation of expression in reference sample[diploid]) for breast, colorectal and lung cancers. Figures presenting CNA data were generated in Prism and assembled in Photoshop.

4.5. Assessing the Methylation Status of SCF Complex Members

The methylation status of each gene was assessed using beta (β) values and corresponding CNA data and extracted from TCGA Firehose Legacy (https://gdac.broadinstitute.org/) for each cancer type. Briefly, β-values were calculated as follows: β = M/M + U, where M > 0 and U > 0 represent the methylated and unmethylated signal intensities measured by the Illumina 27k (ovarian serous cystadenocarcinoma) or 450k BeadChip arrays (all other cancer types), respectively [108]. For each gene, samples were characterized as being hypomethylated ($\beta < 0.2$), partially methylated ($0.2 \leq \beta \leq 0.7$), or hypermethylated ($\beta > 0.7$). All data were accessed no later than 10 November 2021. Figures presenting methylation data were generated in Prism and assembled in Photoshop.

Supplementary Materials: The following are available online at https://www.mdpi.com/article/10.3390/ijms23010084/s1.

Author Contributions: Conceptualization R.C.G., A.C.F., N.M.N. and K.J.M.; formal analyses, R.C.G., A.C.F. and N.M.N.; writing—original draft preparation R.C.G., A.C.F., N.M.N., K.J.M.; writing—review and editing, R.C.G., A.C.F., N.M.N. and K.J.M.; supervision, K.J.M.; funding acquisition, K.J.M. All authors have read and agreed to the published version of the manuscript.

Funding: Operational support is provided by a University of Manitoba Graduate Fellowship (R.C.G.), a Graduate Enhancement of Tri-agency Stipends Award (A.C.F.), a CancerCare Manitoba/Research Manitoba MSc Studentship (N.M.N.), a Natural Sciences and Engineering Research Council of Canada (NSERC) Discovery Grant (K.J.M.; RGPIN: 2018-05007) and a CancerCare Manitoba Foundation Operating Grant (K.J.M.).

Institutional Review Board Statement: Ethical review and approval were waived for this study, due to the use of publicly available datasets from TCGA Research Network (https://www.cancer.gov/tcga) [37] and Oncomine (https://www.oncomine.org) [35,36], accessed on 18 November 2021. No other human specimens or clinical data were utilized.

Informed Consent Statement: All patient data presented in Figures 1–7 are from TCGA Pan-Cancer Atlas and Firehose Legacy datasets made publicly available through cBioPortal (https://cbioportal.org) and Broad Institute (https://gdac.broadinstitute.org/), accessed on 18 November 2021, respectively, and for which informed consent was previously obtained by TCGA.

Data Availability Statement: Patient-related data from Figures 1–7 are based on data available from TCGA Research Network (https://www.cancer.gov/tcga, accessed on 18 November 2021) and accessed online from 20 August to 10 November 2021.

Acknowledgments: We acknowledge that the CancerCare Manitoba Research Institute is located on original lands of Anishinaabeg, Cree, Oji-Cree, Dakota and Dene peoples, and on the homeland of the Métis Nation. We respect the Treaties that were made on these territories and acknowledge the harms and mistakes of the past. We dedicate ourselves to move forward in partnership with Indigenous communities in a spirit of reconciliation and collaboration. We thank members of the McManus laboratory for constructive criticisms during the writing of this review. We also acknowledge the ongoing support of the CancerCare Manitoba Research Institute and the CancerCare Manitoba Foundation.

Conflicts of Interest: The authors declare no conflict of interest. The funders had no role in the design of the study; in the collection, analyses, or interpretation of data; in the writing of the manuscript, or in the decision to publish the results.

References

1. Sung, H.; Ferlay, J.; Siegel, R.L.; Laversanne, M.; Soerjomataram, I.; Jemal, A.; Bray, F. Global cancer statistics 2020: GLOBOCAN estimates of incidence and mortality worldwide for 36 cancers in 185 countries. *CA Cancer J. Clin.* **2021**. [CrossRef] [PubMed]
2. Nakayama, K.; Nagahama, H.; Minamishima, Y.A.; Matsumoto, M.; Nakamichi, I.; Kitagawa, K.; Shirane, M.; Tsunematsu, R.; Tsukiyama, T.; Ishida, N.; et al. Targeted disruption of Skp2 results in accumulation of cyclin E and p27(Kip1), polyploidy and centrosome overduplication. *EMBO J.* **2000**, *19*, 2069–2081. [CrossRef]
3. Marzio, A.; Puccini, J.; Kwon, Y.; Maverakis, N.K.; Arbini, A.; Sung, P.; Bar-Sagi, D.; Pagano, M. The F-Box Domain-Dependent Activity of EMI1 Regulates PARPi Sensitivity in Triple-Negative Breast Cancers. *Mol. Cell* **2019**, *73*, 224–237.e226. [CrossRef] [PubMed]
4. Duda, D.M.; Olszewski, J.L.; Tron, A.E.; Hammel, M.; Lambert, L.J.; Waddell, M.B.; Mittag, T.; DeCaprio, J.A.; Schulman, B.A. Structure of a Glomulin-RBX1-CUL1 Complex: Inhibition of a RING E3 Ligase through Masking of Its E2-Binding Surface. *Mol. Cell* **2012**, *47*, 371–382. [CrossRef] [PubMed]
5. Deshaies, R.J. SCF and Cullin/RING H2-Based Ubiquitin Ligases. *Annu. Rev. Cell Dev. Biol.* **1999**, *15*, 435–467. [CrossRef]
6. Cepeda, D.; Ng, H.F.; Sharifi, H.R.; Mahmoudi, S.; Cerrato, V.S.; Fredlund, E.; Magnusson, K.; Nilsson, H.; Malyukova, A.; Rantala, J.; et al. CDK-mediated activation of the SCFFBXO28 ubiquitin ligase promotes MYC-driven transcription and tumourigenesis and predicts poor survival in breast cancer. *EMBO Mol. Med.* **2013**, *5*, 999–1018. [CrossRef]
7. Kim, S.Y.; Herbst, A.; Tworkowski, K.A.; Salghetti, S.E.; Tansey, W.P. Skp2 regulates Myc protein stability and activity. *Mol. Cell* **2003**, *11*, 1177–1188. [CrossRef]
8. Zhang, Q.; Spears, E.; Boone, D.N.; Li, Z.; Gregory, M.A.; Hanna, S.R. Domain-specific c-Myc ubiquitylation controls c-Myc transcriptional and apoptotic activity. *Proc. Natl. Acad. Sci. USA* **2013**, *110*, 978–983. [CrossRef]
9. Bungsy, M.; Palmer, M.C.L.; Jeusset, L.M.; Neudorf, N.M.; Lichtensztejn, Z.; Nachtigal, M.W.; McManus, K.J. Reduced RBX1 expression induces chromosome instability and promotes cellular transformation in high-grade serous ovarian cancer precursor cells. *Cancer Lett.* **2021**, *500*, 194–207. [CrossRef]
10. Lepage, C.C.; Palmer, M.C.L.; Farrell, A.C.; Neudorf, N.M.; Lichtensztejn, Z.; Nachtigal, M.W.; McManus, K.J. Reduced SKP1 and CUL1 expression underlies increases in Cyclin E1 and chromosome instability in cellular precursors of high-grade serous ovarian cancer. *Br. J. Cancer* **2021**, *124*, 1699–1710. [CrossRef]
11. Palmer, M.C.L.; Neudorf, N.M.; Farrell, A.C.; Razi, T.; Lichtensztein, Z.; McManus, K.J. The F-box protein, FBXO7 is required to maintain chromosome stability in humans. *Hum. Mol. Genet.* 2021; in press. [CrossRef] [PubMed]
12. Thompson, L.L.; Baergen, A.K.; Lichtensztejn, Z.; McManus, K.J. Reduced SKP1 expression induces chromosome instability through aberrant cyclin E1 protein turnover. *Cancers* **2020**, *12*, 531. [CrossRef] [PubMed]
13. Hershko, A.; Ciechanover, A. THE UBIQUITIN SYSTEM. *Annu. Rev. Biochem.* **1998**, *67*, 425–479. [CrossRef] [PubMed]
14. Loda, M.; Cukor, B.; Tam, S.W.; Lavin, P.; Fiorentino, M.; Draetta, G.F.; Jessup, J.M.; Pagano, M. Increased proteasome-dependent degradation of the cyclin-dependent kinase inhibitor p27 in aggressive colorectal carcinomas. *Nat. Med.* **1997**, *3*, 231–234. [CrossRef]
15. Joazeiro, C.A.; Wing, S.S.; Huang, H.; Leverson, J.D.; Hunter, T.; Liu, Y.C. The tyrosine kinase negative regulator c-Cbl as a RING-type, E2-dependent ubiquitin-protein ligase. *Science* **1999**, *286*, 309–312. [CrossRef] [PubMed]
16. Scheffner, M.; Werness, B.A.; Huibregtse, J.M.; Levine, A.J.; Howley, P.M. The E6 oncoprotein encoded by human papillomavirus types 16 and 18 promotes the degradation of p53. *Cell* **1990**, *63*, 1129–1136. [CrossRef]
17. Waterman, H.; Levkowitz, G.; Alroy, I.; Yarden, Y. The RING finger of c-Cbl mediates desensitization of the epidermal growth factor receptor. *J. Biol. Chem.* **1999**, *274*, 22151–22154. [CrossRef]
18. Nakayama, K.I.; Hatakeyama, S.; Nakayama, K. Regulation of the cell cycle at the G1-S transition by proteolysis of cyclin E and p27Kip1. *Biochem. Biophys. Res. Commun.* **2001**, *282*, 853–860. [CrossRef]

19. Bassermann, F.; Eichner, R.; Pagano, M. The ubiquitin proteasome system - Implications for cell cycle control and the targeted treatment of cancer. *Biochim. Biophys. Acta* **2014**, *1843*, 150–162. [CrossRef] [PubMed]
20. Thompson, L.L.; Rutherford, K.A.; Lepage, C.C.; McManus, K.J. The SCF Complex Is Essential to Maintain Genome and Chromosome Stability. *Int. J. Mol. Sci.* **2021**, *22*, 8544. [CrossRef]
21. Kleiger, G.; Mayor, T. Perilous journey: A tour of the ubiquitin-proteasome system. *Trends Cell Biol.* **2014**, *24*, 352–359. [CrossRef]
22. Buetow, L.; Gabrielsen, M.; Huang, D.T. Single-Turnover RING/U-Box E3-Mediated Lysine Discharge Assays. *Methods Mol. Biol.* **2018**, *1844*, 19–31. [CrossRef]
23. Pickart, C.M. Mechanisms Underlying Ubiquitination. *Annu. Rev. Biochem.* **2001**, *70*, 503–533. [CrossRef]
24. Thrower, J.S. Recognition of the polyubiquitin proteolytic signal. *EMBO J.* **2000**, *19*, 94–102. [CrossRef]
25. Tetzlaff, M.T.; Yu, W.; Li, M.; Zhang, P.; Finegold, M.; Mahon, K.; Harper, J.W.; Schwartz, R.J.; Elledge, S.J. Defective cardiovascular development and elevated cyclin E and Notch proteins in mice lacking the Fbw7 F-box protein. *Proc. Natl. Acad. Sci. USA* **2004**, *101*, 3338–3345. [CrossRef]
26. Yada, M.; Hatakeyama, S.; Kamura, T.; Nishiyama, M.; Tsunematsu, R.; Imaki, H.; Ishida, N.; Okumura, F.; Nakayama, K.; Nakayama, K.I. Phosphorylation-dependent degradation of c-Myc is mediated by the F-box protein Fbw7. *EMBO J.* **2004**, *23*, 2116–2125. [CrossRef] [PubMed]
27. Hussain, M.; Lu, Y.; Liu, Y.Q.; Su, K.; Zhang, J.; Liu, J.; Zhou, G.B. Skp1: Implications in cancer and SCF-oriented anti-cancer drug discovery. *Pharmacol. Res.* **2016**, *111*, 34–42. [CrossRef]
28. Petroski, M.D.; Deshaies, R.J. Function and regulation of cullin-RING ubiquitin ligases. *Nat. Rev. Mol. Cell Biol.* **2005**, *6*, 9–20. [CrossRef] [PubMed]
29. Dias, D.C.; Dolios, G.; Wang, R.; Pan, Z.Q. CUL7: A DOC domain-containing cullin selectively binds Skp1·Fbx29 to form an SCF-like complex. *Proc. Natl. Acad. Sci. USA* **2002**, *99*, 16601–16606. [CrossRef] [PubMed]
30. Cardozo, T.; Pagano, M. The SCF ubiquitin ligase: Insights into a molecular machine. *Nat. Rev. Mol. Cell Biol.* **2004**, *5*, 739–751. [CrossRef]
31. Schulman, B.A.; Carrano, A.C.; Jeffrey, P.D.; Bowen, Z.; Kinnucan, E.R.; Finnin, M.S.; Elledge, S.J.; Harper, J.W.; Pagano, M.; Pavletich, N.P. Insights into SCF ubiquitin ligases from the structure of the Skp1-Skp2 complex. *Nature* **2000**, *408*, 381–386. [CrossRef] [PubMed]
32. Yoshida, Y.; Murakami, A.; Tanaka, K. Skp1 stabilizes the conformation of F-box proteins. *Biochem. Biophys. Res. Commun.* **2011**, *410*, 24–28. [CrossRef] [PubMed]
33. Kamura, T.; Koepp, D.M.; Conrad, M.N.; Skowyra, D.; Moreland, R.J.; Iliopoulos, O.; Lane, W.S.; Kaelin, W.G.; Elledge, S.J.; Conaway, R.C.; et al. Rbx1, a component of the VHL tumor suppressor complex and SCF ubiquitin ligase. *Science* **1999**, *284*, 657–661. [CrossRef] [PubMed]
34. Silverman, J.S.; Skaar, J.R.; Pagano, M. SCF ubiquitin ligases in the maintenance of genome stability. *Trends Biochem. Sci.* **2012**, *37*, 66–73. [CrossRef]
35. Cerami, E.; Gao, J.; Dogrusoz, U.; Gross, B.E.; Sumer, S.O.; Aksoy, B.A.; Jacobsen, A.; Byrne, C.J.; Heuer, M.L.; Larsson, E.; et al. The cBio cancer genomics portal: An open platform for exploring multidimensional cancer genomics data. *Cancer Discov.* **2012**, *2*, 401–404. [CrossRef]
36. Gao, J.; Aksoy, B.A.; Dogrusoz, U.; Dresdner, G.; Gross, B.; Sumer, S.O.; Sun, Y.; Jacobsen, A.; Sinha, R.; Larsson, E.; et al. Integrative analysis of complex cancer genomics and clinical profiles using the cBioPortal. *Sci. Signal.* **2013**, *6*, pl1. [CrossRef]
37. Hoadley, K.A.; Yau, C.; Hinoue, T.; Wolf, D.M.; Lazar, A.J.; Drill, E.; Shen, R.; Taylor, A.M.; Cherniack, A.D.; Thorsson, V.; et al. Cell-of-Origin Patterns Dominate the Molecular Classification of 10,000 Tumors from 33 Types of Cancer. *Cell* **2018**, *173*, 291–304.e296. [CrossRef]
38. Liu, H.; Xing, Y.; Yang, S.; Tian, D. Remarkable difference of somatic mutation patterns between oncogenes and tumor suppressor genes. *Oncol. Rep.* **2011**, *26*, 1539–1546. [CrossRef]
39. Sehnal, D.; Bittrich, S.; Deshpande, M.; Svobodová, R.; Berka, K.; Bazgier, V.; Velankar, S.; Burley, S.K.; Koca, J.; Rose, A.S. Mol* Viewer: Modern web app for 3D visualization and analysis of large biomolecular structures. *Nucleic Acis Res.* **2021**, *49*, W431–W437. [CrossRef]
40. Ng, P.C.; Henikoff, S. Predicting deleterious amino acid substitutions. *Genome Res.* **2001**, *11*, 863–874. [CrossRef] [PubMed]
41. Adzhubei, I.; Jordan, D.M.; Sunyaev, S.R. Predicting functional effect of human missense mutations using PolyPhen-2. *Curr. Protoc. Hum. Genet.* **2013**, *76*, 7–20. [CrossRef] [PubMed]
42. Latres, E.; Chiarle, R.; Schulman, B.A.; Pavletich, N.P.; Pellicer, A.; Inghirami, G.; Pagano, M. Role of the F-box protein Skp2 in lymphomagenesis. *Proc. Natl. Acad. Sci. USA* **2001**, *98*, 2515–2520. [CrossRef]
43. Von Der Lehr, N.; Johansson, S.; Wu, S.; Bahram, F.; Castell, A.; Cetinkaya, C.; Hydbring, P.; Weidung, I.; Nakayama, K.; Nakayama, K.I.; et al. The F-box protein Skp2 participates in c-Myc proteosomal degradation and acts as a cofactor for c-Myc-regulated transcription. *Mol. Cell* **2003**, *11*, 1189–1200. [CrossRef]
44. Ng, R.W.M.; Arooz, T.; Yam, C.H.; Chan, I.W.Y.; Lau, A.W.S.; Poon, R.Y.C. Characterization of the cullin and F-box protein partner Skp1. *FEBS Lett.* **1998**, *438*, 183–189. [CrossRef]
45. Xu, C.; Min, J. Structure and function of WD40 domain proteins. *Protein Cell* **2011**, *2*, 202–214. [CrossRef]
46. Strohmaier, H.; Spruck, C.H.; Kaiser, P.; Won, K.A.; Sangfelt, O.; Reed, S.I. Human F-box protein hCdc4 targets cyclin E for proteolysis and is mutated in a breast cancer cell line. *Nature* **2001**, *413*, 316–322. [CrossRef] [PubMed]

47. Koepp, D.M.; Schaefer, L.K.; Ye, X.; Keyomarsi, K.; Chu, C.; Harper, J.W.; Elledge, S.J. Phosphorylation-dependent ubiquitination of cyctin E by the SCFFbw7 ubiquitin ligase. *Science* **2001**, *294*, 173–177. [CrossRef]
48. Guccione, E.; Richard, S. The regulation, functions and clinical relevance of arginine methylation. *Nat. Rev. Mol. Cell Biol.* **2019**, *20*, 642–657. [CrossRef]
49. Blanc, R.S.; Richard, S. Arginine Methylation: The Coming of Age. *Mol. Cell* **2017**, *65*, 8–24. [CrossRef]
50. Siegfried, Z.; Simon, I. DNA methylation and gene expression. *Wiley Interdiscip. Rev. Syst. Biol. Med.* **2010**, *2*, 362–371. [CrossRef]
51. Akhoondi, S.; Sun, D.; von der Lehr, N.; Apostolidou, S.; Klotz, K.; Maljukova, A.; Cepeda, D.; Fiegl, H.; Dafou, D.; Marth, C.; et al. FBXW7/hCDC4 is a general tumor suppressor in human cancer. *Cancer Res.* **2007**, *67*, 9006–9012. [CrossRef] [PubMed]
52. Yeh, C.H.; Bellon, M.; Nicot, C. FBXW7: A critical tumor suppressor of human cancers. *Mol. Cancer* **2018**, *17*, 115. [CrossRef]
53. Bretones, G.; Acosta, J.C.; Caraballo, J.M.; Ferrandiz, N.; Gomez-Casares, M.T.; Albajar, M.; Blanco, R.; Ruiz, P.; Hung, W.C.; Albero, M.P.; et al. SKP2 oncogene is a direct MYC target gene and MYC down-regulates p27(KIP1) through SKP2 in human leukemia cells. *J. Biol. Chem.* **2011**, *286*, 9815–9825. [CrossRef] [PubMed]
54. Li, C.; Du, L.; Ren, Y.; Liu, X.; Jiao, Q.; Cui, D.; Wen, M.; Wang, C.; Wei, G.; Wang, Y.; et al. SKP2 promotes breast cancer tumorigenesis and radiation tolerance through PDCD4 ubiquitination. *J. Exp. Clin. Cancer Res.* **2019**, *38*, 76. [CrossRef] [PubMed]
55. Wang, W.; Ungermannova, D.; Chen, L.; Liu, X. A negatively charged amino acid in Skp2 is required for Skp2-Cks1 interaction and ubiquitination of p27Kip1. *J. Biol. Chem.* **2003**, *278*, 32390–32396. [CrossRef]
56. Jia, L.; Soengas, M.S.; Sun, Y. ROC1/RBX1 E3 ubiquitin ligase silencing suppresses tumor cell growth via sequential induction of G2-M arrest, apoptosis, and senescence. *Cancer Res.* **2009**, *69*, 4974–4982. [CrossRef]
57. Abbastabar, M.; Kheyrollah, M.; Azizian, K.; Bagherlou, N.; Tehrani, S.S.; Maniati, M.; Karimian, A. Multiple functions of p27 in cell cycle, apoptosis, epigenetic modification and transcriptional regulation for the control of cell growth: A double-edged sword protein. *DNA Repair* **2018**, *69*, 63–72. [CrossRef]
58. Gorski, J.W.; Ueland, F.R.; Kolesar, J.M. CCNE1 Amplification as a Predictive Biomarker of Chemotherapy Resistance in Epithelial Ovarian Cancer. *Diagnostics* **2020**, *10*, 279. [CrossRef]
59. Aziz, K.; Limzerwala, J.F.; Sturmlechner, I.; Hurley, E.; Zhang, C.; Jeganathan, K.B.; Nelson, G.; Bronk, S.; Fierro Velasco, R.O.; van Deursen, E.J.; et al. Ccne1 Overexpression Causes Chromosome Instability in Liver Cells and Liver Tumor Development in Mice. *Gastroenterology* **2019**, *157*, 210–226.e212. [CrossRef]
60. Dhanasekaran, R.; Deutzmann, A.; Mahauad-Fernandez, W.D.; Hansen, A.S.; Gouw, A.M.; Felsher, D.W. The MYC oncogene—The grand orchestrator of cancer growth and immune evasion. *Nat. Rev. Clin. Oncol.* **2021**, *19*, 23–36. [CrossRef]
61. Dong, Y.; Tu, R.; Liu, H.; Qing, G. Regulation of cancer cell metabolism: Oncogenic MYC in the driver's seat. *Signal Transduct. Target. Ther.* **2020**, *5*, 124. [CrossRef] [PubMed]
62. Ecker, A.; Simma, O.; Hoelbl, A.; Kenner, L.; Beug, H.; Moriggl, R.; Sexl, V. The dark and the bright side of Stat3: Proto-oncogene and tumor-suppressor. *Front. Biosci.* **2009**, *14*, 2944–2958. [CrossRef] [PubMed]
63. Nesbit, C.E.; Tersak, J.M.; Prochownik, E.V. MYC oncogenes and human neoplastic disease. *Oncogene* **1999**, *18*, 3004–3016. [CrossRef]
64. Orlicky, S.; Tang, X.; Willems, A.; Tyers, M.; Sicheri, F. Structural basis for phosphodependent substrate selection and orientation by the SCFCdc4 ubiquitin ligase. *Cell* **2003**, *112*, 243–256. [CrossRef]
65. Postow, L.; Funabiki, H. An SCF complex containing Fbxl12 mediates DNA damage-induced Ku80 ubiquitylation. *Cell Cycle* **2013**, *12*, 587–595. [CrossRef]
66. Etemadmoghadam, D.; Weir, B.A.; Au-Yeung, G.; Alsop, K.; Mitchell, G.; George, J.; Australian Ovarian Cancer Study, G.; Davis, S.; D'Andrea, A.D.; Simpson, K.; et al. Synthetic lethality between CCNE1 amplification and loss of BRCA1. *Proc. Natl. Acad. Sci. USA* **2013**, *110*, 19489–19494. [CrossRef]
67. Karst, A.M.; Jones, P.M.; Vena, N.; Ligon, A.H.; Liu, J.F.; Hirsch, M.S.; Etemadmoghadam, D.; Bowtell, D.D.; Drapkin, R. Cyclin E1 deregulation occurs early in secretory cell transformation to promote formation of fallopian tube-derived high-grade serous ovarian cancers. *Cancer Res.* **2014**, *74*, 1141–1152. [CrossRef]
68. Rajagopalan, H.; Jallepalli, P.V.; Rago, C.; Velculescu, V.E.; Kinzler, K.W.; Vogelstein, B.; Lengauer, C. Inactivation of hCDC4 can cause chromosomal instability. *Nature* **2004**, *428*, 77–81. [CrossRef]
69. Nakayama, K.; Nagahama, H.; Minamishima, Y.A.; Miyake, S.; Ishida, N.; Hatakeyama, S.; Kitagawa, M.; Iemura, S.; Natsume, T.; Nakayama, K.I. Skp2-mediated degradation of p27 regulates progression into mitosis. *Dev. Cell* **2004**, *6*, 661–672. [CrossRef]
70. Hanahan, D.; Weinberg, R.A. Hallmarks of cancer: The next generation. *Cell* **2011**, *144*, 646–674. [CrossRef]
71. Nowak, M.A.; Komarova, N.L.; Sengupta, A.; Jallepalli, P.V.; Shih Ie, M.; Vogelstein, B.; Lengauer, C. The role of chromosomal instability in tumor initiation. *Proc. Natl. Acad. Sci. USA* **2002**, *99*, 16226–16231. [CrossRef] [PubMed]
72. Vishwakarma, R.; McManus, K.J. Chromosome Instability; Implications in Cancer Development, Progression, and Clinical Outcomes. *Cancers* **2020**, *12*, 824. [CrossRef]
73. Gao, C.; Su, Y.; Koeman, J.; Haak, E.; Dykema, K.; Essenberg, C.; Hudson, E.; Petillo, D.; Khoo, S.K.; Vande Woude, G.F. Chromosome instability drives phenotypic switching to metastasis. *Proc. Natl. Acad. Sci. USA* **2016**, *113*, 14793–14798. [CrossRef] [PubMed]
74. Lee, A.J.; Endesfelder, D.; Rowan, A.J.; Walther, A.; Birkbak, N.J.; Futreal, P.A.; Downward, J.; Szallasi, Z.; Tomlinson, I.P.; Howell, M.; et al. Chromosomal instability confers intrinsic multidrug resistance. *Cancer Res.* **2011**, *71*, 1858–1870. [CrossRef]

75. Carter, S.L.; Eklund, A.C.; Kohane, I.S.; Harris, L.N.; Szallasi, Z. A signature of chromosomal instability inferred from gene expression profiles predicts clinical outcome in multiple human cancers. *Nat. Genet.* **2006**, *38*, 1043–1048. [CrossRef] [PubMed]
76. Lepage, C.C.; Morden, C.R.; Palmer, M.C.L.; Nachtigal, M.W.; McManus, K.J. Detecting Chromosome Instability in Cancer: Approaches to Resolve Cell-to-Cell Heterogeneity. *Cancers* **2019**, *11*, 226. [CrossRef]
77. Thompson, L.L.; McManus, K.J. A novel multiplexed, image-based approach to detect phenotypes that underlie chromosome instability in human cells. *PLoS ONE* **2015**, *10*, e0123200. [CrossRef] [PubMed]
78. Thompson, L.L.; Jeusset, L.M.; Lepage, C.C.; McManus, K.J. Evolving Therapeutic Strategies to Exploit Chromosome Instability in Cancer. *Cancers* **2017**, *9*, 151. [CrossRef]
79. Wang, W.; Qiu, J.; Liu, Z.; Zeng, Y.; Fan, J.; Liu, Y.; Guo, Y. Overexpression of RING box protein-1 (RBX1) associated with poor prognosis of non-muscle-invasive bladder transitional cell carcinoma. *J. Surg. Oncol.* **2013**, *107*, 758–761. [CrossRef]
80. Kunishige, T.; Migita, K.; Matsumoto, S.; Wakatsuki, K.; Nakade, H.; Miyao, S.; Kuniyasu, H.; Sho, M. Ring box protein-1 is associated with a poor prognosis and tumor progression in esophageal cancer. *Oncol. Lett.* **2020**, *20*, 2919–2927. [CrossRef]
81. Yang, D.; Zhao, Y.; Liu, J.; Sun, Y.; Jia, L. Protective autophagy induced by RBX1/ROC1 knockdown or CRL inactivation via modulating the DEPTOR-MTOR axis. *Autophagy* **2012**, *8*, 1856–1858. [CrossRef]
82. Liu, Y.Q.; Wang, X.L.; Cheng, X.; Lu, Y.Z.; Wang, G.Z.; Li, X.C.; Zhang, J.; Wen, Z.S.; Huang, Z.L.; Gao, Q.L.; et al. Skp1 in lung cancer: Clinical significance and therapeutic efficacy of its small molecule inhibitors. *Oncotarget* **2015**, *6*, 34953–34967. [CrossRef] [PubMed]
83. Tian, C.; Lang, T.; Qiu, J.; Han, K.; Zhou, L.; Min, D.; Zhang, Z.; Qi, D. SKP1 promotes YAP-mediated colorectal cancer stemness via suppressing RASSF1. *Cancer Cell. Int.* **2020**, *20*, 579. [CrossRef] [PubMed]
84. Mao, S.Y.; Xiong, D.B.; Huang, T.B.; Zheng, J.H.; Yao, X.D. Expression of CUL1 correlates with tumour-grade and recurrence in urothelial carcinoma. *ANZ J. Surg.* **2017**, *87*, 624–629. [CrossRef] [PubMed]
85. Chen, G.; Li, G. Increased Cul1 expression promotes melanoma cell proliferation through regulating p27 expression. *Int. J. Oncol.* **2010**, *37*, 1339–1344. [CrossRef] [PubMed]
86. Bai, J.; Zhou, Y.; Chen, G.; Zeng, J.; Ding, J.; Tan, Y.; Zhou, J.; Li, G. Overexpression of Cullin1 is associated with poor prognosis of patients with gastric cancer. *Hum. Pathol.* **2011**, *42*, 375–383. [CrossRef]
87. Hershko, D.D. Oncogenic properties and prognostic implications of the ubiquitin ligase Skp2 in cancer. *Cancer* **2008**, *112*, 1415–1424. [CrossRef] [PubMed]
88. Asmamaw, M.D.; Liu, Y.; Zheng, Y.C.; Shi, X.J.; Liu, H.M. Skp2 in the ubiquitin-proteasome system: A comprehensive review. *Med. Res. Rev.* **2020**, *40*, 1920–1949. [CrossRef]
89. Cai, Z.; Moten, A.; Peng, D.; Hsu, C.C.; Pan, B.S.; Manne, R.; Li, H.Y.; Lin, H.K. The Skp2 Pathway: A Critical Target for Cancer Therapy. *Semin. Cancer Biol.* **2020**, *67*, 16–33. [CrossRef]
90. Petersen, I.; Kotb, W.F.; Friedrich, K.H.; Schluns, K.; Bocking, A.; Dietel, M. Core classification of lung cancer: Correlating nuclear size and mitoses with ploidy and clinicopathological parameters. *Lung Cancer* **2009**, *65*, 312–318. [CrossRef]
91. Zeimet, A.G.; Fiegl, H.; Goebel, G.; Kopp, F.; Allasia, C.; Reimer, D.; Steppan, I.; Mueller-Holzner, E.; Ehrlich, M.; Marth, C. DNA ploidy, nuclear size, proliferation index and DNA-hypomethylation in ovarian cancer. *Gynecol. Oncol.* **2011**, *121*, 24–31. [CrossRef]
92. Geigl, J.B.; Obenauf, A.C.; Schwarzbraun, T.; Speicher, M.R. Defining 'chromosomal instability'. *Trends Genet.* **2008**, *24*, 64–69. [CrossRef]
93. Barber, T.D.; McManus, K.; Yuen, K.W.; Reis, M.; Parmigiani, G.; Shen, D.; Barrett, I.; Nouhi, Y.; Spencer, F.; Markowitz, S.; et al. Chromatid cohesion defects may underlie chromosome instability in human colorectal cancers. *Proc. Natl. Acad. Sci. USA* **2008**, *105*, 3443–3448. [CrossRef] [PubMed]
94. Galindo-Moreno, M.; Giraldez, S.; Limon-Mortes, M.C.; Belmonte-Fernandez, A.; Reed, S.I.; Saez, C.; Japon, M.A.; Tortolero, M.; Romero, F. SCF(FBXW7)-mediated degradation of p53 promotes cell recovery after UV-induced DNA damage. *FASEB J.* **2019**, *33*, 11420–11430. [CrossRef]
95. Orr, B.; Compton, D.A. A double-edged sword: How oncogenes and tumor suppressor genes can contribute to chromosomal instability. *Front. Oncol.* **2013**, *3*, 1–14. [CrossRef] [PubMed]
96. Vaidyanathan, S.; Cato, K.; Tang, L.; Pavey, S.; Haass, N.K.; Gabrielli, B.G.; Duijf, P.H. In vivo overexpression of Emi1 promotes chromosome instability and tumorigenesis. *Oncogene* **2016**, *35*, 5446–5455. [CrossRef]
97. Erlanson, M.; Landberg, G. Prognostic implications of p27 and cyclin E protein contents in malignant lymphomas. *Leuk. Lymphoma* **2001**, *40*, 461–470. [CrossRef]
98. Takada, M.; Zhang, W.; Suzuki, A.; Kuroda, T.S.; Yu, Z.; Inuzuka, H.; Gao, D.; Wan, L.; Zhuang, M.; Hu, L.; et al. FBW7 Loss Promotes Chromosomal Instability and Tumorigenesis via Cyclin E1/CDK2-Mediated Phosphorylation of CENP-A. *Cancer Res.* **2017**, *77*, 4881–4893. [CrossRef] [PubMed]
99. Cisyk, A.L.; Penner-Goeke, S.; Lichtensztejn, Z.; Nugent, Z.; Wightman, R.H.; Singh, H.; McManus, K.J. Characterizing the prevalence of chromosome instability in interval colorectal cancer. *Neoplasia* **2015**, *17*, 306–316. [CrossRef] [PubMed]
100. Penner-Goeke, S.; Lichtensztejn, Z.; Neufeld, M.; Ali, J.L.; Altman, A.D.; Nachtigal, M.W.; McManus, K.J. The temporal dynamics of chromosome instability in ovarian cancer cell lines and primary patient samples. *PLoS Genet.* **2017**, *13*, e1006707. [CrossRef]
101. Morden, C.R.; Farrell, A.C.; Sliwowski, M.; Lichtensztejn, Z.; Altman, A.D.; Nachtigal, M.W.; McManus, K.J. Chromosome instability is prevalent and dynamic in high-grade serous ovarian cancer patient samples. *Gynecol. Oncol.* **2021**, *161*, 769–778. [CrossRef] [PubMed]

102. Bonifácio, V.D.B. Ovarian Cancer Biomarkers: Moving Forward in Early Detection. *Adv. Exp. Med. Biol.* **2020**, *1219*, 355–363. [CrossRef]
103. Moore, J.S.; Aulet, T.H. Colorectal Cancer Screening. *Surg. Clin. N. Am.* **2017**, *97*, 487–502. [CrossRef]
104. Vaser, R.; Adusumalli, S.; Leng, S.N.; Sikic, M.; Ng, P.C. SIFT missense predictions for genomes. *Nat. Protoc.* **2016**, *11*, 1–9. [CrossRef]
105. Adzhubei, I.A.; Schmidt, S.; Peshkin, L.; Ramensky, V.E.; Gerasimova, A.; Bork, P.; Kondrashov, A.S.; Sunyaev, S.R. A method and server for predicting damaging missense mutations. *Nat. Methods* **2010**, *7*, 248–249. [CrossRef]
106. Zheng, N.; Schulman, B.A.; Song, L.; Miller, J.J.; Jeffrey, P.D.; Wang, P.; Chu, C.; Koepp, D.M.; Elledge, S.J.; Pagano, M.; et al. Structure of the Cul1-Rbx1-Skp1-F boxSkp2 SCF ubiquitin ligase complex. *Nature* **2002**, *416*, 703–709. [CrossRef] [PubMed]
107. Hao, B.; Oehlmann, S.; Sowa, M.E.; Harper, J.W.; Pavletich, N.P. Structure of a Fbw7-Skp1-Cyclin E Complex: Multisite-Phosphorylated Substrate Recognition by SCF Ubiquitin Ligases. *Mol. Cell* **2007**, *26*, 131–143. [CrossRef] [PubMed]
108. Weinhold, L.; Wahl, S.; Pechlivanis, S.; Hoffmann, P.; Schmid, M. A statistical model for the analysis of beta values in DNA methylation studies. *BMC Bioinform.* **2016**, *17*, 480. [CrossRef]

International Journal of
Molecular Sciences

MDPI

Article

Synergistic Effects of Weighted Genetic Risk Scores and Resistin and sST2 Levels on the Prognostication of Long-Term Outcomes in Patients with Coronary Artery Disease

Hsin-Hua Chou [1,2], Lung-An Hsu [3], Jyh-Ming Jimmy Juang [4,5], Fu-Tien Chiang [4,5,6], Ming-Sheng Teng [7], Semon Wu [7,8] and Yu-Lin Ko [1,2,7,*]

1 Division of Cardiology, Department of Internal Medicine, Taipei Tzu Chi Hospital, Buddhist Tzu Chi Medical Foundation, New Taipei City 23142, Taiwan; chouhhtw@gmail.com
2 School of Medicine, Tzu Chi University, Hualien 97004, Taiwan
3 Cardiovascular Division, Department of Internal Medicine, Chang Gung Memorial Hospital, Chang Gung University College of Medicine, Taoyuan 33305, Taiwan; hsula@adm.cgmh.org.tw
4 Cardiovascular Center and Division of Cardiology, Department of Internal Medicine, National Taiwan University Hospital, Taipei 10002, Taiwan; jjmjuang@ntu.edu.tw (J.-M.J.J.); futienc@ntuh.gov.tw (F.-T.C.)
5 College of Medicine, National Taiwan University, Taipei 10051, Taiwan
6 Cardiovascular Center and Division of Cardiology, Fu-Jen Catholic University Hospital, New Taipei City 24352, Taiwan
7 Department of Research, Taipei Tzu Chi Hospital, Buddhist Tzu Chi Medical Foundation, New Taipei City 23142, Taiwan; vincent@tzuchi.com.tw (M.-S.T.); semonwu@yahoo.com.tw (S.W.)
8 Department of Life Science, Chinese Culture University, Taipei 11114, Taiwan
* Correspondence: yulinkotw@yahoo.com.tw; Tel.: +886-2-6628-9779 (ext. 5355); Fax: +886-2-6628-9009

Abstract: Resistin and soluble suppression of tumorigenicity 2 (sST2) are useful predictors in patients with coronary artery disease (CAD). Their serum levels are significantly attributed to variations in *RETN* and *IL1RL1* loci. We investigated candidate variants in the *RETN locus* for resistin levels and those in the *IL1RL1* locus for sST2 levels and evaluated the prognostication of these two biomarkers and the corresponding variants for long-term outcomes in the patients with CAD. We included 4652, 557, and 512 Chinese participants from the Taiwan Biobank (TWB), cardiovascular health examination (CH), and CAD cohorts, respectively. Candidate variants in *RETN* and *IL1RL1* were investigated using whole-genome sequence (WGS) and genome-wide association study (GWAS) data in the TWB cohort. The weighted genetic risk scores (WGRS) of *RETN* and *IL1RL1* with resistin and sST2 levels were calculated. Kaplan–Meier curves were used to analyze the prognostication of resistin and sST2 levels, WGRS of *RETN* and *IL1RL1*, and their combinations. Three *RETN* variants (rs3219175, rs370006313, and rs3745368) and two *IL1RL1* variants (rs10183388 and rs4142132) were independently associated with resistin and sST2 levels as per the WGS and GWAS data in the TWB cohort and were further validated in the CH and CAD cohorts. In combination, these variants explained 53.7% and 28.0% of the variation in resistin and sST2 levels, respectively. In the CAD cohort, higher resistin and sST2 levels predicted higher rates of all-cause mortality and major adverse cardiac events (MACEs) during long-term follow-up, but WGRS of *RETN* and *IL1RL1* variants had no impact on these outcomes. A synergistic effect of certain combinations of biomarkers with *RETN* and *IL1RL1* variants was found on the prognostication of long-term outcomes: Patients with high resistin levels/low *RETN* WGRS and those with high sST2 levels/low *IL1RL1* WGRS had significantly higher all-cause mortality and MACEs rates, and those with both these combinations had the poorest outcomes. Both higher resistin and sST2 levels, but not *RETN* and *IL1RL1* variants, predict poor long-term outcomes in patients with CAD. Furthermore, combining resistin and sST2 levels with the WGRS of *RETN* and *IL1RL1* genotyping exerts a synergistic effect on the prognostication of CAD outcomes. Future studies including a large sample size of participants with different ethnic populations are needed to verify this finding.

Citation: Chou, H.-H.; Hsu, L.-A.; Juang, J.-M.J.; Chiang, F.-T.; Teng, M.-S.; Wu, S.; Ko, Y.-L. Synergistic Effects of Weighted Genetic Risk Scores and Resistin and sST2 Levels on the Prognostication of Long-Term Outcomes in Patients with Coronary Artery Disease. *Int. J. Mol. Sci.* **2022**, 23, 4292. https://doi.org/10.3390/ijms23084292

Academic Editor: Elixabet Lopez-Lopez

Received: 20 March 2022
Accepted: 11 April 2022
Published: 13 April 2022

Keywords: resistin; soluble suppression of tumorigenicity 2; weighted genetic risk score; Taiwan Biobank; coronary artery disease; all-cause mortality; major adverse cardiac events

1. Introduction

Despite advancements in guideline-directed medical therapy, cardiovascular diseases remain the leading cause of disease burden and mortality worldwide. The global prevalence of cardiovascular disease nearly doubled from 271 million in 1990 to 523 million in 2019, and the number of cardiovascular deaths steadily increased from 12.1 million in 1990 to 18.6 million in 2019, half of which were due to ischemic heart disease [1]. To provide new insights into cardiovascular disease pathophysiology and to improve patient prognosis, investigators in the past two decades have focused on the association of numerous adipokines/cytokines with cardiovascular disease. Among these, resistin and suppression of tumorigenicity 2 (ST2) have attracted increasing interest in the past 10 years.

Resistin, belonging to a family of cysteine-rich proteins known as resistin-like molecules, was originally shown to be synthesized only by adipocytes and was demonstrated to induce insulin resistance in mice [2]. In humans, however, this protein appears to be expressed mainly by monocyte-derived macrophages [3]. As a proinflammatory adipokine in humans, resistin markedly upregulates the expression of inflammatory cytokines and cellular adhesive molecules [4,5]. Resistin appears to mediate the pathogenesis of atherosclerosis by promoting endothelial dysfunction, vascular smooth muscle cell proliferation, arterial inflammation, and foam cell formation [6,7]. It is predictive of atherosclerosis and poor clinical outcomes in patients with coronary artery disease (CAD), ischemic stroke, and congestive heart failure [7,8].

Approximately 70% of the observed variation in circulating resistin levels may be attributable to genetic factors [9], and several candidate genetic loci for resistin levels have been identified in different ethnic populations [9–12]. Genetic polymorphisms around the *RETN* locus are also related to resistin levels and various metabolic phenotypes, with discrepant results [11–16].

The ST2 receptor, a member of the interleukin (IL)-1 receptor family encoded by the IL-1 receptor-like 1 (*IL1RL1*) gene, is expressed as a membrane-bound receptor variant form (ST2L) and a truncated soluble form (sST2) [17]. IL-33, a biomechanically induced protein predominantly synthesized by cardiac fibroblasts, is the functional ligand of ST2L [18,19]. The IL-33/ST2 signaling pathway is upregulated in cardiomyocytes and fibroblasts in response to mechanical stimulation or injury and is cardioprotective in terms of reducing myocardial fibrosis, preventing cardiomyocyte hypertrophy, reducing apoptosis, and improving myocardial function [19,20]. By contrast, sST2 binds to IL-33 and acts as a "decoy" receptor for IL-33 to inhibit IL-33/ST2L signaling [19]. An elevated sST2 level is an independent predictor of subsequent mortality in patients with heart failure, acute myocardial infarction, and stable CAD [21–25].

IL1RL1 variations can affect sST2 levels. In the genome-wide association study (GWAS) of the Framingham offspring cohort, up to 45% of the variation in sST2 levels not explained by clinical variables was attributed to genetic factors [26]. The most significant single nucleotide polymorphism (SNP) is rs950880, accounting for 12% of the individual variability in circulating sST2 levels. SNPs in the *IL1RL1* gene have also been linked to the severity of several immune and inflammatory diseases [27]. However, the effects of *IL1RL1* on predicting the outcome of cardiovascular disease remain unclear.

Our recent study indicated that individuals with the rs950880 AA genotype tended to have lower sST2 levels [28]; this genotype was an independent predictor of all-cause mortality in patients with CAD and lower-extremity arterial disease, and patients with high sST2 levels and the rs950880 AA genotype had the lowest survival rate. However, whether other biomarker levels can predict long-term outcomes in patients with CAD when combined with level-determining genotypes remains unknown. The Taiwan Biobank

(TWB) conducted a large-scale population-based cohort study on 30–70-year-old volunteers with no history of cancer [29]. The genetic determinants of resistin and sST2 levels were derived from a regional association plot analysis using whole-genome sequence (WGS) data in a subgroup of 859 participants from the TWB cohort and from GWAS in 5000 participants from the TWB cohort. The weighted genetic risk scores (WGRS) of *RETN* and *IL1RL1* with resistin and sST2 levels were calculated using the data of independent level-determining genotypes. We hypothesized that WGRS of *RETN* and *IL1RL1* combined with both biomarker levels may better predict the long-term outcomes of patients with CAD.

2. Results

2.1. WGS Revealed Candidate SNPs in RETN and IL1RL1 Gene Loci

Given the previously reported *RETN* gene as the candidate locus for resistin levels, we first performed a regional association plot study with conditional analysis using data from 859 participants from the TWB cohort and the significance of resistin levels with 509 SNPs at positions between 7.715 and 7.755 Mb on chromosome 19p13.2 around the *RETN* gene region was assessed. Our data revealed that 17 SNPs exceeded the genome-wide significance threshold ($p < 5 \times 10^{-8}$), with rs3219175 being the lead SNP ($p = 4.24 \times 10^{-72}$) (Supplementary Figure S1A). To clarify whether the association of other *RETN* SNPs was independent of the lead SNP, we performed a stepwise conditional analysis. With adjustment for rs3219175, rs370006313 in the regional plot at the *RETN* locus became more significant with resistin levels ($p = 6.56 \times 10^{-68}$, Supplementary Figure S1B). Furthermore, with adjustment for both rs3219175 and rs370006313, rs3745368 remained significant ($p = 1.24 \times 10^{-27}$, Supplementary Figure S1C). With adjustment for all the three SNPs, none of the SNPs in the regional plot near the *RETN* locus exhibited significance at $p < 0.01$ (Supplementary Figure S1D), indicating that in this chromosomal region, variation in resistin concentrations was mainly explained by at least three signals

We further evaluated the candidate gene variants for sST2 levels in the *IL1RL1* gene region using a regional association plot study with conditional analysis. In brief, 307 SNPs at positions between 102.915 and 102.985 Mb on chromosome 2q12.1 around the *IL1RL1* gene region were analyzed, and 242 SNPs exceeded the genome-wide significance threshold ($p < 5 \times 10^{-8}$), with rs6543115 being the lead SNP ($p = 2.35 \times 10^{-85}$) (Supplementary Figure S2A). With adjustment for rs6543115, rs1420091 in the regional plot at the *IL1RL1* locus became more significant with sST2 levels ($p = 3.57 \times 10^{-15}$, Supplementary Figure S2B). With adjustment for the two aforementioned SNPs, none of the SNPs initially showed genome-wide significance in the regional plot near the *IL1RL1* locus exhibited significance at $p < 0.01$ (Supplementary Figure S2C).

2.2. GWAS and Replication Genotyping Results for Resistin and sST2 Levels

We then performed GWAS using the data of 4652 participants from the TWB cohort on a TWB genotype array for resistin levels. Because of the absence of the three lead aforementioned *RETN* SNPs (i.e., rs3219175, rs370006313, and rs3745368) in the TWB genotype array, we performed genotyping of the three *RETN* SNPs using TaqMan SNP Genotyping Assays. GWAS with conditional analysis indicated that the genome-wide significance threshold was exceeded only on chromosome 19p13.2, where *RETN* is located, and serial conditional analysis revealed the lead SNPs to be rs3219175, rs370006313, and rs3745368 ($p < 1.00 \times 10^{-307}$, $p = 2.99 \times 10^{-241}$, and $p = 1.21 \times 10^{-126}$, respectively) (Figure 1).

We then performed GWAS using the data of 4652 participants from the TWB cohort on a TWB genotype array for sST2 levels. GWAS with conditional analysis indicated that the genome-wide significance threshold was exceeded only on chromosome 2q12.1, where *IL1RL1* is located, and serial conditional analysis revealed that the lead SNPs were rs10183388 and rs4142132 ($p < 1.00 \times 10^{-307}$ and $p = 8.12 \times 10^{-51}$, respectively) (Figure 2). Further analysis revealed that rs10183388 and rs4142132 were in nearly complete linkage disequilibrium with the *IL1RL1* lead SNPs rs6543115 and rs1420091, respectively, in the

WGS study (r^2 = 0.98 and 0.96, respectively) (Supplementary Figure S3). The three lead *RETN* SNPs explained 53.7% of the variation in resistin levels, and the two lead *IL1RL1* SNPs explained 28.0% of the variation in sST2 levels (Supplementary Table S1). The SNP table of *RETN* and *IL1RL1* is provided in Supplementary Table S2.

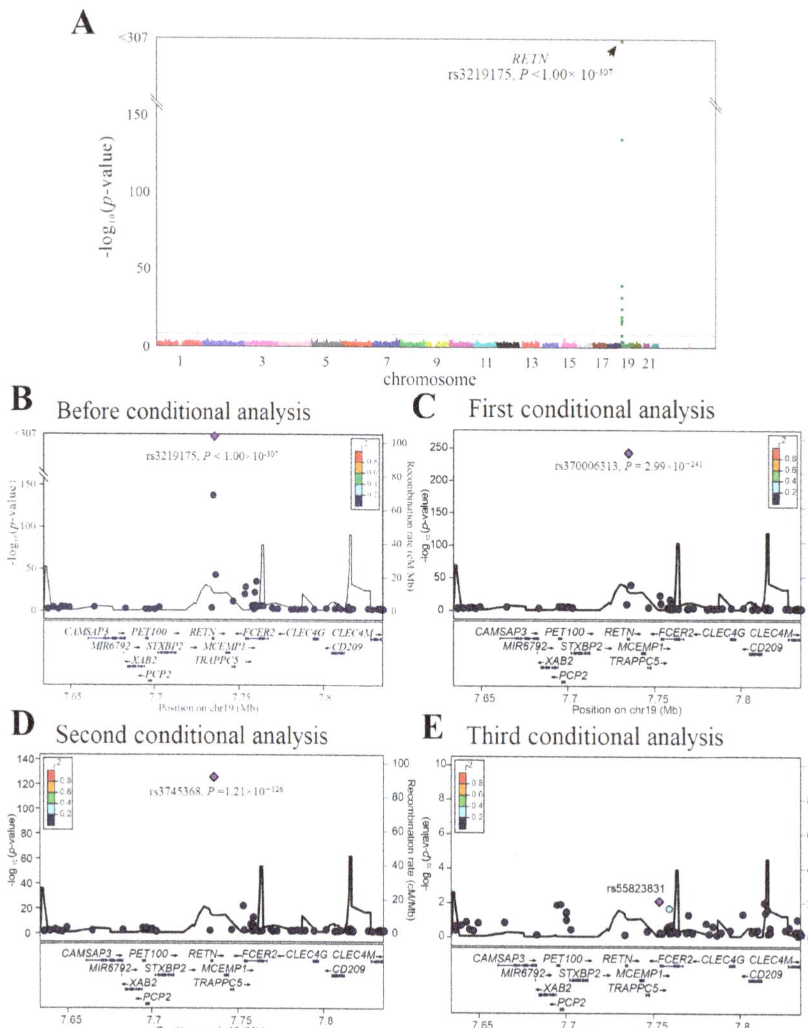

Figure 1. Conditional analysis of *RETN* candidate SNPs using GWAS data in the TWB cohort. (**A**) Manhattan plots for resistin levels from the genome-wide association study of 4652 Taiwan Biobank participants depict the only one peak above genome-wide significance on chromosome 19p13.2, where the *RETN* gene is located (arrow). (**B**) Before conditional analysis, regional association plots for resistin level surrounding the *RETN* locus show rs3219175 as the lead SNP. (**C**) After the first conditional analysis adjusting the rs3219175 genotypes, rs370006313 in the regional plot at the *RETN* locus becomes a significant association with resistin levels. (**D**) After the second conditional analysis adjusting for both rs3219175 and rs370006313 genotypes, rs3745368 is significantly associated with resistin levels. (**E**) After the third conditional analysis adjusting for the aforementioned SNPs, no more single SNP is found to be genome-wide significantly associated with resistin levels.

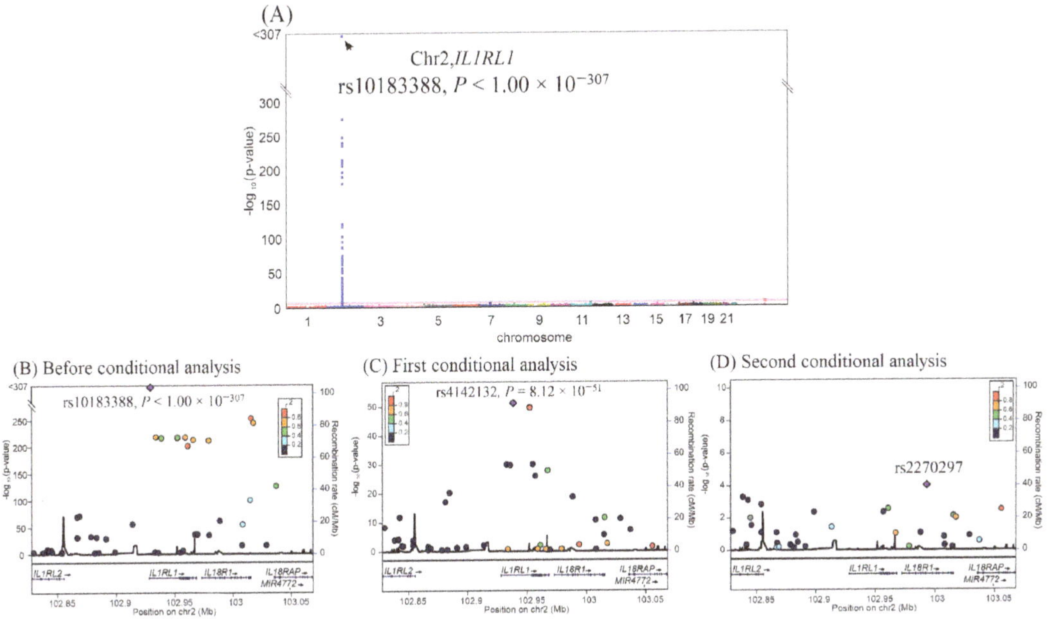

Figure 2. Conditional analysis of *IL1RL1* candidate SNPs using GWAS data in the TWB cohort. (**A**) Manhattan plots for sST2 levels from the genome-wide association study of 4652 Taiwan Biobank participants depict the only one peak above genome-wide significance on chromosome 2q12.1, where the *IL1RL1* gene is located (arrow). (**B**) Before conditional analysis, regional association plots for sST2 levels around the *IL1RL1* locus show rs10183388 as the lead SNP. (**C**) After the first conditional analysis adjusting for rs10183388 genotype, rs4142131 in the regional plot at *the IL1RL1* locus becomes a significant association with sST2 levels. (**D**) After the second conditional analysis adjusting for both rs10183388 and rs4142131, no more single SNP is found to be associated with sST2 levels significantly.

For the validation study, we analyzed the three lead *RETN* SNPs for resistin levels and two lead *IL1RL1* SNPs for sST2 levels in the cardiovascular health examination cohort (CH cohort). As shown in Table 1, the three *RETN* SNPs were significantly associated with resistin levels (rs3219175, rs370006313, and rs3745368 genotypes, $p = 6.77 \times 10^{-41}$, $p = 1.36 \times 10^{-12}$, and $p = 2.70 \times 10^{-5}$, respectively), and the two *IL1RL1* SNPs were significantly associated with sST2 levels (rs10183388 and rs4142132 genotypes, $p = 9.11 \times 10^{-16}$ and $p = 2.73 \times 10^{-8}$, respectively).

We evaluated the association of the *RETN* and *IL1RL1* SNPs with resistin and sST2 levels in the patients with CAD (Table 1), and the results revealed significant associations, except for rs370006313 genotypes for resistin levels (rs3219175, rs370006313, and rs3745368 genotypes for resistin levels, $p = 4.89 \times 10^{-15}$, $p = 0.055$, and $p = 0.001$, respectively, and rs10183388 and rs4142132 genotypes for sST2 levels, $p = 4.49 \times 10^{-8}$ and $p = 8.00 \times 10^{-6}$, respectively).

2.3. Baseline Characteristics of TWB, CH, and CAD Cohorts

The baseline characteristics of the participants in the TWB, CH, and CAD cohorts are provided in Table 2. The participants in the CAD cohort were older, with male predominance, and they tended to have a higher prevalence of smoking, obesity, hypertension, diabetes mellitus, and dyslipidemia. Compared with the TWB and CH cohorts, the CAD cohort had higher fasting glucose, triglyceride, aspartate aminotransferase (AST), and uric acid levels; higher leukocyte counts; lower total cholesterol, high-density lipoprotein (HDL)-cholesterol, and low-density lipoprotein (LDL)-cholesterol levels; and lower hematocrit, platelet counts, and estimated glomerular filtration rate (eGFR). The patients with CAD also had higher resistin levels but lower sST2 levels.

Table 1. Associations of *RETN* SNPs with resistin levels and *IL1RL1* SNPs with sST2 levels in the study populations.

Population	Genotypes	MAF	MM	Mm	mm	β	*p*1 Value	*p*2 Value
	RETN SNPs and resistin levels							
TWB cohort (N = 4652)	rs3219175	0.188	10.1 (7.7–13.0) (3058)	19.6 (15.8–24.6) (1442)	32.4 (27.2–41.1) (152)	0.274	$<1.00 \times 10^{-307}$	$<1.00 \times 10^{-307}$
	rs370006313	0.009	12.3 (8.8 –18.5) (4570)	65.3 (49.5–89.0) (81)	43.1 (1)	0.676	2.70×10^{-135}	8.04×10^{-136}
	rs3745368	0.150	13.4 (9.7–19.8) (3377)	10.3 (7.3–16.3) (1151)	6.4 (4.5–10.1) (124)	−0.099	5.90×10^{-42}	8.69×10^{-42}
	RETN WGRS					1.062	$<1.00 \times 10^{-307}$	$<1.00 \times 10^{-307}$
CH cohort (N = 557)	rs3219175	0.203	12.40 (9.15–15.95) (361)	22.50 (17.48–29.75) (166)	39.30 (29.10–49.08) (30)	0.234	3.36×10^{-41}	6.77×10^{-41}
	rs370006313	0.006	14.8 (10.3–22.6) (550)	58.3 (54.7–86.5) (7)	-	0.685	2.42×10^{-12}	1.36×10^{-12}
	rs3745368	0.129	16.0 (11.20 –24.0) (413)	12.10 (8.2–18.8) (134)	7.90 (7.1–36.6) (5)	−0.100	2.80×10^{-5}	2.70×10^{-5}
	RETN WGRS					1.034	2.61×10^{-65}	1.98×10^{-65}
CAD cohort (N = 512)	rs3219175	0.194	11.1 (7.5–21.7) (333)	21.5 (13.1–39.4) (159)	27.0 (18.4–42.3) (20)	0.211	4.44×10^{-15}	4.89×10^{-15}
	rs370006313	0.008	15.2 (8.7–26.7) (504)	40.9 (16.7–52.9) (8)	-	0.263	0.035	0.055
	rs3745368	0.159	16.3 (9.3–30.3) (365)	12.4 (7.2–22.5) (131)	13.4 (6.4–24.3) (16)	−0.093	0.001	0.001
	RETN WGRS					1.041	1.69×10^{-19}	1.84×10^{-19}
	IL1RL1 SNPs and sST2 level							
TWB cohort (N = 4652)	rs10183388	0.456	4.9 (3.6–6.5) (1370)	7.6 (5.9–10.0) (2352)	9.9 (7.5–12.9) (957)	0.157	6.16×10^{-294}	$<1.00 \times 10^{-307}$
	rs4142132	0.491	5.2 (3.9–6.8) (1180)	7.4 (5.6–9.8) (2375)	9.3 (7.1–12.5) (1097)	0.125	1.23×10^{-176}	1.92×10^{-212}
	IL1RL1-WGRS					0.520	1.15×10^{-246}	3.71×10^{-296}
CH cohort (N = 557)	rs10183388	0.435	7.4 (5.3–9.5) (179)	10.2 (8.0–13.2) (271)	13.2 (8.8–18.3) (107)	0.130	2.93×10^{-14}	9.11×10^{-16}
	rs4142132	0.480	7.6 (5.8–9.5) (154)	9.9 (7.5–13.2) (271)	11.7 (8.2–16.6) (132)	0.090	1.48×10^{-7}	2.73×10^{-8}
	IL1RL1-WGRS					0.547	7.67×10^{-12}	4.56×10^{-13}
CAD cohort (N = 512)	rs10183388	0.457	6.1 (3.4–8.7) (162)	6.4 (4.6–9.3) (234)	8.6 (6.0–12.4) (116)	0.092	2.64×10^{-8}	4.49×10^{-8}
	rs4142132	0.496	6.0 (3.7–8.8) (133)	6.4 (4.4–9.3) (250)	8.3 (5.9–12.2) (129)	0.077	5.00×10^{-6}	8.00×10^{-6}
	IL1RL1-WGRS					0.533	1.83×10^{-7}	3.01×10^{-7}

TWB, Taiwan Biobank; CH, cardiovascular healthy examination; CAD, cardiovascular disease; MAF, minor allele frequency; M, major allele; m, minor allele; sST2, soluble suppression of tumorigenicity 2; WGRS, weighted genetic risk score. *p*1: unadjusted, *p*2: adjusted for age, sex, body mass index, and current smoking status.

Table 2. Baseline characteristics between TWB, CH, and CAD groups.

	TWB (N = 4652)	CH (N = 557)	CAD (N = 512)
Age	48.6 ± 11.0	46.0 ± 9.9 ^{b}	65.5 ± 11.1 ^{b,c}
Male sex §	2067 (44.4%)	290 (52.1%)	413 (80.7%)
Smoking §	861 (18.5%)	107 (9.2%)	126 (24.6%)
Obesity §	1692 (36.4%)	225 (40.4%)	301 (58.8%)
Hypertension §	826 (17.8%)	110 (19.7%)	400 (78.1%)
Diabetes mellitus §	264 (5.7%)	28 (5.0%)	228 (44.5%)
Dyslipidemia §	2197 (47.2%)	297 (53.3%)	329 (64.3%)
BMI (kgw/m^2) §	24.2 ± 3.6	24.4 ± 3.4	26.1 ± 4.0 ^{b,c}
Fasting glucose (mmol/L)	5.29 ± 1.04	5.41 ± 1.35	6.30 ± 2.04 ^{b,c}
Cholesterol (mmol/L)	5.03 ± 0.93	5.14 ± 0.94 ^{a}	4.63 ± 1.01 ^{b,c}
Triglyceride (mmol/L) *	1.05 (0.73–1.54)	1.32 (0.89–1.88) ^{b}	1.40 (1.01–2.04) ^{b}
HDL-C (mmol/L)	1.41 ± 0.34	1.42 ± 0.36	1.02 ± 0.40 ^{b,c}
LDL-C (mmol/L)	3.14 ± 0.82	3.00 ± 0.84 ^{b}	2.61 ± 0.81 ^{b,c}
AST (μkat/L)	0.41 ± 0.25	-	0.43 ± 0.27 ^{a}
Uric acid (μmol/L)	331.9 ± 87.4	379.0 ± 95.8 ^{b}	387.2 ± 105.7 ^{b}
Creatinin (μmol/L)	65.4 ± 17.7	87.5 ± 42.4 ^{b}	122.9 ± 125.5 ^{b,c}
eGFR (mL/min/1.73 m^2)	95.0 ± 20.3	84.2 ± 20.6 ^{b}	65.7 ± 22.9 ^{b,c}
Leukocyte count (10^3/uL)	6.02 ± 1.57	-	6.67 ± 2.12 ^{b}
Hematocrit (%)	43.3 ± 4.58	-	40.8 ± 5.38 ^{b}
Platelet counts (10^3/uL)	239.8 ± 56.6	-	214.5 ± 62.9 ^{b}
Resistin (ng/mL) *	12.44 (8.86–18.95)	14.90 (10.35–22.9) ^{b}	15.39 (8.68–26.98) ^{b}
sST2 (ng/mL) *	7.12 (5.14–9.84)	9.33 (7.09–12.94) ^{b}	6.76 (4.49–9.63) ^{c}

Continuous data are presented as the mean ± SD or median (interquartile range), and categorical data are presented as numbers (%). Abbreviations: AST, aspartate aminotransferase; BMI, body mass index; CAD, cardiovascular disease; CH, cardiovascular healthy examination; eGFR, estimated glomerular filtration rate; HDL, high-density lipoprotein cholesterol; LDL-C, low-density lipoprotein cholesterol; TWB, Taiwan Biobank. AST, leukocyte count, hematocrit, and platelet counts were not available in the CH cohort. * Data with skew distribution are logarithmically transformed before statistical testing to meet the assumption of normal distribution. § $p < 0.01$ for $\chi 2$ test. ^{a} $p < 0.05$ vs. Q1 by the Bonferroni method. ^{b} $p < 0.001$ vs. Q1 by Bonferroni method. ^{c} $p < 0.001$ vs. Q2 by the Bonferroni method.

2.4. Correlations of Resistin and sST2 Levels with Clinical, Metabolic, Biochemical, Hematological Parameters, and WGRS of SNPs in All Cohorts

The correlations of resistin and sST2 levels with clinical, metabolic, biochemical, and hematological parameters and WGRS of SNPs are summarized in Supplementary Tables S3–S5. Resistin levels were strongly correlated with *RETN* WGRS in all three cohorts. Resistin levels were also significantly correlated with the lipid profile, eGFR, leukocyte count, and platelet count in the TWB cohort and with eGFR, leukocyte count, and hematocrit in the CAD cohort. Similarly, sST2 levels were strongly correlated with *IL1RL1* WGRS in all three cohorts. sST2 levels were significantly correlated with fasting glucose, lipid profile, AST, uric acid, eGFR, and leukocyte count in the TWB cohort and with eGFR and leukocyte count in the CAD cohort.

2.5. Association of Resistin Levels and RETN SNPs with Long-Term Outcomes for the Patients with CAD

In the CAD cohort, the follow-up duration was 1017 ± 324 days; 32 patients died and 58 patients developed major adverse cardiac events (MACEs). Kaplan–Meier survival analysis indicated that the patients with higher resistin levels had significantly higher rates of all-cause mortality and MACEs (Figure 3A,D, $p = 3.10 \times 10^{-4}$ and $p = 0.007$, respectively). The patients with lower *RETN* WGSR also had a higher rate of all-cause mortality, but *RETN* WGRS did not predict the MACEs rate (Figure 3B,E, $p = 0.042$ and $p = 0.228$, respectively). When the patients with CAD were further divided into four subgroups according to the resistin levels and *RETN* WGRS, the combination of high resistin levels and low *RETN* WGRS was a strong predictor of all-cause mortality and MACEs (Figure 3C,F, $p = 4.0 \times 10^{-6}$ and $p = 0.001$, respectively). The results of Cox regression analysis of all-cause mortality and MACEs between the groups stratified by the resistin levels and *RETN* WGRS are presented in Supplementary Table S6.

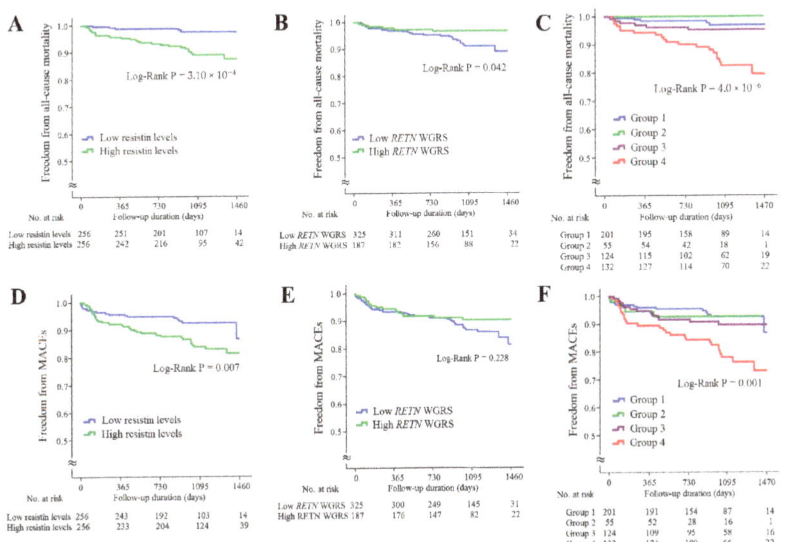

Figure 3. Kaplan–Meier curve analysis of resistin levels and *RETN* WGRS with long-term outcome in the patients with CAD. (**A–C**) Freedom from all-cause mortality in patients stratified by resistin levels, *RETN* WGRS, and combination of resistin levels and *RETN* WGRS. (**D–F**) Freedom from MACEs in patients stratified by resistin levels, *RETN* WGRS, and combination of resistin levels and *RETN* WGRS. WGRS, weighted genetic risk scores; MACEs, major adverse cardiac events. Group 1: low resistin levels/low *RETN* WGRS; Group 2: low resistin levels/high *RETN* WGRS; Group 3: high resistin levels/high *RETN* WGRS; Group 4: high resistin levels/low *RETN* WGRS.

2.6. Associations of sST2 Levels and IL1RL1 SNPs with Long-Term Outcome for the Patients with CAD

The associations of sST2 levels and *IL1RL1* SNPs with long-term outcomes for the patients with CAD are provided in Figure 4. The patients with higher sST2 levels had significantly higher rates of all-cause mortality and MACEs (Figure 4A,D, $p = 0.008$ and 0.009, respectively). However, the *IL1RL1* WGRS did not predict all-cause mortality and MACEs (Figure 4B,E, $p = 0.331$ and 0.971, respectively). When these patients were further divided into four subgroups according to the sST2 levels and *IL1RL1* WGRS, the combination of high sST2 levels and low *IL1RL1* WGRS was a strong predictor of all-cause mortality and MACEs (Figure 4C,F, $p = 0.001$ and 0.005, respectively). The results of Cox regression analysis of all-cause mortality and MACEs between the groups stratified by the sST2 levels and *IL1RL1* WGRS are provided in Supplementary Table S7.

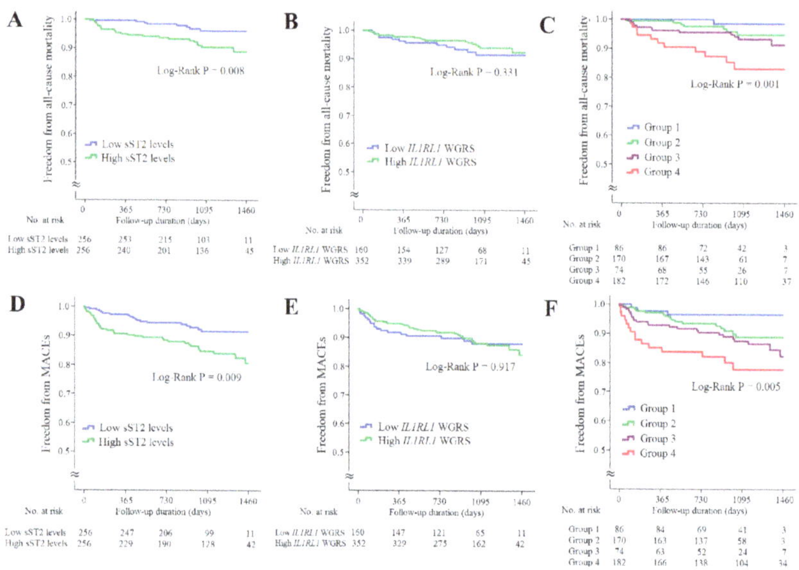

Figure 4. Kaplan–Meier curve analysis of sST2 levels and *IL1RL1* WGRS with long-term outcomes in the patients with CAD. (**A–C**) Freedom from all-cause mortality in patients stratified by sST2 levels, *IL1RL1* WGRS, and combination of sST2 levels and *IL1RL1* WGRS. (**D–F**) Freedom from MACEs in patients stratified by sST2 levels, *IL1RL1* WGRS, and combination of sST2 levels and *IL1RL1* WGRS. WGRS, weighted genetic risk scores; MACEs, major adverse cardiac events. Group 1: low sST2 levels/low *IL1RL1* WGRS; Group 2: low sST2 levels/high *IL1RL1* WGRS; Group 3: high sST2 levels/high *IL1RL1* WGRS; Group 4: high sST2 levels/low *IL1RL1* WGRS.

2.7. Synergistic Effects of WGRS with Resistin and sST2 on Predicting Long-Term Outcomes of the Patients with CAD

The patients with CAD were further divided into three subgroups according to the presence of high resistin levels/low *RETN* WGRS or high sST2 levels/low *IL1RL1* WGRS (Figure 5). The patients with either high resistin levels/low *RETN* WGRS or high sST2 levels/low *IL1RL1* WGRS had significantly higher rates of all-cause mortality and MACEs, and the patients with both high resistin levels/low *RETN* WGRS and high sST2 levels/low *IL1RL1* WGRS had the worst outcomes (Figure 5A,B, $p = 3.12 \times 10^{-11}$ for all-cause mortality and $p = 4.00 \times 10^{-6}$ for MACEs). The results of Cox regression analysis of all-cause mortality and MACEs between the groups stratified by the presence of high resistin levels/low *RETN* WGRS and high sST2 levels/low *IL1RL1* WGRS are provided in Table 3. The patients with either high resistin levels/low *RETN* WGRS or high sST2 levels/low *IL1RL1* WGRS and the patients with both high resistin levels/low *RETN* WGRS and high sST2 levels/low *IL1RL1* WGRS had significantly higher rates of all-cause mortality and MACEs after adjustment for baseline characteristics and diseases such as hypertension, diabetes mellitus, or hyperlipidemia. However, predictive power was attenuated after further adjustment for uric acid level, eGFR, and inflammatory markers, including C-reactive protein (CRP), chemerin, and growth differentiation factor (GDF)-15 levels. Furthermore, significantly higher levels of inflammatory biomarkers and lower eGFR were observed in the patients with high resistin levels/low *RETN* WGRS or high sST2 levels/low *IL1RL1* WGRS (Supplementary Table S8).

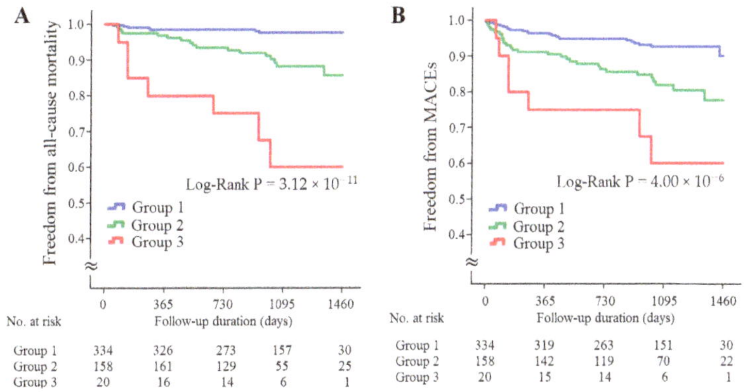

Figure 5. Kaplan–Meier curve analysis of combining resistin and sST2 levels with corresponding WGRS in predicting freedom from all-cause mortality (**A**) and freedom from MACEs (**B**) in the patients with CAD. WGRS, weighted genetic risk scores; MACEs, major adverse cardiac events. Group 1: patients without high resistin levels/low *RETN* WGRS and high sST2 levels/low *IL1RL1* WGRS; Group 2: patients with either high resistin levels/low *RETN* WGRS or high sST2 levels/low *IL1RL1* WGRS; Group 3, patients with both high resistin levels/low *RETN* WGRS and high sST2 levels/low *IL1RL1* WGRS.

Table 3. Cox regression analysis of all-cause mortality and MACEs rate between the groups stratified by the presence of high resistin level/low *RETN* WGRS and high sST2 levels/low *IL1RL1* WGRS.

	Group 1 *	Group 2 *		Group 3 *	
		All-cause mortality			
Patient numbers	334	158		20	
Number of events	7	17		7	
		HR (95% CI)	*p*-value	HR (95% CI)	*p*-value
Model 1	Reference	5.04 (2.09–12.15)	3.22×10^{-4}	19.59 (6.86–55.91)	2.71×10^{-8}
Model 2	Reference	5.04 (2.08–12.22)	3.48×10^{-4}	15.21 (5.24–44.13)	5.48×10^{-7}
Model 3	Reference	5.14 (2.10–12.55)	3.29×10^{-4}	11.59 (3.82–35.17)	1.50×10^{-5}
Model 4	Reference	3.78 (1.08–13.23)	0.037	21.76 (4.50–105.11)	1.27×10^{-4}
		MACEs			
Patient numbers	334	158		20	
Number of events	23	28		7	
		HR (95% CI)	*p*-value	HR (95% CI)	*p*-value
Model 1	Reference	2.62 (1.51–4.55)	0.001	5.88 (2.52–13.71)	4.1×10^{-5}
Model 2	Reference	2.70 (1.55–4.69)	4.50×10^{-4}	5.42 (2.30–12.78)	1.11×10^{-4}
Model 3	Reference	2.74 (1.57–4.78)	4.10×10^{-4}	4.87 (2.06–11.56)	3.24×10^{-4}
Model 4	Reference	1.72 (0.82–3.64)	0.153	4.03 (1.29–12.59)	0.017

* Group 1, patients without high resistin levels/low *RETN* WGRS and high sST2 levels/low *IL1RL1* WGRS; Group 2, patients with either high resistin levels/low *RETN* WGRS or high sST2 levels/low *IL1RL1* WGRS; Group 3, patients with both high resistin levels/low *RETN* WGRS and high sST2 levels/low *IL1RL1* WGRS. Abbreviations: HR, hazard ratio; CI, confidence interval. Model 1, Unadjusted. Model 2, Adjusted for age, sex, BMI, and smoking status. Model 3: Adjusted for age, sex, BMI, smoking status, diabetes mellitus, hypertension, and dyslipidemia. Model 4: Adjusted for age, sex, BMI, smoking status, diabetes mellitus, hypertension, dyslipidemia, uric acid level, estimated glomerular filtration rate, CRP levels, chemerin levels, and GDF-15 levels.

3. Discussion

In this study, we performed two GWAS analyses for 4652 participants in the TWB cohort as well as regional association plot analyses for a subgroup of 859 participants with WGS data, aiming to elucidate the genetic basis of resistin and sST2 levels in the Chinese population. Three *RETN* variants (rs3219175, rs370006313, and rs3745368) and two *IL1RL1* variants (rs10183388 and rs4142132) were found to be independently associated with resistin and sST2 levels. In combination, these variants explained 53.7% and 28.0% of the variation in resistin and sST2 levels, respectively. These results were confirmed in validation studies in the CH and CAD cohorts. To the best of our knowledge, the three *RETN* variants were the strongest genetic determinants of resistin levels compared with the results of other GWAS analyses in ethnic population studies [13–15,30]. We further evaluated the effect of the combination of biomarker levels and genetic variants on predicting long-term outcomes in the patients with CAD. Both higher resistin and sST2 levels predicted higher rates of all-cause mortality and MACEs in the patients with CAD during long-term follow-up, but not WGRS of *RETN* and *IL1RL1* variants, except a borderline significance of low *RENT* WGRS on all-cause mortality. Notably, we observed a synergistic effect of the combination of biomarkers with WGRS of *RETN* and *IL1RL1* on the prediction of the long-term outcomes in the patients with CAD. The patients with high resistin levels/low *RETN* WGRS and the patients with high sST2 levels/low *IL1RL1* WGRS had significantly higher rates of all-cause mortality and MACEs. Furthermore, the patients with both high resistin levels/low *RETN* WGRS and high sST2 levels/low *IL1RL1* WGRS had the poorest outcomes over long-term follow-up.

3.1. Novel Variants Associated with Resistin Levels

GWAS of the data of 4652 participants in the TWB cohort revealed the *RETN* gene region as the only locus associated with resistin levels, with rs3219175 as the lead SNP ($p < 1.00 \times 10^{-307}$). This result confirmed the data from East Asian [11,12,16,30] and African populations [14] but was different from those reported in Caucasian populations, in which *TYW3/CRYZ* and *NDST4* loci, but not *RETN* variants, showed a genome-wide significant association with circulating resistin levels [10]. We used the GWAS data to analyze the *RETN* gene locus and found three *RETN* SNPs, namely rs3219175, rs370006313, and rs3745368, to be independently associated with resistin levels, which contributed to 53.7% of the variation in resistin levels in the TWB cohort. The data were replicated in the CH cohort (41.1% of variations). Notably, minor allele frequencies (MAFs) of two SNPs, rs3219175 and rs3745368, were significantly lower in the Caucasian population than that in East Asian and African populations [11–15]. By contrast, the MAF of the rs370006313 variant was 0.9% in our cohorts. According to the pubmed.gov website, the MAF of rs370006313 was 0% for 1006 European participants and 1322 African participants from the 1000-Genome project. Furthermore, the *RETN* –420C > G (rs1862513) genotypes, initially considered candidate SNPs for resistin levels [9,11,12,16], were not found to be independent determinants of resistin levels in our study population.

These results indicate ethnic heterogeneity not only at the level of gene loci but also at the level of each gene variant. rs370006313 is a novel SNP that is localized within the minimal promoter of human *RETN*, and it has not been previously reported to be associated with resistin levels. Although rs370006313 is a rare variant in the Taiwanese population, the resistin level increased by 4.7 times in the heterozygous state; thus, this SNP contributed to a 13.8% variation in resistin level in the relatively healthy population. Further research is warranted to elucidate the functional significance of this variant.

3.2. Genetic Determinants of sST2 Levels in Taiwan

In the GWAS analysis, our data revealed that chromosome 2q12.1, where the *IL1RL1* gene is located, was the only gene locus associated with sST2 levels. Further GWAS analysis revealed that the two lead *IL1RL1* SNPs, rs10183388 and rs4142131, were independently associated with sST2 levels and contributed to 28% of the variation; thus, these SNPs are strong genetic determinants of sST2 levels in the Chinese population. Consistently, the Framingham offspring cohort study reported that all candidate SNPs associated with the sST2 concentration were on chromosome 2q12.1 and that the 11 "independent" genome-wide significant SNPs across the *IL1RL1* locus accounted for 36% of the heritability of sST2 levels, with rs950880 as the lead SNP [26]. In the TWB cohort, GWAS analysis also revealed a significant association of rs950880 with sST2 levels ($p = 1.73 \times 10^{-216}$). Notably, this association disappeared after serial conditional analysis with rs10183388 and rs4142131, suggesting ethnic heterogeneity in *IL1RL1* variants.

3.3. Association of Resistin Levels with Cardiometabolic Phenotypes and Long-Term Outcomes in the Patients with CAD

As a proinflammatory adipokine, resistin is associated with several cardiometabolic phenotypes, including obesity, diabetes, and hyperlipidemia [31,32]. Elevated resistin levels are also associated with impaired renal function and several inflammatory biomarkers [33,34]. Resistin may also contribute to cholesterol and triglyceride accumulation in macrophages, arterial inflammation, endothelial dysfunction, and angiogenesis [31], leading to accelerated atherogenesis and CAD. In line with these findings, our study also found a strong association of resistin levels with BMI, diabetes mellitus, dyslipidemia, impaired renal function, and elevated leukocyte counts in the TWB cohort. Furthermore, elevated resistin levels were a strong predictor of poor long-term outcomes in the patients with CAD, and the participants with higher resistin levels had significantly higher rates of all-cause mortality and MACEs.

3.4. Association of sST2 Levels with Cardiometabolic Phenotypes and Long-Term Outcome in the Patients with CAD

sST2, a protein secreted by cultured myocytes subjected to mechanical strain, is a well-known biomarker of severe heart failure and a strong predictor of mortality [21,22]. IL-33, the ligand of sST2, is also induced and released by stretched myocytes. In patients presenting to the emergency department with myocardial infarction with ST elevation and dyspnea, sST2 levels are strongly predictive of heart failure and mortality [23]. In patients with stable CAD, increased sST2 levels also predict long-term MACEs and all-cause mortality [24,25]. Recently, sST2 levels have also been found to be a novel biomarker and clinical predictor of metabolic syndrome. In patients without CAD or heart failure, Zong et al. reported that increased sST2 levels were significantly associated with a higher prevalence of hypertension, diabetes mellitus, hypertriglyceridemia, and lower HDL cholesterol levels [35]. Consistently, in the present study, elevated sST2 levels were associated with a higher prevalence of diabetes mellitus, higher BMI, total cholesterol, and lower HDL cholesterol levels in the TWB cohort. Furthermore, elevated sST2 levels were significantly associated with higher AST, uric acid, leukocyte counts, and lower eGFR. For the patients with CAD, elevated sST2 levels predicted poor long-term clinical outcomes. The patients with higher sST2 levels had higher rates of all-cause mortality and MACEs during the follow-up period.

3.5. RENT and IL1RL1 Variants on Long-Term Outcomes in the Patients with CAD

The results of the association between *RENT* variants and CAD risk were inconsistent in previous studies. The association of *RETN* –420C > G with the risk of CAD was not significant in the Caucasian population [36,37]. By contrast, Tang et al. evaluated the association of *RETN* –420C > G with the presence of CAD and reported a 62% increased risk of CAD in participants with variant genotypes (CG and GG) [38]. Together, these findings suggest the ethnic heterogeneity of *RETN* variants with the risk of CAD. Importantly, none of these studies evaluated the prognostication ability of *RETN* variants for CAD outcomes. In the present study, the patients with CAD with low *RETN* WGRS had an increased all-cause mortality rate during the 4-year follow-up, but low *RETN* WGRS did not predict a high MACEs rate.

The evidence of the association between *IL1RL1* variants and the risk of CAD remains limited. In a case-control association analysis of 4521 individuals with CAD and 4809 controls in the Chinese Han population, Tu et al. reported a strong association of the *IL1RL1* SNP rs11685424 with the risk of CAD and disease severity [39]. Our previous study reported that among patients with CAD and lower-extremity arterial disease, those with the *IL1RL1* SNP rs950880 AA genotype tended to have lower sST2 levels and a lower survival rate [28]. The present study further evaluated the effect of *IL1RL1* variants on the prediction of long-term outcomes in the patients with CAD using the WGS and GWAS data from the TWB cohort. However, the WGRS of the two *IL1RL1* lead SNPs did not predict the outcomes of the patients with CAD, including all-cause mortality and MACEs.

3.6. Synergistic Effects of Genetic Variants and Resistin and sST2 Levels on Predicting Long-Term Outcomes in the Patients with CAD

When we combined the biomarkers with the WGRS of *RETN* and *IL1RL1*, we observed a synergistic effect on predicting long-term outcomes in the patients with CAD. The patients with high resistin levels/low *RETN* WGRS had the highest all-cause mortality and MACEs rates during the follow-up period (6.2 times for all-cause mortality and 3.1 times for MACEs compared with patients with low resistin levels/low *RETN* WGRS, Figure 3 and Supplementary Table S6). In agreement with this finding, the patients with high sST2 levels/low *IL1RL1* WGRS also had the poorest prognosis during the follow-up period (14.1 times for all-cause mortality and 6.0 times for MACEs compared with the patients with low sST2 level/low *IL1RL1* WGRS, Figure 4 and Supplementary Table S7).

We further analyzed the synergistic effects of combining resistin levels with *RETN* WGRS and sST2 levels with *IL1RL1* WGRS in the prognostication of long-term outcomes in the patients with CAD. The patients with high resistin levels/low *RETN* WGRS or high sST2 levels/low *IL1RL1* WGRS had 5.0 times higher all-cause mortality and 2.6 times higher MACEs rates than those without these presentations (Table 3). Furthermore, the patients with both high resistin levels/low *RETN* WGRS and high sST2 levels/low *IL1RL1* WGRS had the poorest outcomes (19.6 times for all-cause mortality and 5.9 times for the MACEs). These results were further adjusted for traditional cardiovascular risk factors, including age, sex, BMI, smoking status, diabetes mellitus, hypertension, and dyslipidemia, and adjusted for known predictors of poorer cardiovascular outcomes, such as uric acid levels [40], eGFR [41], and inflammatory biomarkers. These findings remained unchanged after adjustment for traditional cardiovascular risk factors but were attenuated after further adjustment of uric acid levels, eGFR, and inflammatory markers such as CRP, chemerin, and GDF-15 levels. Furthermore, significantly higher levels of CRP, chemerin, and GDF-15 and lower eGFR were observed in the patients with high resistin levels/low *RETN* WGRS and/or high sST2 levels/low *IL1RL1* WGRS (Supplementary Table S8). These findings indicate that inflammation and impaired renal function may explain the poorer outcomes in the patients with high biomarker levels and low WGRS for the corresponding genetic variants.

3.7. Study Limitations

Although we had enrolled a large sample size with available WGS and GWAS data in the TWB cohort, the sample sizes in the validation group (CH cohort) and study group (CAD cohort) were relatively small. However, the candidate SNPs in the *RETN* and *IL1RL1* genes were consistent in the three groups, providing strong evidence of the lead SNPs for *RETN* and *IL1RL1* in the Chinese population. Second, the event rates of all-cause mortality and MACEs were low in the CAD cohort. Nevertheless, we found a significant prediction power of resistin and sST2 levels for outcomes in patients with CAD, as well as a synergistic effect when the resistin and sST2 levels were combined with the WGRS of *RETN* and *IL1RL1*. Future studies should include a large sample size of patients with CAD to verify this finding. Finally, because we only included individuals of Han Chinese ethnicity, our results cannot be extended to other ethnic groups.

4. Materials and Methods

4.1. Study Populations

4.1.1. TWB Cohort

The study cohort for the GWAS comprised 5000 participants from the TWB cohort, and WGS data were available for 859 of them. The data were collected from recruitment centers across Taiwan between 2008 and 2015. The inclusion criteria were participants without a history of cancer, stroke, CAD, or systemic disease. All participants self-reported having Han Chinese ethnicity. After participants were excluded based on the exclusion criteria, 4652 participants remained and were included in the GWAS (Supplementary Figure S4). Supplementary Table S9 shows the definitions of hypertension, diabetes mellitus, hyperlipidemia, and current smoking status. The Research Ethics Committee of Taipei Tzu Chi Hospital (approval number: 05-X04-007), Buddhist Tzu Chi Medical Foundation, and Ethics and Governance Council of the Taiwan Biobank (approval number: TWBR10507-02 and TWBR10611-03) approved our study. Written informed consent was obtained from all participants before participation.

4.1.2. CH Cohort

The validation group was recruited during routine cardiovascular health examinations from October 2003 to September 2005 at Chang Gung Memorial Hospital and comprised 617 Han Chinese participants (327 men with a mean age of 45.2 $\pm$ 10.5 years and 290 women with a mean age of 46.8 $\pm$ 10.1 years), who responded to a questionnaire on their medical history and lifestyle characteristics. A total of 60 participants were excluded from the current study, and 557 participants were enrolled in the analysis (Supplementary Figure S4). All of the participants provided written informed consent, and the study was approved by the Ethics Committee of Chang Gung Memorial Hospital and the Ethics Committee of Taipei Tzu Chi Hospital, Buddhist Tzu Chi Medical Foundation.

4.1.3. CAD Cohort

In the CAD cohort, 565 patients who received coronary angiography, had at least 50% stenosis of one major coronary artery, and had available blood samples for DNA and biomarker analyses were recruited between July 2010 and September 2013 from National Taiwan University Hospital. Among these, 53 patients were excluded from the current study, and 512 patients were included in the analysis (Supplementary Figure S4). All clinical data were obtained from the patients' medical records. The primary enpoint was all-cause mortality. The secondary endpoint was MACEs, including the composite endpoints of all-cause mortality, hospitalization for heart failure, nonfatal myocardial infarction, or nonfatal stroke. Seven patients who were lost to follow-up after enrollment were contacted by telephone before the end of the study. Three of these patients had died, and the cause of death was provided by the relatives. All of the participants provided written informed consent, and the study was approved by the Research Ethics Committee of National Taiwan University Hospital.

4.2. Laboratory Examination

We examined the following clinical phenotypes: body height, body weight, body mass index (BMI), and systolic, mean, and diastolic blood pressure. In addition, we collected the following biochemical data: lipid profile, including total cholesterol, HDL cholesterol, LDL cholesterol, and triglyceride levels; fasting plasma glucose; and liver and renal functional test-related parameters such as serum creatinine, eGFR, uric acid, and AST. Hematological parameters included white and red blood cell counts, platelet counts, and hematocrit. Circulating levels of resistin, sST2, and inflammatory markers such as chemerin and GDF-15 were measured using commercially available enzyme-linked immunosorbent assay kits (R&D, Minneapolis, MN, USA). Circulating plasma levels of CRP were measured using the particle-enhanced turbidimetric immunoassay technique (Siemens Healthcare Diagnostics, Camberley, UK). The increase in turbidity that accompanies aggregation is proportional to the CRP concentration. Overall, the intra- and interassay coefficients of variation were 1.2–9.5% (Supplementary Table S10).

4.3. Genomic DNA Extraction and Genotyping

DNA of the participants was isolated from blood samples using a QIAamp DNA blood kit following the manufacturer's instructions (Qiagen, Valencia, CA, USA). SNP genotyping was conducted using custom TWB chips and was run on the Axiom Genome-Wide Array Plate System (Affymetrix, Santa Clara, CA, USA). Genotyping for *RETN* rs3219175, rs370006313, and rs3745368 genotypes in the participants from the TWB cohort, the participants from the CH cohort, and the patients with CAD as well as for *IL1RL1* rs10183388 and rs4142132 genotypes were performed using TaqMan SNP Genotyping Assays (Applied Biosystems, Foster City, CA, USA). For quality control purposes, approximately 10% of the samples were re-genotyped blindly, and identical results were obtained.

4.4. TWB Whole-Genome Sequencing: RETN and IL1RL1 Gene Region

The WGS data of 859 participants from the TWB cohort were evaluated using an ultra-fast whole-genome secondary analysis on Illumina sequencing platforms [42] (Illumina HiSeq 2500/4000). The resulting reads were aligned to the hg19 reference genome with iSAAC 01.13.10.21. iSAAC Variant Caller 2.0.17 was used to perform SNP and insertion-deletion variant discovery and genotyping [42]. An in-house protocol written in shell script was performed to combine 880 vcf files. A union table of all detected variants in 880 vcf files was used for further analysis. Two gene loci linked to resistin and sST2 levels, according to the GWAS results, were included in the analysis, namely 20 Kb around the *RETN* and *IL1RL1* gene regions. The association between SNPs and resistin and sST2 levels was then analyzed using the GWAS method.

4.5. GWAS Analysis

The Axiom Genome-Wide CHB 1 Array Plate (Affymetrix) designed by the Taiwan Biomarker Study Group is a TWB genotype array used for GWAS analysis. In this genotyping platform, SNPs with minor allele frequencies of $\geq$5% in a set of 1950 samples were selected from Taiwan Han Chinese populations previously genotyped at the National Center of Genome Medicine of the Academia Sinica, Taipei, Taiwan [43]. Each genomic DNA was genotyped on the Axiom TWB genome-wide array comprising 642,832 SNPs with assistance from the National Center of Genome Medicine of Academia Sinica. All the samples in the analysis had a call rate of $\geq$97%. For SNP quality control, SNP call rate < 97%, minor allele frequency <0.01, and violation of Hardy–Weinberg equilibrium ($p < 10^{-6}$) were criteria for exclusion from subsequent analyses. In total, 4652 participants and 614,821 SNPs were included in the GWAS analysis after quality control.

4.6. Statistical Analysis

Before analysis, resistin and sST2 levels were logarithmically transformed to adhere to a normality assumption. A generalized linear model was used to analyze resistin and sST2 levels in relation to the investigated genotypes and confounders. We assumed the genetic effect to be additive after adjustment for age, sex, BMI, and current smoking status. The software package PLINK was used to conduct genome-wide scans, and $p < 5 \times 10^{-8}$ was considered genome-wide significant. For GWAS, a conditional analysis was conducted to assess the residual association with all remaining SNPs after adjustment for the most strongly associated SNP at a locus by adding the SNP as a covariate into the regression model.

The baseline characteristics of all participants were evaluated using analysis of variance. Continuous variables are expressed as mean $\pm$ standard deviation or median and interquartile ranges when the distribution is strongly skewed. Differences in categorical data distribution were examined using a chi-square test or a chi-square test for trends. The Bonferroni method was used for post-hoc analysis after ANOVA. The genetic risk score was calculated using the weighted method, which assumed each SNP to be independently associated with resistin or sST2 levels (i.e., no interaction between each SNP) [44]. We assumed an additive effect of risk alleles for each SNP and applied linear weighting of 0, 1, or 2 to genotypes containing a corresponding number of risk alleles. WGRS were calculated by multiplying the estimated beta-coefficient of each SNP by the number of corresponding risk alleles (0, 1, or 2). Pearson's correlation coefficients were used to examine the relationship of resistin and sST2 levels with clinical and biochemical factors in addition to WGRS. Each variable with a significant association with resistin and sST2 was entered into the multivariate linear regression model with the stepwise method to identify the independent correlates of resistin and sST2.

The rates of freedom from primary and secondary endpoints in the patients with CAD, stratified by the levels and resistin and sST2 and the WGRS of *RETN* and *IL1RL1* genotyping, were assessed using Kaplan–Meier curves and compared using the long-rank test. The outcomes were further analyzed by combining WGRS according to *RETN* and *IL1RL1* genotyping with both biomarker levels. Cox regression analysis was used to determine the hazard ratio of primary and secondary endpoints in each group. IBM SPSS Statistics version 24.0 (IBM Corp., Armonk, NY, USA) was used to perform all calculations, with two-sided $p < 0.05$ set as the statistically significant level.

5. Conclusions

In the GWAS of the TWB cohort, we found that three *RETN* lead SNPs (rs3219175, rs370006313, and rs3745368) were strongly associated with resistin levels and that two *IL1RL1* lead SNPs (rs10183388 and rs4142132) were significantly associated with sST2 levels in the Chinese population. These results were validated in the CH cohort and CAD cohort. Serum resistin and sST2 levels were significantly associated with cardiometabolic risk factors in the TWB and CAD cohorts and were strong predictors of poor clinical outcomes in patients with CAD. Furthermore, a synergistic effect was noted when combining resistin and sST2 levels with the WGRS of *RETN* and *IL1RL1* in the prognostication of CAD outcomes. Patients with high resistin levels/low *RETN* WGRS and high sST2 levels/low *IL1RL1* WGRS had the poorest outcome during long-term follow-up. This study provides insights into the effects of biomarkers and corresponding genetic variants on the prognosis of long-term clinical outcomes in patients with CAD.

Supplementary Materials: The following supporting information can be downloaded at: https://www.mdpi.com/article/10.3390/ijms23084292/s1.

Author Contributions: Conceptualization, Y.-L.K. and H.-H.C.; methodology, Y.-L.K., H.-H.C., M.-S.T. and S.W.; formal analysis, H.-H.C., Y.-L.K., M.-S.T. and S.W.; resources, L.-A.H., J.-M.J.J. and F.-T.C.; data curation, L.-A.H., J.-M.J.J. and F.-T.C.; writing—original draft preparation, H.-H.C.; writing—review and editing, Y.-L.K.; visualization, Y.-L.K.; project administration, Y.-L.K.; supervision, Y.-L.K.; funding acquisition, H.-H.C. and Y.-L.K. All authors have read and agreed to the published version of the manuscript.

Funding: This study was supported by grants from Buddhist Tzu Chi Medical Foundation Academic Advancement (TCMF-EP 111-02, TCMF-A 107-01-15), grants from the Ministry of Science and Technology (MOST 108-2314-B-303-026-MY3) and Taipei Tzu Chi Hospital, Buddhist Tzu Chi Medical Foundation (TCRD-TPE-MOST-109-05) to Y.-L.K., and Taipei Tzu Chi Hospital, Buddhist Tzu Chi Medical Foundation (TCRD-TPE-109-RT-1) to H.-H.C.

Institutional Review Board Statement: The study was conducted in accordance with the Declaration of Helsinki. The Research Ethics Committee of Taipei Tzu Chi Hospital (approval number: 05-X04-007), Buddhist Tzu Chi Medical Foundation, had approved the present study. The TWB study cohort had been approved by the Ethics and Governance Council of the Taiwan Biobank (approval number: TWBR10507-02 and TWBR10611-03). The Cardiovascular Healthy examination cohort had been approved by the Ethics Committee of Chang Gung Memorial Hospital, and the CAD study cohort was approved by the Research Ethics Committee of National Taiwan University Hospital. Written informed consent was obtained from all participants before participation.

Informed Consent Statement: Informed consent was obtained from all subjects involved in the study.

Data Availability Statement: The data presented in this study are available on request from the corresponding author.

Acknowledgments: We greatly appreciate technical support from the Core Laboratory of the Taipei Tzu Chi Hospital, Buddhist Tzu Chi Medical Foundation, and expert statistical analysis assistance from Tsung-Han Hsieh.

Conflicts of Interest: The authors declare no conflict of interest.

References

1. Roth, G.A.; Mensah, G.A.; Johnson, C.O.; Addolorato, G.; Ammirati, E.; Baddour, L.M.; Barengo, N.C.; Beaton, A.Z.; Benjamin, E.J.; Benziger, C.P.; et al. Global Burden of Cardiovascular Diseases and Risk Factors, 1990–2019: Update From the GBD 2019 Study. *J. Am. Coll. Cardiol.* **2020**, *76*, 2982–3021. [CrossRef]
2. Steppan, C.M.; Bailey, S.T.; Bhat, S.; Brown, E.J.; Banerjee, R.R.; Wright, C.M.; Patel, H.R.; Ahima, R.S.; Lazar, M.A. The hormone resistin links obesity to diabetes. *Nature* **2001**, *409*, 307–312. [CrossRef] [PubMed]
3. Patel, L.; Buckels, A.C.; Kinghorn, I.J.; Murdock, P.R.; Holbrook, J.D.; Plumpton, C.; Macphee, C.H.; Smith, S.A. Resistin is expressed in human macrophages and directly regulated by PPAR gamma activators. *Biochem. Biophys. Res. Commun.* **2003**, *300*, 472–476. [CrossRef]
4. Bokarewa, M.; Nagaev, I.; Dahlberg, L.; Smith, U.; Tarkowski, A. Resistin, an adipokine with potent proinflammatory properties. *J. Immunol.* **2005**, *174*, 5789–5795. [CrossRef]
5. Silswal, N.; Singh, A.K.; Aruna, B.; Mukhopadhyay, S.; Ghosh, S.; Ehtesham, N.Z. Human resistin stimulates the pro-inflammatory cytokines TNF-alpha and IL-12 in macrophages by NF-kappaB-dependent pathway. *Biochem. Biophys. Res. Commun.* **2005**, *334*, 1092–1101. [CrossRef] [PubMed]
6. Cho, Y.; Lee, S.E.; Lee, H.C.; Hur, J.; Lee, S.; Youn, S.W.; Lee, J.; Lee, H.J.; Lee, T.K.; Park, J.; et al. Adipokine resistin is a key player to modulate monocytes, endothelial cells, and smooth muscle cells, leading to progression of atherosclerosis in rabbit carotid artery. *J. Am. Coll. Cardiol.* **2011**, *57*, 99–109. [CrossRef] [PubMed]
7. Park, H.K.; Ahima, R.S. Resistin in rodents and humans. *Diabetes Metab. J.* **2013**, *37*, 404–414. [CrossRef] [PubMed]
8. Schwartz, D.R.; Lazar, M.A. Human resistin: Found in translation from mouse to man. *Trends Endocrinol. Metab.* **2011**, *22*, 259–265. [CrossRef]
9. Menzaghi, C.; Coco, A.; Salvemini, L.; Thompson, R.; De Cosmo, S.; Doria, A.; Trischitta, V. Heritability of serum resistin and its genetic correlation with insulin resistance-related features in nondiabetic Caucasians. *J. Clin. Endocrinol. Metab.* **2006**, *91*, 2792–2795. [CrossRef]
10. Qi, Q.; Menzaghi, C.; Smith, S.; Liang, L.; de Rekeneire, N.; Garcia, M.E.; Lohman, K.K.; Miljkovic, I.; Strotmeyer, E.S.; Cummings, S.R.; et al. Genome-wide association analysis identifies TYW3/CRYZ and NDST4 loci associated with circulating resistin levels. *Hum. Mol. Genet.* **2012**, *21*, 4774–4780. [CrossRef]
11. Onuma, H.; Tabara, Y.; Kawamura, R.; Tanaka, T.; Ohashi, J.; Nishida, W.; Takata, Y.; Ochi, M.; Yamada, K.; Kawamoto, R.; et al. A at single nucleotide polymorphism-358 is required for G at -420 to confer the highest plasma resistin in the general Japanese population. *PLoS ONE* **2010**, *5*, e9718. [CrossRef]
12. Osawa, H.; Yamada, K.; Onuma, H.; Murakami, A.; Ochi, M.; Kawata, H.; Nishimiya, T.; Niiya, T.; Shimizu, I.; Nishida, W.; et al. The G/G genotype of a resistin single-nucleotide polymorphism at -420 increases type 2 diabetes mellitus susceptibility by inducing promoter activity through specific binding of Sp1/3. *Am. J. Hum. Genet.* **2004**, *75*, 678–686. [CrossRef]
13. Hivert, M.F.; Manning, A.K.; McAteer, J.B.; Dupuis, J.; Fox, C.S.; Cupples, L.A.; Meigs, J.B.; Florez, J.C. Association of variants in RETN with plasma resistin levels and diabetes-related traits in the Framingham Offspring Study. *Diabetes* **2009**, *58*, 750–756. [CrossRef] [PubMed]
14. Meeks, K.A.C.; Doumatey, A.P.; Bentley, A.R.; Gouveia, M.H.; Chen, G.; Zhou, J.; Lei, L.; Adeyemo, A.A.; Rotimi, C.N. Genetics of Circulating Resistin Level, a Biomarker for Cardiovascular Diseases, Is Informed by Mendelian Randomization and the Unique Characteristics of African Genomes. *Circ. Genom. Precis. Med.* **2020**, *13*, 488–503. [CrossRef] [PubMed]
15. Asano, H.; Izawa, H.; Nagata, K.; Nakatochi, M.; Kobayashi, M.; Hirashiki, A.; Shintani, S.; Nishizawa, T.; Tanimura, D.; Naruse, K.; et al. Plasma resistin concentration determined by common variants in the resistin gene and associated with metabolic traits in an aged Japanese population. *Diabetologia* **2010**, *53*, 234–246. [CrossRef]
16. Cho, Y.M.; Youn, B.S.; Chung, S.S.; Kim, K.W.; Lee, H.K.; Yu, K.Y.; Park, H.J.; Shin, H.D.; Park, K.S. Common genetic polymorphisms in the promoter of resistin gene are major determinants of plasma resistin concentrations in humans. *Diabetologia* **2004**, *47*, 559–565. [CrossRef]
17. Iwahana, H.; Yanagisawa, K.; Ito-Kosaka, A.; Kuroiwa, K.; Tago, K.; Komatsu, N.; Katashima, R.; Itakura, M.; Tominaga, S. Different promoter usage and multiple transcription initiation sites of the interleukin-1 receptor-related human ST2 gene in UT-7 and TM12 cells. *Eur. J. Biochem.* **1999**, *264*, 397–406. [CrossRef]

18. Schmitz, J.; Owyang, A.; Oldham, E.; Song, Y.; Murphy, E.; McClanahan, T.K.; Zurawski, G.; Moshrefi, M.; Qin, J.; Li, X.; et al. IL-33, an interleukin-1-like cytokine that signals via the IL-1 receptor-related protein ST2 and induces T helper type 2-associated cytokines. *Immunity* **2005**, *23*, 479–490. [CrossRef]
19. Sanada, S.; Hakuno, D.; Higgins, L.J.; Schreiter, E.R.; McKenzie, A.N.; Lee, R.T. IL-33 and ST2 comprise a critical biomechanically induced and cardioprotective signaling system. *J. Clin. Investig.* **2007**, *117*, 1538–1549. [CrossRef] [PubMed]
20. Seki, K.; Sanada, S.; Kudinova, A.Y.; Steinhauser, M.L.; Handa, V.; Gannon, J.; Lee, R.T. Interleukin-33 prevents apoptosis and improves survival after experimental myocardial infarction through ST2 signaling. *Circ. Heart Fail.* **2009**, *2*, 684–691. [CrossRef] [PubMed]
21. Weinberg, E.O.; Shimpo, M.; Hurwitz, S.; Tominaga, S.; Rouleau, J.L.; Lee, R.T. Identification of serum soluble ST2 receptor as a novel heart failure biomarker. *Circulation* **2003**, *107*, 721–726. [CrossRef] [PubMed]
22. Januzzi, J.L., Jr.; Peacock, W.F.; Maisel, A.S.; Chae, C.U.; Jesse, R.L.; Baggish, A.L.; O'Donoghue, M.; Sakhuja, R.; Chen, A.A.; van Kimmenade, R.R.; et al. Measurement of the interleukin family member ST2 in patients with acute dyspnea: Results from the PRIDE (Pro-Brain Natriuretic Peptide Investigation of Dyspnea in the Emergency Department) study. *J. Am. Coll. Cardiol.* **2007**, *50*, 607–613. [CrossRef]
23. Shimpo, M.; Morrow, D.A.; Weinberg, E.O.; Sabatine, M.S.; Murphy, S.A.; Antman, E.M.; Lee, R.T. Serum levels of the interleukin-1 receptor family member ST2 predict mortality and clinical outcome in acute myocardial infarction. *Circulation* **2004**, *109*, 2186–2190. [CrossRef] [PubMed]
24. Dieplinger, B.; Egger, M.; Haltmayer, M.; Kleber, M.E.; Scharnagl, H.; Silbernagel, G.; de Boer, R.A.; Maerz, W.; Mueller, T. Increased soluble ST2 predicts long-term mortality in patients with stable coronary artery disease: Results from the Ludwigshafen risk and cardiovascular health study. *Clin. Chem.* **2014**, *60*, 530–540. [CrossRef] [PubMed]
25. Li, M.; Duan, L.; Cai, Y.; Hao, B.; Chen, J.; Li, H.; Liu, H. Prognostic value of soluble suppression of tumorigenesis-2 (sST2) for cardiovascular events in coronary artery disease patients with and without diabetes mellitus. *Cardiovasc. Diabetol.* **2021**, *20*, 49. [CrossRef] [PubMed]
26. Ho, J.E.; Chen, W.Y.; Chen, M.H.; Larson, M.G.; McCabe, E.L.; Cheng, S.; Ghorbani, A.; Coglianese, E.; Emilsson, V.; Johnson, A.D.; et al. Common genetic variation at the IL1RL1 locus regulates IL-33/ST2 signaling. *J. Clin. Investig.* **2013**, *123*, 4208–4218. [CrossRef]
27. Akhabir, L.; Sandford, A. Genetics of interleukin 1 receptor-like 1 in immune and inflammatory diseases. *Curr. Genomics.* **2010**, *11*, 591–606. [CrossRef] [PubMed]
28. Lin, J.F.; Wu, S.; Juang, J.J.; Chiang, F.T.; Hsu, L.A.; Teng, M.S.; Cheng, S.T.; Huang, H.L.; Sun, Y.C.; Liu, P.Y.; et al. IL1RL1 single nucleotide polymorphism predicts sST2 level and mortality in coronary and peripheral artery disease. *Atherosclerosis* **2017**, *257*, 71–77. [CrossRef] [PubMed]
29. Fan, C.T.; Lin, J.C.; Lee, C.H. Taiwan Biobank: A project aiming to aid Taiwan's transition into a biomedical island. *Pharmacogenomics* **2008**, *9*, 235–246. [CrossRef]
30. Chung, C.M.; Lin, T.H.; Chen, J.W.; Leu, H.B.; Yin, W.H.; Ho, H.Y.; Sheu, S.H.; Tsai, W.C.; Chen, J.H.; Lin, S.J.; et al. Common quantitative trait locus downstream of RETN gene identified by genome-wide association study is associated with risk of type 2 diabetes mellitus in Han Chinese: A Mendelian randomization effect. *Diabetes Metab. Res. Rev.* **2014**, *30*, 232–240. [CrossRef] [PubMed]
31. Filkova, M.; Haluzik, M.; Gay, S.; Senolt, L. The role of resistin as a regulator of inflammation: Implications for various human pathologies. *Clin. Immunol.* **2009**, *133*, 157–170. [CrossRef] [PubMed]
32. Figarska, S.M.; Gustafsson, S.; Sundstrom, J.; Arnlov, J.; Malarstig, A.; Elmstahl, S.; Fall, T.; Lind, L.; Ingelsson, E. Associations of Circulating Protein Levels With Lipid Fractions in the General Population. *Arter. Thromb. Vasc. Biol.* **2018**, *38*, 2505–2518. [CrossRef] [PubMed]
33. Axelsson, J.; Bergsten, A.; Qureshi, A.R.; Heimburger, O.; Barany, P.; Lonnqvist, F.; Lindholm, B.; Nordfors, L.; Alvestrand, A.; Stenvinkel, P. Elevated resistin levels in chronic kidney disease are associated with decreased glomerular filtration rate and inflammation, but not with insulin resistance. *Kidney Int.* **2006**, *69*, 596–604. [CrossRef] [PubMed]
34. Reilly, M.P.; Lehrke, M.; Wolfe, M.L.; Rohatgi, A.; Lazar, M.A.; Rader, D.J. Resistin is an inflammatory marker of atherosclerosis in humans. *Circulation* **2005**, *111*, 932–939. [CrossRef]
35. Zong, X.; Fan, Q.; Zhang, H.; Yang, Q.; Xie, H.; Chen, Q.; Zhang, R.; Tao, R. Soluble ST2 levels for predicting the presence and severity of metabolic syndrome. *Endocr. Connect.* **2021**, *10*, 336–344. [CrossRef]
36. Hoffmann, M.M.; Pilz, S.; Weihrauch, G.; Seelhorst, U.; Wellnitz, B.; Winkelmann, B.R.; Boehm, B.O.; Marz, W. Effect of the resistin -420C > G polymorphism on cardiovascular disease and mortality. *Clin. Endocrinol.* **2008**, *69*, 344–345. [CrossRef]
37. Norata, G.D.; Ongari, M.; Garlaschelli, K.; Tibolla, G.; Grigore, L.; Raselli, S.; Vettoretti, S.; Baragetti, I.; Noto, D.; Cefalu, A.B.; et al. Effect of the -420C/G variant of the resistin gene promoter on metabolic syndrome, obesity, myocardial infarction and kidney dysfunction. *J. Intern. Med.* **2007**, *262*, 104–112. [CrossRef]
38. Tang, N.P.; Wang, L.S.; Yang, L.; Zhou, B.; Gu, H.J.; Sun, Q.M.; Cong, R.H.; Zhu, H.J.; Wang, B. A polymorphism in the resistin gene promoter and the risk of coronary artery disease in a Chinese population. *Clin. Endocrinol.* **2008**, *68*, 82–87. [CrossRef]
39. Tu, X.; Nie, S.; Liao, Y.; Zhang, H.; Fan, Q.; Xu, C.; Bai, Y.; Wang, F.; Ren, X.; Tang, T.; et al. The IL-33-ST2L pathway is associated with coronary artery disease in a Chinese Han population. *Am. J. Hum. Genet.* **2013**, *93*, 652–660. [CrossRef]

40. Maloberti, A.; Biolcati, M.; Ruzzenenti, G.; Giani, V.; Leidi, F.; Monticelli, M.; Algeri, M.; Scarpellini, S.; Nava, S.; Soriano, F.; et al. The Role of Uric Acid in Acute and Chronic Coronary Syndromes. *J. Clin. Med.* **2021**, *10*, 4750. [CrossRef]
41. Chronic Kidney Disease Prognosis Consortium; Matsushita, K.; van der Velde, M.; Astor, B.C.; Woodward, M.; Levey, A.S.; de Jong, P.E.; Coresh, J.; Gansevoort, R.T. Association of estimated glomerular filtration rate and albuminuria with all-cause and cardiovascular mortality in general population cohorts: A collaborative meta-analysis. *Lancet* **2010**, *375*, 2073–2081.
42. Raczy, C.; Petrovski, R.; Saunders, C.T.; Chorny, I.; Kruglyak, S.; Margulies, E.H.; Chuang, H.Y.; Kallberg, M.; Kumar, S.A.; Liao, A.; et al. Isaac: Ultra-fast whole-genome secondary analysis on Illumina sequencing platforms. *Bioinformatics* **2013**, *29*, 2041–2043. [CrossRef] [PubMed]
43. Chen, C.H.; Yang, J.H.; Chiang, C.W.K.; Hsiung, C.N.; Wu, P.E.; Chang, L.C.; Chu, H.W.; Chang, J.; Song, I.W.; Yang, S.L.; et al. Population structure of Han Chinese in the modern Taiwanese population based on 10,000 participants in the Taiwan Biobank project. *Hum. Mol. Genet.* **2016**, *25*, 5321–5331. [CrossRef] [PubMed]
44. Wu, S.; Teng, M.S.; Er, L.K.; Hsiao, W.Y.; Hsu, L.A.; Yeh, C.H.; Lin, J.F.; Lin, Y.Y.; Su, C.W.; Ko, Y.L. Association between NF-kappaB Pathway Gene Variants and sICAM1 Levels in Taiwanese. *PLoS ONE* **2017**, *12*, e0169516.

International Journal of
Molecular Sciences

MDPI

Article

Differential Genetic and Epigenetic Effects of the *KLF14* Gene on Body Shape Indices and Metabolic Traits

Semon Wu [1], Lung-An Hsu [2], Ming-Sheng Teng [3], Hsin-Hua Chou [4,5] and Yu-Lin Ko [3,4,5,*]

1 Department of Life Science, Chinese Culture University, Taipei 11114, Taiwan; semonwu@yahoo.com.tw
2 The First Cardiovascular Division, Department of Internal Medicine, Chang Gung Memorial Hospital and Chang Gung University College of Medicine, Taoyuan 33305, Taiwan; hsula@adm.cgmh.org.tw
3 Department of Research, Taipei Tzu Chi Hospital, Buddhist Tzu Chi Medical Foundation, New Taipei City 23142, Taiwan; vincent@tzuchi.com.tw
4 The Division of Cardiology, Department of Internal Medicine and Cardiovascular Center, Taipei Tzu Chi Hospital, Buddhist Tzu Chi Medical Foundation, New Taipei City 23142, Taiwan; chouhhtw@gmail.com
5 School of Medicine, Tzu Chi University, Hualien 97004, Taiwan
* Correspondence: yulinkotw@yahoo.com.tw; Tel.: +886-2-6628-9779 (ext. 5355); Fax: +886-2-6628-9009

Abstract: The *KLF14* gene is a key metabolic transcriptional transregulator with monoallelic maternal expression. *KLF14* variants are only associated with adipose tissue gene expression, and *KLF14* promoter methylation is strongly associated with age. This study investigated whether age, sex, and obesity mediate the effects of *KLF14* variants and DNA methylation status on body shape indices and metabolic traits. In total, the data of 78,742 and 1636 participants from the Taiwan Biobank were included in the regional plot association analysis for *KLF14* variants and *KLF14* methylation, respectively. Regional plot association studies revealed that the *KLF14* rs4731702 variant and the nearby strong linkage disequilibrium polymorphisms were the lead variants for lipid profiles, blood pressure status, insulin resistance surrogate markers, and metabolic syndrome mainly in female participants and for body shape indices mainly in obese women. Significant age-dependent associations between *KLF14* promoter methylation levels and body shape indices, and metabolic traits were also noted predominantly in female participants. *KLF14* variants and *KLF14* hypermethylation status were associated with metabolically healthy and unhealthy phenotypes, respectively, in obese individuals, and only the *KLF14* variants demonstrated a significant association with both higher adiposity and lower cardiometabolic risk in the same allele, revealing uncoupled excessive adiposity from its cardiometabolic comorbidities, especially in obese women. Variations of *KLF14* are associated with body shape indices, metabolic traits, insulin resistance, and metabolically healthy status. Differential genetic and epigenetic effects of *KLF14* are age-, sex- and obesity-dependent. These results provided a personalized reference for the management of cardiometabolic diseases in precision medicine.

Keywords: *KLF14*; body shape indices; metabolic traits; differential effect; genetic variants; DNA methylation

Citation: Wu, S.; Hsu, L.-A.; Teng, M.-S.; Chou, H.-H.; Ko, Y.-L. Differential Genetic and Epigenetic Effects of the *KLF14* Gene on Body Shape Indices and Metabolic Traits. *Int. J. Mol. Sci.* **2022**, *23*, 4165. https://doi.org/10.3390/ijms23084165

Academic Editor: Elixabet Lopez-Lopez

Received: 9 March 2022
Accepted: 6 April 2022
Published: 9 April 2022

1. Introduction

Krüpple-like factors (KLFs), belonging to the SP/KLF family, are a group of evolutionarily conserved transcription factors with C-terminal zinc finger domains that are crucial for the recognition of and binding to the GT/GC-rich or CACCC box cis-regulatory sites in gene promoters and enhancers [1,2]. KLFs play a critical role in homeostasis maintenance by regulating the gene expression in many organ systems, such as cardiovascular, renal, digestive, respiratory, immunological, and hematopoietic systems [3–6]. KLFs also regulate cell signaling pathways controlling cell proliferation, apoptosis, migration, differentiation, and other physiological activities [4]. As a member of the KLF family, KLF14 is induced by TGF-β in intraembryonic and ectodermal tissues [4,6] and forms a corepressor transcription complex with Sin3A and HDAC2 to inhibit the expression of TGF-II receptors. KLF14 is only expressed in mammalian tissues, including the muscle, brain, fat, and liver, and it has monoallelic maternal expression in all tissues of both humans and mice [7]. In the livers of high-fat-diet–fed mice and db/db obese mice, KLF14 activates the PI3K/Akt signaling pathway to increase insulin sensitivity [8]. It also mediates lipid signaling, and KLF14-mediated metabolic phenotypes are attributed to the regulation of lipid signaling through the dysregulation of sphingokinase 1, a critical lipid signaling molecule [5,9,10]. With the lack of *Klf14*-mediated apoA-I promoter activation, hepatic-specific *Klf14* deletions in $Apoe^{-/-}$ mice cause marked acceleration of atherosclerotic lesion development in combination with decreased high-density lipoprotein (HDL) cholesterol levels, and cholesterol efflux, whereas Klf14 activation reduces atherosclerosis [11]. Together, these findings indicate that epigenetic regulation–induced *KLF14* loss-of-function may trigger the onset of metabolic diseases. Furthermore, KLF14 is a master metabolic transcriptional regulator that mediates adipogenesis, insulin signaling, lipid metabolism, inflammatory and immune responses, and cell proliferation and differentiation, thus acting as a potential novel therapeutic target to reduce the risk of atherosclerotic cardiovascular disease [12].

The *KLF14* gene is a single exon imprinted gene localized in the human chromosome 7q32.2, with a total length of 1059 bp; it encodes 323 amino acids, and only the allele inherited from the mother is expressed [13]. Using chromatin-state maps, a 1.6-kb enhancer region approximately 5 kb upstream of the *KLF14* transcription start site has been identified in adipose tissue [14]. Genome-wide association studies (GWASs) have identified *KLF14* variants to be associated with a multitude of metabolic pathologies, such as insulin resistance (IR), diabetes mellitus, coronary artery disease, ischemic stroke, and myocardial infarction [14–17]. Lead single-nucleotide polymorphisms (SNPs) in GWASs map to a 3–48-kb region upstream of *KLF14*, and these mapped genetic variants are also associated with the abundance of the *KLF14* transcript only in adipose tissue [14,18]. The associations are usually stronger in women than in men. Thus, the potential interactive effect of sex and obesity on the association between *KLF14* variants and cardiometabolic phenotypes required further elucidation.

Aging is an important risk factor for chronic metabolic and inflammatory disorders such as atherosclerosis, cancer, and type 2 diabetes mellitus [19–21]. CpG sites located in *KLF14* are associated with aging-related alteration of hypermethylation in whole blood samples, thus providing an accurate estimate of chronological age and serving as age-related epigenetic biomarkers [22–24]. In addition, the age-related DNA methylation level of *KLF14* may be a risk marker for diabetes mellitus [25]. Recently, extensive studies have also focused on the genetic loci of metabolically healthy phenotype in obese individuals and of genetic loci that uncouple excessive adiposity from its cardiometabolic comorbidities [26,27]. The Taiwan Biobank (TWB) population-based cohort study enrolled >100,000 volunteers aged 30–70 years with no history of cancer [28,29]. In the current study, we included participants from the TWB cohort study to elucidate the effects of age, sex, and obesity on the association of *KLF14* variants and methylation status with conventional and allometric body shape indices, IR surrogate marker levels, various metabolic traits, and metabolic syndrome. We also assessed whether *KLF14* variants and methylation levels are markers of metabolically healthy obese phenotype and whether these variations uncouple excessive adiposity from the adverse metabolic traits in Taiwan.

2. Results

2.1. Regional Plot Association Studies for Genetic Variants at Positions between 130.3 and 130.5 Mb on Chromosome 7q32.2 and Study Phenotypes

The flowchart of participant enrollment is presented in Figure 1. We performed regional plot association studies in 78,742 TWB participants to examine the association between genetic variants at positions between 130.3 and 130.5 Mb on chromosome 7q32.2 and the studied phenotypes. Lead SNPs with genome-wide significance were noted for the HDL cholesterol level (rs4731702, $p = 6.69 \times 10^{-14}$), triglyceride level (rs13240528,

$p = 1.46 \times 10^{-16}$), metabolic syndrome (rs3996352, $p = 1.30 \times 10^{-13}$), mean blood pressure (BP) (rs1364422, $p = 3.59 \times 10^{-8}$), and hip index (HI) (rs35057928, $p = 3.48 \times 10^{-14}$), whereas associations with $p < 1.41 \times 10^{-4}$ were noted for body mass index (BMI) (rs3996352, $p = 1.58 \times 10^{-6}$), a body shape index (ABSI) (rs34072724, $p = 2.14 \times 10^{-6}$), hip circumference (rs35057928, $p = 3.54 \times 10^{-7}$), and waist circumference (rs34072724, $p = 1.41 \times 10^{-5}$) (Table 1 and Supplementary Figures S1–S10). Except for rs1364422, which is the lead SNP for mean BP, all other lead SNPs were in nearly complete LD with rs4731702 ($r^2 > 0.95$) (Supplementary Figure S11). Our data revealed that the lead SNP for each phenotype was located upstream of the *KLF14* gene region, revealing pleiotropic effects on this gene locus.

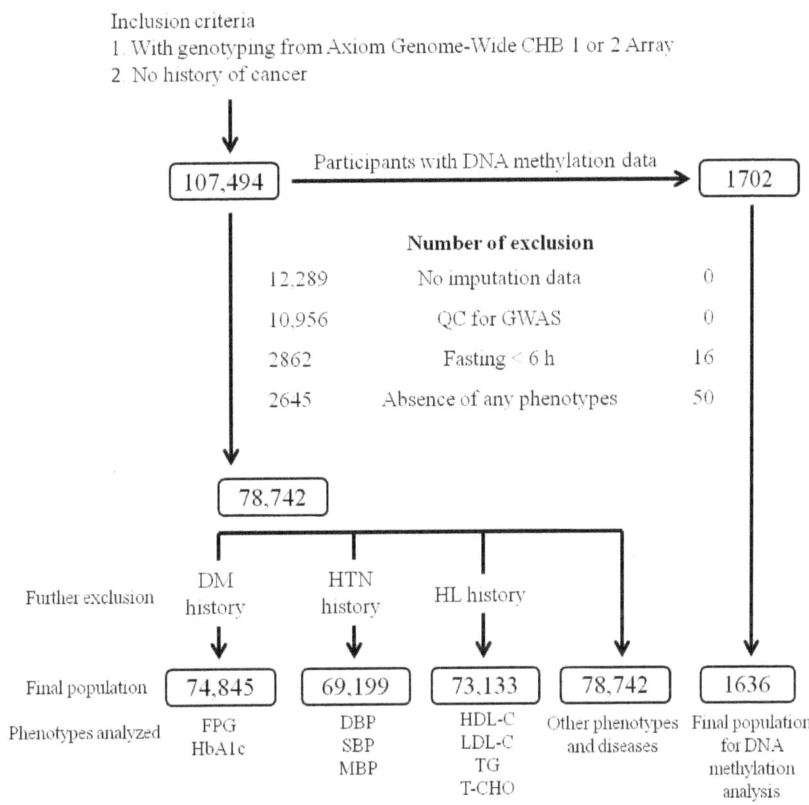

Figure 1. Study inclusion and exclusion criteria flowchart. This flowchart presents the inclusion and exclusion criteria used to screen for Taiwan Biobank (TWB) participants. GWAS—genome-wide association study; QC—quality control; HL—hyperlipidemia, HTN—hypertension; DM—diabetes mellitus; FPG—fasting plasma glucose; HbA1c—glycated hemoglobin; SBP—systolic blood pressure; DBP—diastolic blood pressure; MBP—mean blood pressure; LDL-C:—low-density lipoprotein cholesterol; HDL-C—high-density lipoprotein cholesterol; TG—triglyceride; T-CHO—total cholesterol. Other phenotypes include age, body mass index (BMI), waist circumference, hip circumference, waist–hip ratio, a body shape index (ABSI), waist–hip index (WHI), hip index (HI), the product of triglyceride and fasting plasma glucose (the TyG index), TyG with adiposity status (TyG-BMI), and TyG with waist circumference (TyG-WC), current smoking status, and metabolic syndrome.

Table 1. Lead single nucleotide polymorphisms (SNPs) for various phenotypes at the *KLF14* gene region.

Phenotypes	Lead SNPs	Position	Ref/Alt	MAF	LD *	*p* Value
HDL cholesterol (mmol/L)	rs4731702	130748625	T/C	0.3117	1.0	6.69×10^{-14}
Triglyceride (mmol/L)	rs13240528	130760369	A/C	0.3097	0.955	1.46×10^{-16}
Mean blood pressure (mmHg)	rs1364422	130761222	T/C	0.4218	0.283	3.59×10^{-8}
Metabolic syndrome (%)	rs3996352	130760175	G/A	0.3098	0.955	1.3×10^{-13}
Body mass index (kg/m^2)	rs3996352	130760175	G/A	0.3098	0.955	1.58×10^{-6}
Waist circumference (cm)	rs34072724	130747710	A/G	0.3118	1.0	1.41×10^{-5}
A body shape index	rs34072724	130747710	A/G	0.3118	1.0	2.14×10^{-6}
Hip circumference (cm)	rs35057928	130749767	C/T	0.3104	0.978	3.54×10^{-7}
Hip index	rs35057928	130749767	C/T	0.3104	0.978	3.48×10^{-14}
Body fat percentage (%)	rs34084575	130751643	A/-	0.3123	1.0	0.0002

HDL—high-density lipoprotein; Ref—Reference allele; Alt—Alternate allele; MAF—Minor allele frequency; * LD—Linkage disequilibrium between rs4731702 and lead SNPs.

2.2. Association of the KLF14 rs4731702 Genotype with Parameters of Body Shape Indices, IR Surrogate Markers, and Metabolic Traits

Because of nearly complete LD between lead *KLF14* polymorphisms, we selected *rs4731702*, a polymorphism associated with adipocyte size and body composition traits on *KLF14* [14], for further genotype–phenotype analysis of the various clinical phenotypes and laboratory parameters of 78,742 participants. By using an additive model, after adjustment for age, sex, BMI, and smoking status, we observed genome-wide significant associations of *rs4731702* with HI, HDL cholesterol, and triglyceride levels and metabolic syndrome, whereas associations with $p < 2.65 \times 10^{-4}$ were noted for hip circumference, waist circumference, body fat percentage (BFP), BMI, ABSI, systolic BP, diastolic BP, and mean BP. The products of triglyceride (TG) and fasting plasma glucose (the TyG index), TyG and body mass index (TyG-BMI), TyG and waist circumference (TyG-WC), and hypertension (Supplementary Table S1) influence type 2 diabetes risk via a female-specific effect on adipocyte size and body composition.

2.3. Interactive Effects of Sex and Obesity on the Association between the KLF14 rs4731702 Genotype and Various Phenotypes

We investigated whether sex and obesity affect the association between the *rs4731702* genotype and the studied phenotypes. As presented in Table 2, for female participants, genome-wide significant associations were found between the *rs4731702* genotype and hip circumference, HI, HDL cholesterol and triglyceride levels, TyG index, TyG-BMI, and metabolic syndrome. Significant associations with $p < 2.65 \times 10^{-4}$ were found between the *rs4731702* genotype and waist circumference, BMI, ABSI, diastolic and mean BP, TyG-WC, and diabetes mellitus in women. However, no significant associations were found between the *rs4731702* genotype and the studied phenotypes in men. The results of the two-sample *t* tests revealed that sex affected the association between the *KLF14* rs4731702 genotype and waist circumference, hip circumference, ABSI, HI, HDL cholesterol, and triglyceride levels (Table 2). These results suggest that the association between the *rs4731702* genotype and several phenotypes is sex dependent. We further evaluated the genotype–phenotype associations of *rs4731702* in obese and nonobese participants. Differential associations according to adiposity status were noted only in body shape indices, including waist and hip circumferences and ABSI, with significant associations occurring only in obese participants but not in nonobese participants (Supplementary Table S2). We then divided the participants into four groups according to sex and obesity. No obvious associations were noted between the *rs4731702* genotype and various phenotypes in men, whether obese or nonobese (Supplementary Table S3). In women, differential associations with adiposity status were noted between the *rs4731702* genotype and body shape indices but not between the *rs4731702* genotype and metabolic traits (Table 3).

Table 2. Association between *KLF14* rs4731702 genotype and body shape indices and metabolic traits according to sex.

Clinical and Laboratory Parameters	Male (n = 28,483)				Female (n = 50,259)			
	Median (IQR)	Beta	SE	p Value *	Median (IQR)	Beta	SE	p Value *
Age (years)	51.0 (40.0–59.0)	−0.0034	0.1001	0.9733	51.0 (41.0–58.0)	0.0311	0.0705	0.6596
Body shape indices								
Body height (cm)	169.5 (165.5–173.5)	0.0041	0.0524	0.9377	157.5 (153.5–161.0)	−0.0984	0.0360	0.0063
Body weight (kg)	71.7 (65.1–79.2)	0.0108	0.0457	0.8131	56.9 (51.7–63.4)	−0.0719	0.0272	0.0081
Hip circumference (cm)	97.0 (93.0–101.3)	0.0175	0.0328	0.5927	94.0 (90.0–99.0)	0.1483	0.0261	1.40×10^{-8} ¶
Waist circumference (cm)	87.0 (82.0–93.0)	0.0295	0.0395	0.4549	79.6 (74.0–86.0)	0.1718	0.0368	2.98×10^{-6} ¶
Waist–hip ratio	0.90 (0.86–0.94)	0.0001	0.0004	0.8457	0.84 (0.80–0.89)	0.0003	0.0004	0.4090
Body fat percentage (%)	22.9 (19.6–26.1)	0.0424	0.0253	0.0939	31.6 (27.6–35.8)	0.0540	0.0153	4.17×10^{-4}
Body mass index (kg/m^2)	25.0 (23.0–27.3)	0.0701	0.0315	0.0262	23.0 (21.0–25.5)	0.0995	0.0251	7.19×10^{-5}
ABSI	7.8 (7.6–8.1)	0.0015	0.0032	0.6524	7.8 (7.5–8.2)	0.0177	0.0034	2.34×10^{-7} ¶¶
WHI	4.0 (3.9–4.1)	0.0002	0.0017	0.9198	3.8 (3.7–4.0)	0.0014	0.0018	0.4300
HI	14.6 (14.3–14.9)	0.0022	0.0041	0.5900	15.5 (15.1–15.8)	0.0303	0.0036	2.16×10^{-17} ¶¶
Blood pressure and heart rate								
Mean heart rate † (/min)	34.5 (32.0–38.0)	−0.0969	0.0952	0.3088	35.0 (32.5–38.0)	0.0109	0.0620	0.8599
Systolic BP † (mmHg)	121.0 (111.5–131.3)	−0.4146	0.1412	0.0033	111.0 (102.0–123.0)	−0.3622	0.1026	4.15×10^{-4}
Diastolic BP † (mmHg)	76.5 (70.0–83.0)	−0.3409	0.0952	3.45×10^{-4}	69.0 (62.7–76.0)	−0.2618	0.0659	7.10×10^{-5}
Mean BP † (mmHg)	91.2 (84.3–98.7)	−0.3654	0.1028	3.78×10^{-4}	83.1 (76.3–91.3)	−0.2953	0.0731	5.40×10^{-5}
Lipid profiles								
Total cholesterol § (mmol/L)	190.0 (169.0–213.0)	0.0000	0.0007	0.9856	194.0 (172.0–218.0)	0.0004	0.0005	0.4241
HDL cholesterol § (mmol/L)	47.0 (40.0–54.0)	0.0002	0.0009	0.7872	57.0 (49.0–66.0)	0.0058	0.0006	9.26×10^{-20} ¶¶
LDL cholesterol § (mmol/L)	121.0 (101.0–142.0)	0.0004	0.0011	0.6975	118.0 (98.0–140.0)	−0.0008	0.0008	0.3344
Triglyceride § (mmol/L)	106.0 (74.0–155.0)	−0.0030	0.0021	0.1659	82.0 (59.0–118.0)	−0.0127	0.0014	4.47×10^{-19} ¶¶
Glucose metabolism								
Fasting plasma glucose ‡ (mmol/L)	94.0 (89.0–99.0)	−0.1248	0.1576	0.4284	80.0 (86.0–95.0)	−0.3007	0.0938	0.0013
HbA1c ‡ (%)	5.6 (5.4–5.9)	−0.0062	0.0063	0.3223	5.6 (5.4–5.8)	−0.0129	0.0037	5.53×10^{-4}
Insulin resistance surrogate markers								
TyG index §,‡	10080.0 (6887.0–14952.0)	−161.99	117.87	0.1694	7384.0 (5200.0–10878.0)	−332.47	54.04	7.71×10^{-10}
TyG-BMI §,‡ ($\times 10^3$)	252.5 (164.3–393.0)	−4483	3138	0.1531	170.1 (113.4–266.2)	−8072	1385	5.59×10^{-9}
TyG-WC §,‡ ($\times 10^3$)	879.6 (578.6–1353.6)	−14,439	10905	0.1855	585.6 (395.9–900.1)	−25,264	4760	1.12×10^{-7}
Atherosclerotic risk factors								
Diabetes mellitus (%)	12.13%	−0.0362	0.0293	0.2161	7.73%	−0.0998	0.0270	2.23×10^{-4}
Hypertension (%)	30.56%	−0.0665	0.0217	0.002[illegible]	17.22%	−0.0510	0.0198	0.0101
Metabolic syndrome (%)	26.52%	−0.0640	0.0239	0.0073	19.34%	−0.1414	0.0200	1.47×10^{-12}

Participants were analyzed with the exclusion of those with a history of † hypertension, ‡ diabetes mellitus, and § hyperlipidemia. Abbreviations as in Figure 1; SE; —standard error; IQR—Inter Quartile range. Data are presented as median (interquartile range). * p—adjusted for age, BMI, and current smoking; Age—adjusted for BMI and current smoking; and BMI—adjusted for age and smoking. Significance was defined as a p value of $<0.05/(131 + 27) = 3.16 \times 10^{-4}$. t-test: ** $p < 0.001$, * $p < 0.01$.

Table 3. Association between *KLF14* rs4731702 genotype and body shape indices and metabolic traits according to obesity in females.

Clinical and Laboratory Parameters	Female, Non-Obese (n = 35,677)				Female, Obese (n = 14,582)			
	Median (IQR)	Beta	SE	p Value *	Median (IQR)	Beta	SE	p Value *
Age (years)	50.0 (41.0–58.0)	0.0035	0.0834	0.9667	52.0 (43.0–59.0)	0.1340	0.1275	0.2933
Body-shape indices								
Body height (cm)	157.5 (154.0–161.5)	−0.1199	0.0427	0.0049	156.5 (153.0–160.5)	−0.0496	0.0671	0.4597
Body weight (kg)	53.9 (50.0–57.9)	−0.0817	0.0293	0.0053	67.5 (63.0–73.3)	−0.0464	0.0597	0.4371
Hip circumference (cm)	92.0 (89.0–95.0)	0.0829	0.0290	0.0043	101.0 (98.0–105.5)	0.3001	0.0551	5.23×10^{-8} **
Waist circumference (cm)	76.5 (72.0–81.0)	0.0506	0.0417	0.2251	89.0 (84.5–95.0)	0.4659	0.0747	4.49×10^{-10} **
Waist hip ratio	0.8 (0.8–0.9)	−0.0002	0.0005	0.6526	0.8 (0.8–0.9)	0.0018	0.0008	0.0198
Body fat percentage (%)	29.4 (26.2–32.2)	0.0421	0.0171	0.0138	38.4 (36.4–41.3)	0.0998	0.0281	3.83×10^{-4}
Body mass index (kg/m^2)	21.8 (20.3–23.2)	0.0225	0.0156	0.1486	27.2 (25.9–29.3)	0.1106	0.0364	0.0024
ABSI	7.8 (7.5–8.2)	0.0077	0.0041	0.0593	7.8 (7.5–8.1)	0.0411	0.0062	2.64×10^{-11} *
WHI	3.8 (3.7–4.0)	−0.0010	0.0021	0.6383	3.8 (3.7–4.0)	0.0076	0.0033	0.0209
HI	15.6 (15.2–15.9)	0.0213	0.0040	1.06×10^{-7}	15.3 (14.9–15.6)	0.0492	0.0072	6.94×10^{-12} **
Blood pressure and heart rate								
Mean heart rate † (/min)	35.0 (32.5–38.0)	0.1158	0.0721	0.1079	35.5 (32.5–38.5)	−0.2848	0.1210	0.0186
Systolic BP † (mmHg)	109.0 (100.5–120.0)	−0.2944	0.1177	0.0124	118.0 (108.7–130.0)	−0.5344	0.2080	0.0102
Diastolic BP † (mmHg)	67.5 (61.5–74.0)	−0.2066	0.0758	0.0064	73.0 (67.0–80.0)	−0.4107	0.1328	0.0020
Mean BP † (mmHg)	81.3 (75.0–89.0)	−0.2359	0.0840	0.0050	88.3 (81.3–96.2)	−0.4519	0.1475	0.0022
Lipid profiles								
Total cholesterol § (mmol/L)	193.0 (171.0–217.0)	0.0007	0.0006	0.2729	199.0 (176.0–222.0)	−0.0001	0.0010	0.9301
HDL cholesterol § (mmol/L)	59.0 (51.0–68.0)	0.0052	0.0008	3.07×10^{-12}	51.0 (45.0–59.0)	0.0070	0.0012	4.94×10^{-9}
LDL cholesterol § (mmol/L)	115.0 (96.0–136.0)	−0.0008	0.0009	0.3903	126.0 (106.0–146.0)	−0.0004	0.0015	0.7926
Triglyceride § (mmol/L)	74.0 (55.0–105.0)	−0.0110	0.0016	1.87×10^{-11}	107.0 (77.0–150.0)	−0.0165	0.0028	6.58×10^{-9}
Glucose metabolism								
Fasting plasma glucose ‡ (mmol/L)	89.0 (85.0–94.0)	−0.1831	0.0952	0.0543	93.0 (88.0–99.0)	−0.5934	0.2276	0.0091
HbA1c ‡ (%)	5.5 (5.3–5.8)	−0.0100	0.0038	0.0090	5.7 (5.5–6.0)	−0.0204	0.0090	0.0235
Insulin resistance surrogate markers								
TyG index §,‡	6650.0 (4823.0–9546.0)	−256.27	55.32	4.00×10^{-6}	9879.0 (6942.0–14,234.5)	−519.90	131.25	7.50×10^{-5}
TyG-BMI §,‡ ($\times 10^3$)	143.6 (101.2–211.2)	−5670	1272	8.00×10^{-6}	274.2 (190.3–404.5)	−14211	3747	1.50×10^{-4}
TyG-WC §,‡ ($\times 10^3$)	506.5 (357.2–746.2)	−19,217	4678	4.00×10^{-5}	887.4 (612.8–1302.6)	−40,422	12118	8.53×10^{-4}
Atherosclerotic risk factors								
Diabetes mellitus (%)	4.78%	−0.1062	0.0396	0.0074	14.97%	−0.0874	0.0367	0.0173
Hypertension (%)	12.29%	−0.0640	0.0265	0.0159	28.29%	−0.0302	0.0298	0.3112
Metabolic syndrome (%)	9.96%	−0.1254	0.0296	2.23×10^{-5}	42.30%	−0.1442	0.0269	8.02×10^{-8}

Abbreviations, adjusted conditions, and subjects recruited for analysis as in Figure 1 and Table 2. p *—adjusted for age, BMI, and current smoking; Age—adjusted for BMI and current smoking; and BMI—adjusted for age and smoking. Significance was defined as a p value of $<0.05/(131 + 27) = 3.16 \times 10^{-4}$. t-test: ** $p < 0.001$, * $p < 0.01$. † hypertension, ‡ diabetes mellitus, and § hyperlipidemia.

2.4. Association between KLF14 Promoter Methylation Status and KLF14 Variants and Age

We examined whether *KLF14* variants are associated with the DNA methylation status of the *KLF14* promoter region. We observed no significant association between *KLF14* variants and methylation levels in 32 *KLF14* methylation sites (Supplementary Table S4). By contrast, after adjustment for sex, BMI, and smoking, genome-wide significant associations were observed between age and 15 *KLF14* promoter DNA methylation sites, with hypermethylation associated with increasing age and with the minimum p value of 3.83×10^{-253} for *KLF14* cg08097417 (Figure 2). Both male and female participants exhibited a highly significant association between age and *KLF14* promoter methylation status (Supplementary Table S4).

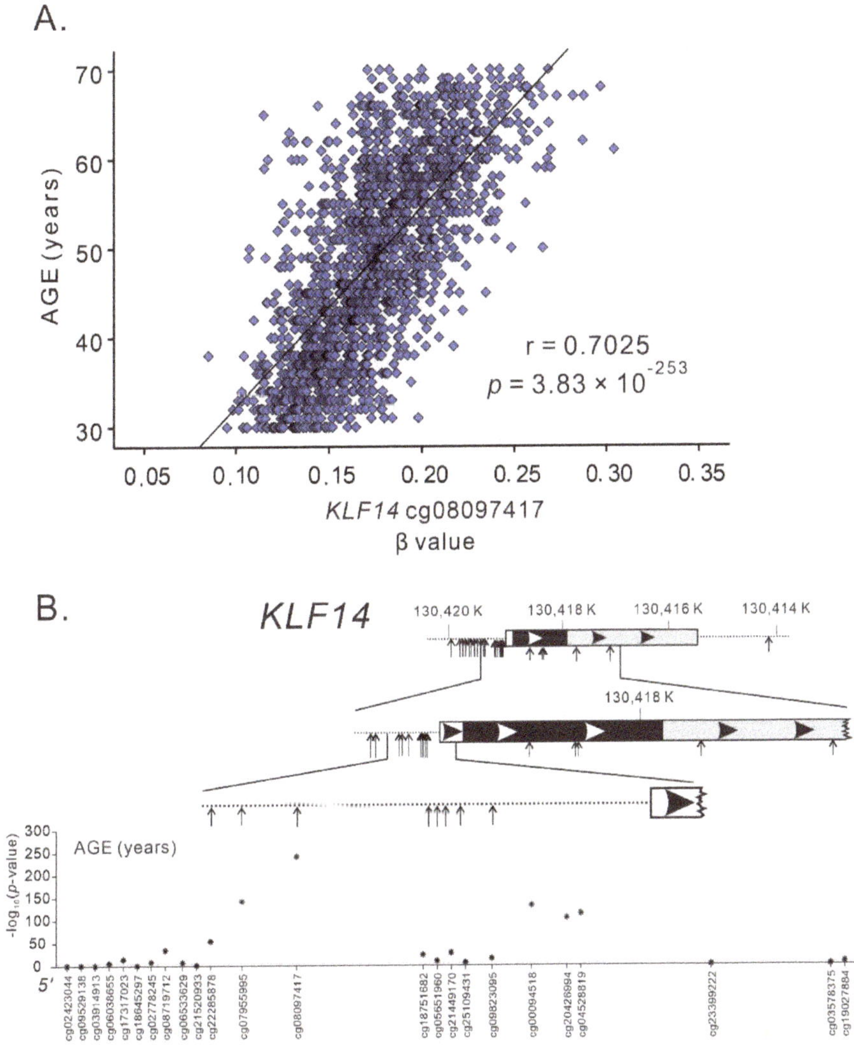

Figure 2. *KLF14* methylation and age: (**A**) Association between *KLF14* promoter methylation cg08097417 level and chronologic age and (**B**) Genomic structure of *KLF14* and the association between *KLF14* methylation status and age.

2.5. Associations between KLF14 Methylation Status at the Promoter Region and Various Phenotypes Are Mediated by Chronologic Age

We evaluated the association between a *KLF14* promoter methylation site, cg08097417, which displayed the strongest association with age and various phenotypes. After adjustment for sex, BMI, and current smoking, our data revealed genome-wide significant positive associations between cg08097417 methylation and various body shape indices and metabolic traits, such as waist circumference, waist–hip ratio, ABSI, waist–hip index (WHI), systolic and mean BP, total cholesterol levels, and hypertension. Negative associations were observed in cg08097417 methylation with body height, body weight, and hip circumference, and positive associations with $p < 2.65 \times 10^{-4}$ between cg08097417 methylation and diastolic BP, low-density lipoprotein (LDL) cholesterol, triglyceride, and hemoglobin A1c (HbA1c) levels and diabetes mellitus, which are in the same direction with age-related adiposity and metabolic changes (Table 4). In addition, after further adjustment for age, significance levels markedly decreased, and all the associations became nonsignificant, suggesting that the associations are predominantly mediated by chronologic age (Table 4). We further extended the analysis to other *KLF14* promoter methylation sites with regional plot association studies. The results revealed similar association trends between the *KLF14* promoter methylation levels and body shape indices and metabolic traits, with subsidence of significant associations after adjustment for age (Figure 3).

2.6. Associations between KLF14 Methylation Status at the Promoter Region and Various Phenotypes: Subgroup Analysis with Sex and Obesity

Subgroup analysis revealed an interactive effect of sex on the associations between cg08097417 methylation levels and several metabolic traits, including systolic BP, total and LDL cholesterol and triglyceride levels, and HbA1c, with the association occurring predominantly in the female sex (Supplementary Table S5 and Figure 4). By contrast, no interactive effect was noted on the associations between cg08097417 methylation levels and all study phenotypes according to adiposity status (Supplementary Table S6).

Table 4. Association of the cg08097417 methylation site and age with body shape indices and metabolic traits.

Clinical and Laboratory Parameters	Age (n = 1636)					KLF14 cg08097417 (n = 1636)			
	Median (IQR)	Beta	SE	p Value *	p Value #	Beta	SE	p Value *	p Value ##
Age (years)	49.0 (40.0–58.0)	-	-	-	-	0.0153	0.0004	7.71×10^{-243}	
Body shape indices									
Body height (cm)	163.0 (157.5–170.5)	−0.5529	0.0440	1.23×10^{-34}	2.67×10^{-17}	−0.0088	0.0010	9.13×10^{-19}	0.6149
Body weight (kg)	64.6 (55.7–73.5)	−0.6675	0.0527	4.16×10^{-35}	4.42×10^{-18}	−0.0104	0.0012	1.88×10^{-18}	0.7655
Hip circumference (cm)	96.1 (92.0–101.0)	−0.5762	0.0709	8.73×10^{-16}	4.00×10^{-6}	−0.0106	0.0016	1.28×10^{-11}	0.1022
Waist circumference (cm)	83.5 (76.5–90.0)	0.4137	0.0516	2.05×10^{-15}	1.98×10^{-7}	0.0068	0.0011	2.30×10^{-9}	0.5210
Waist–hip ratio	0.86 (0.82–0.91)	66.4649	4.7723	9.62×10^{-42}	5.31×10^{-18}	1.1323	0.1062	1.08×10^{-25}	0.1063
Body fat percentage (%)	26.9 (22.6–32.3)	0.1568	0.0996	0.1155	0.3052	0.0026	0.0022	0.2334	0.8981
Body mass index (kg/m^2)	23.9 (21.8–26.4)	−0.0126	0.0780	0.8720	0.7680	0.0001	0.0017	0.9349	0.7908
ABSI	7.8 (7.5–8.1)	6.5158	0.5522	6.83×10^{-31}	3.23×10^{-15}	0.1040	0.0123	4.64×10^{-17}	0.5717
WHI	3.9 (3.7–4.1)	14.3854	1.0618	1.05×10^{-39}	2.62×10^{-17}	0.2438	0.0236	2.96×10^{-24}	0.1340
HI	15.1 (14.6–15.6)	−1.0974	0.5654	0.0524	0.8798	−0.0324	0.0123	0.0083	0.0764
Blood pressure and heart rate									
Mean heart rate † (/min)	34.5 (32.0–37.5)	−0.0671	0.0318	0.0352	0.2659	−0.0013	0.0007	0.0560	0.5506
Systolic BP † (mmHg)	113.0 (104.0–124.0)	0.2357	0.0184	1.45×10^{-35}	1.98×10^{-13}	0.0043	0.0004	7.02×10^{-25}	0.0126
Diastolic BP † (mmHg)	71.0 (64.0–79.0)	0.1596	0.0291	5.02×10^{-8}	0.0087	0.0033	0.0006	2.93×10^{-7}	0.0610
Mean BP † (mmHg)	85.0 (78.2–93.3)	0.2432	0.0264	9.40×10^{-20}	6.21×10^{-7}	0.0046	0.0006	2.34×10^{-15}	0.0212
Lipid profiles									
Total cholesterol § (mmol/L)	193.0 (171.0–217.0)	24.5694	3.4825	2.61×10^{-12}	3.00×10^{-5}	0.4394	0.0759	8.53×10^{-9}	0.2152
HDL cholesterol § (mmol/L)	53.0 (45.0–63.0)	5.4550	3.0549	0.0744	0.2430	0.0933	0.0663	0.1595	0.8485
LDL cholesterol § (mmol/L)	120.0 (100.0–141.0)	9.7590	2.3010	2.40×10^{-5}	0.0211	0.1838	0.0500	0.0002	0.3064
Triglyceride § (mmol/L)	89.0 (63.0–131.3)	7.2982	1.2616	8.80×10^{-9}	0.0001	0.1187	0.0275	1.70×10^{-7}	0.6777
Glucose metabolism									
Fasting plasma glucose ‡ (mmol/L)	92.0 (87.0–97.0)	0.0851	0.0173	1.00×10^{-6}	0.0003	0.0012	0.0004	0.0015	0.7686
HbA1c ‡ (%)	5.6 (5.4–5.8)	3.3823	0.4459	5.62×10^{-14}	1.55×10^{-9}	0.0449	0.0098	5.00×10^{-6}	0.3136
Insulin resistance surrogate markers									
TyG index §‡	8239.0 (5670.0–12,240.0)	8.20×10^{-5}	2.70×10^{-5}	0.0026	0.1158	2.00×10^{-6}	5.98×10^{-7}	0.0070	0.4128
TyG-BMI §‡ ($\times 10^3$)	199.3 (127.5–314.5)	2.00×10^{-6}	1.00×10^{-6}	0.0187	0.2553	4.89×10^{-8}	2.22×10^{-8}	0.0280	0.4433
TyG-WC §‡ ($\times 10^3$)	693.3 (440.5–1069.5)	8.56×10^{-7}	2.98×10^{-7}	0.0042	0.1554	1.72×10^{-8}	6.52×10^{-9}	0.0083	0.3771
Atherosclerotic risk factors									
Diabetes mellitus (%)	4.83%	7.7733	1.2499	6.33×10^{-10}	5.00×10^{-6}	2.2274	0.5205	1.90×10^{-5}	0.9187
Hypertension (%)	18.95%	7.8538	0.6828	1.7×10^{-29}	2.58×10^{-10}	3.0710	0.3183	4.99×10^{-22}	0.0041
Metabolic syndrome (%)	18.89%	4.2314	0.7435	1.49×10^{-23}	3.10×10^{-5}	1.3859	0.3097	8.00×10^{-6}	0.9530

Abbreviations, adjusted condition, and subjects recruited for analysis as in Figure 1 and Table 2. * p—adjusted for age, sex, BMI, and current smoking. # p—adjusted for age, sex, BMI, current smoking, and cg08097417; ## p—adjusted for age, sex, BMI, current smoking, and age. Significance between age and studied phenotypes was defined as a p value of $<0.05/(24 + 26) = 0.001$. Significance between cg08097417 and studied phenotypes was defined as a p value of $<0.05/(24 + 27) = 9.80 \times 10^{-4}$. † hypertension, ‡ diabetes mellitus, and § hyperlipidemia.

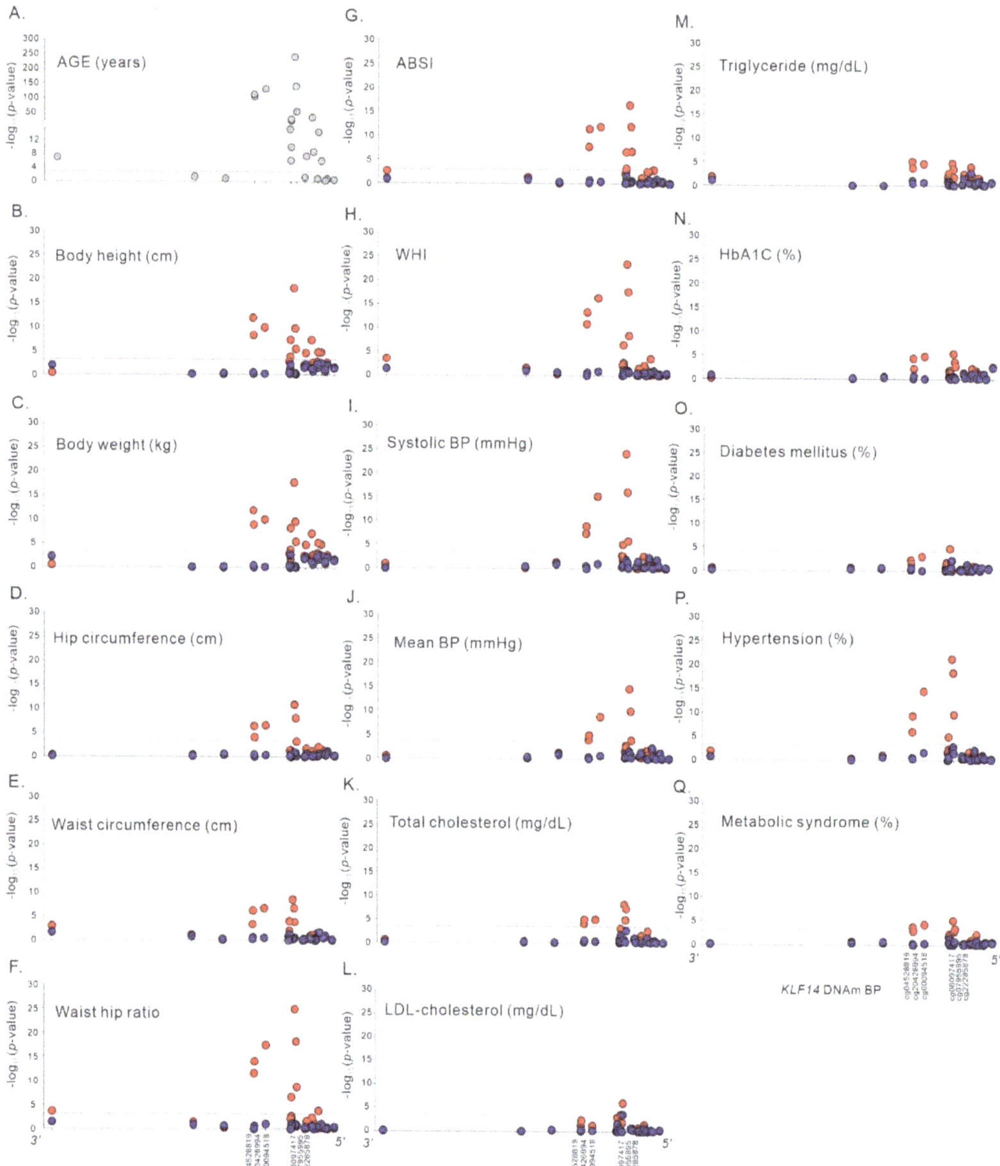

Figure 3. Regional plot association analysis between*KLF14* methylation status and the studied phenotypes (**A**) Age, (**B**) Body height, (**C**) Body weight, (**D**) Hip circumference, (**E**) Waist circumference, (**F**) Waist hip ratio, (**G**) ABSI, (**H**) WHI, (**I**) Systolic BP, (**J**) Mean BP, (**K**) Total cholesterol, (**L**) LDL-cholesterol, (**M**) Triglyceride, (**N**) HbA1c, (**O**) Diabetes mellitus, (**P**) Hypertension, (**Q**) Metabolic syndrome, with (blue circle) or without (red circle) adjustment for age. The results showed subsidence of significant associations after adjustment for age.

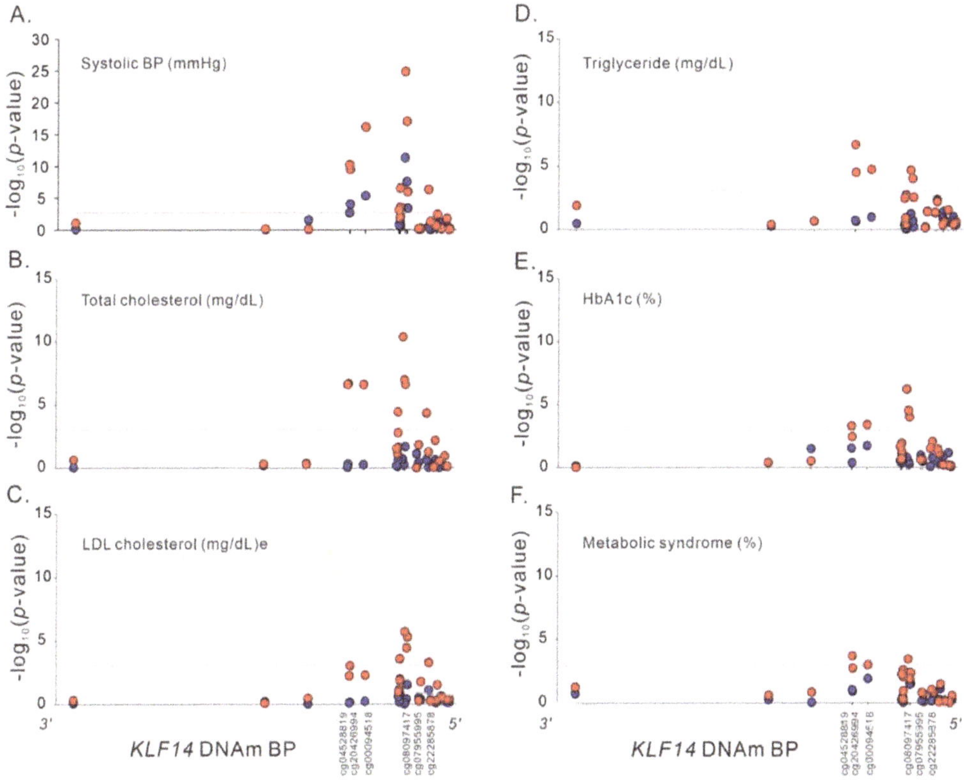

Figure 4. Regional plot association analysis between the studied phenotypes (**A**) Systolic BP, (**B**) Total cholesterol, (**C**) LDL cholesterol, (**D**) Triglyceride, (**E**) HbA1c, (**F**) Metabolic syndrome, and DNA methylation status according to sex, with red circles representing female participants and blue circles representing male participants.

2.7. KLF14 rs4731702 Variant as a Marker of Metabolically Healthy Obese Phenotype with the Adiposity-Increasing Allele Associated with a Favorable Cardiometabolic Risk Profile

We further evaluated whether the *KLF14* rs4731702 genotypes and *KLF14* promoter methylation levels are involved in the uncoupling of adiposity from its adverse metabolic risk profiles using the standardized effect size analysis in the Radar plot. From the Radar plots of different subgroups of participants, the *KLF14* rs4731702 C allele was the most strongly and positively associated with HI, ABSI, hip and waist circumferences, BMI, and BFP, but no association with waist–hip ratio and WHI was found (Figure 5A–C). By contrast, the *KLF14* rs4731702 C allele was negatively associated with nearly all metabolic traits, positively associated with HDL cholesterol levels, and had no association with total and LDL cholesterol levels. These results were provided by analyses for total participants, female participants, and obese female participants (Figure 5A–C). By contrast, cg08097417 hypermethylation showed a positive association with both body shape indices and metabolic traits, except for body height, hip circumference, and HI (Figure 5D). In the analysis conducted using the criteria of Park et al. [27], our data indicated that the rs4731702 C allele has a lower risk of metabolically unhealthy status compared with metabolically healthy status in obese participants (odds ratio: 0.88, 95% confidence interval [CI]: 0.85–0.92, $p = 1.92 \times 10^{-11}$), whereas hypermethylation of the cg08097417 methylation site was associated with a higher risk of metabolically unhealthy status in obese participants (odds ratio 4.04, 95% CI 1.95–8.41, $p = 1.81 \times 10^{-4}$) (Supplementary Table S7).

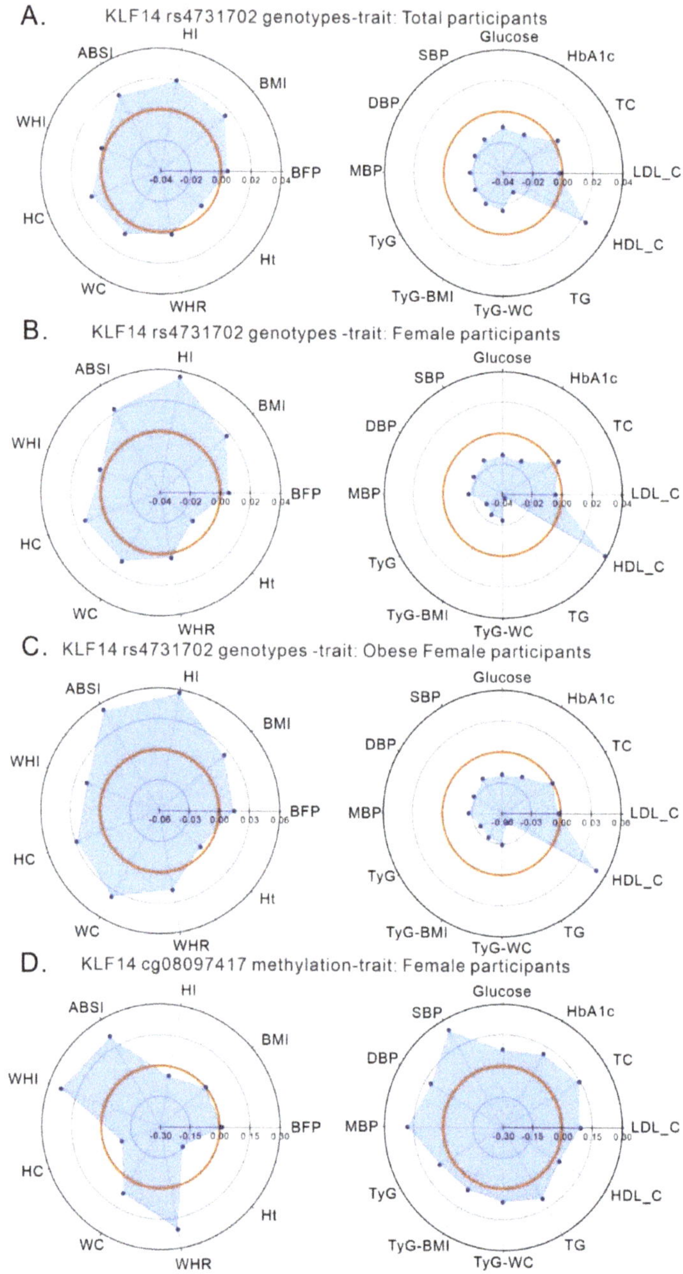

Figure 5. *KLF14* rs4731702 genotype (**A–C**) and *KLF14* promoter methylation site cg08097417 (**D**) and their associations with body shape indices (left panel) and metabolic traits (right panel) in the Taiwan Biobank population. Standardized coefficients (beta) are shown in Radar plots for total participants (**A**), female participants (**B**,**D**), and obese female participants (**C**). The middle of the three concentric circles is labeled "0," representing no association between the *KLF14* variant and methylation status and trait. Points falling outside the middle octagon or dodecagon represent positive variant–trait or methylation–trait associations, whereas those inside it represent negative variant–trait or methylation–trait associations.

3. Discussion

This study investigated the differential genetic and epigenetic effects of *KLF14* on the association between body shape indices and metabolic traits. Several novel results were identified. First, we observed an independent association of genetic and epigenetic effects of *KLF14* on both multiple body shape indices and metabolic traits. Second, the association of *KLF14* variants with body shape indices and metabolic traits occurred predominantly in the female sex, and the association of *KLF14* variants with body shape indices occurred predominantly in obese participants. Thus, obese women were the only subgroup associated with both body shape indices and metabolic traits. Third, *KLF14* methylation is a marker of age in forensic medicine for both sexes. In addition to diabetes mellitus, *KLF14* promoter methylation status showed a significant age-dependent association with various body shape indices and metabolic traits, and the associations with lipid profiles, HbA1c, and metabolic syndrome occurred predominantly in female participants. Finally, differential genetic and epigenetic effects were noted in cross-phenotype associations using the adiposity–cardiometabolic trait pairs. *KLF14* rs4731702 variant, but not methylation status, served as a marker involved in the uncoupling of adiposity from its adverse metabolic risk profiles predominantly in subgroups of obese female participants. To the best of our knowledge, this is the first report investigating the mediated and interactive effects of age, sex, and obesity on the association of genetic and epigenetic *KLF14* variations with body shape indices and metabolic traits. These results support the critical role of *KLF14* as an age-, sex-, and obesity-specific key transcriptional regulator affecting a large transregulatory network of metabolic traits and adiposity status, and the results also provided evidence for the differential genetic and epigenetic effects of *KLF14* on the risk of cardiometabolic disorders.

3.1. Association between KLF14 Variants and Metabolic Traits: The Role of Sex and Obesity

By including the *KLF14* rs4731702 genotypes in the analysis, Small et al. [11] demonstrated colocalization of the GWAS signal for metabolic disorders and the expression quantitative trait locus (eQTL) for nearly 400 genes in trans to the *KLF14* locus. The results revealed *KLF14* to be one of the largest trans-eQTL hotspots in the human genome [30]. Despite the nearly ubiquitous expression of KLF14 in many tissues [31], *KLF14* variants regulate gene expression only in adipose tissue [14]. *KLF14* is also an imprinted gene with monoallelic maternal expression [14]. Our data indicated a significant association of *KLF14* variants with various metabolic traits, such as lipid profiles, BP status, IR surrogate marker levels, and metabolic syndrome, but only in female participants. By contrast, no interactive effect for such associations according to adiposity status was noted in our study population.

3.2. Association of KLF14 Variants with Adiposity Indices, including Body Shape and Body Fat Distribution

Body shape plays a critical role in influencing cardiometabolic complications of obesity, with the abdominal size showing positive associations and gluteofemoral size showing inverse associations with complications [32]. Conventional body shape indices, such as WHR and waist and hip circumferences, are strongly associated with BMI, and even after adjustment for BMI, waist and hip circumferences are strongly associated with height [14]. Allometric body shape indices, such as ABSI and HI, are alternatives to body shape indices and are independent of body size and general obesity [33,34]. In analogy to ABSI as an allometric index of waist circumference, WHI was created as an allometric index of WHR [35]. ABSI has been shown to be significantly associated with cardiometabolic risk factors and mortality [33,36]. ABSI also achieves better mortality risk stratification than other body shape indices, which are correlated with BMI [37]. Similar to our results with sexual dimorphism, GWAS of allometric body shape indices in the UK Biobank Resources of 406,697 participants with British ancestry indicated a genome-wide significant association of *KLF14* variants with HI in women but not in men [35]. We further revealed the interactive effect of obesity on the association between *KLF14* variants and body shape indices. In subgroup analysis, the association between *KLF14* variants and body shape indices occurred only in obese women, showing the critical role of both sex and obesity in associations between *KLF14* variants and body shape indices.

3.3. Epigenetic Effects of KLF14 on Adiposity Indices and Metabolic Traits Occur Only in the Female Sex and Are Dependent on Age

Aging generally induces global hypomethylation with a genome-wide decrease in the average DNA methylation level. By contrast, aging-related hypermethylation events occurred in 13% of the CpG sites among the millions of sites in the genome [38]. *KLF14* has a large CpG island that spans almost its entire open reading frame. By analyzing KLF14 in offspring from a DNA methyltransferase 3a conditional knockout mouse, *KLF14* expression was detected within the hypermethylated region inherited from the mother, which is a typical feature of the KLF family [7]. Using pathway analyses of adipose tissue from 190 individuals, altered DNA methylation in 1050 genes, including *KLF14*, has been reported as epigenetic markers of chronological age in blood [39]. DNA methylation in *KLF14* is linearly correlated with human age with an increase in the methylation level of the *KLF14* promoter region in multiple tissues in mice, such as the whole blood, adipose tissue, kidney, spleen, lung, and colon, which is significantly associated with inflammatory marker levels and with decreased *KLF14* downstream gene methylation in the whole blood, the adipose tissue, and the spleen [40]. More importantly, DNA methylation loci in *KLF14* is an age-related epigenetic biomarker; it can be used as an age prediction model to predict chronological age and may be considered an important factor in the development of aging-associated diseases, such as cancer, diabetes mellitus, and cardiovascular diseases [38,41].

Our data revealed the hypermethylation of the *KLF14* promoter region with age in both sexes. We further demonstrated significant associations of *KLF14* promoter methylation with body shape indices and metabolic traits, and the association occurred predominantly in the female sex, which was similar to the results of *KLF14* variants. However, no significant association between *KLF14* variants and *KLF14* promoter methylation was noted. These results are consistent with those of a previous study reporting a significant association between the *KLF14* rs4731702 genotype and *KLF14* gene expression and DNA methylation in adipose tissue, but not in whole blood. The finding suggested that as a blood biomarker, *KLF14* exerts both genetic and epigenetic effects on body shape indices and metabolic traits independently [14]. The association of *KLF14* methylation with various phenotypes was abolished after adjustment for age, revealing an age-dependent effect. Thus, *KLF14* methylation can serve not only as an epigenetic biomarker of age but also as an epigenetic biomarker of age-related changes in adiposity status, body shape indices, and metabolic phenotypes.

3.4. KLF14 rs4731702 Variant as a Marker of Metabolically Healthy Obese Phenotype That Uncouples Excessive Adiposity from Its Comorbidities

Obesity is typically associated with chronic, low-grade inflammation and a constellation of metabolic abnormalities, which are important risk factors for type 2 diabetes and cardiovascular diseases [42,43]. For a given fat mass, individuals who are obese may not develop metabolic comorbidities, which was termed metabolically healthy obesity (MHO). The mechanism of the discrepancy has been suggested to be related to disproportionate adipose tissue distribution in metabolically unhealthy obesity (MUO) with visceral adiposity [44,45] or an impaired ability to expand subcutaneous fat in the lower part of the body. Pathway analyses have suggested that pathophysiological mechanisms for the differential status of metabolically healthy or unhealthy individuals may involve adipogenesis, angiogenesis, transcriptional regulation, and IR [46]. Cumulative evidence from several large-scale prospective studies suggests that MHO individuals have higher risks than healthy normal-weight individuals of many cardiometabolic outcomes and cardiovascular diseases, such as hypertension, IR, diabetes mellitus, and cerebrovascular disease [47–56]; yet, their risk is lower than that of MUO individuals [46].

Recently, GWASs have identified genetic variants that are associated with increased adiposity in conjunction with a healthy metabolic profile [46]. GWAS investigating genetic determinants of insulin concentrations in individuals without diabetes highlighted that many identified loci showed associations with markers suggestive of clinical IR, such as high triglyceride and low HDL concentrations [57]. Favorable adiposity alleles were also associated with a higher subcutaneous fat mass and lower liver fat content, supporting the hypothesis that storing excess triglycerides in metabolically low-risk depots has beneficial effects on metabolism [57]. Waist-specific and hip-specific polygenic risk scores, which are related to visceral and subcutaneous abdominal fat and gluteofemoral and leg fat, respectively, increased the risks of an adverse metabolic profile, type 2 diabetes, and coronary artery disease [58]. In a genome-wide discovery of genetic loci that uncouple excess adiposity from its comorbidities, Huang et al. [26] identified 62 loci with the same allele significantly associated with both higher adiposity and lower cardiometabolic risk. These loci are enriched for genes expressed in adipose tissue and for regulatory variants influencing nearby genes affecting adipocyte differentiation. In a Korean study examining the genetic determinants of MHO and metabolically unhealthy normal weight phenotypes, the genes common to both models are related to lipid metabolism, and those associated with MHO are related to insulin or glucose metabolism [27]. Our data indicated an inverse association between the rs4731702 genotypes and multiple body shape indices when compared with the association with metabolic traits and IR surrogate markers. These results supported *KLF14* variants as a marker for uncoupling excessive adiposity from its cardiometabolic risk profiles, which occurred predominantly in obese women. By contrast, *KLF14* hypermethylation is associated with a metabolically unhealthy obese phenotype coupling of adiposity with its comorbidities. Interestingly, Cannataro et al. [59] found that microRNAs may modify promoter by methylation on targeted genes and regulate their expression, which affect body shape under Ketogenic Diet. Therefore, we definitely consider investigating the microRNA candidates that regulate KLF14 associated body shape in the near future.

3.5. Limitation

This study has limitations. First, ethnic genetic heterogeneity has been noted in the genetic association studies, and our data may not be applied to other ethnic populations. Second, several regional plot association studies did not reach genome-wide significance for the associations. Further studies with larger sample size and using different ethnic populations may help to strengthen the validity of our analysis.

4. Participants and Methods

4.1. TWB Participants

We included 107,494 TWB participants with no history of cancer who had undergone genotyping from Axiom Genome-Wide CHB 1 or 2 Array (Affymetrix, Santa Clara, CA, USA). Of them, 28,752 were excluded because of the absence of imputation data (12,289 participants) and after quality control (QC) for the GWAS with identity by descent (IBD) > 0.187 revealing relative pairs of second-degree relatives or closer (10,956 participants). Participants with fasting <6 h (2862 participants) and the absence of any of the study phenotypes (2645 participants) were also excluded. When examining the lipid profile, blood pressure status, and glucose metabolism parameters, the participants with a history of hyperlipidemia, hypertension, and diabetes mellitus were also excluded. The flowchart of participant enrollment is presented in Figure 1. The definitions of hypertension, diabetes mellitus, obesity, current smoking, and metabolic syndrome are provided in Supplementary Methods. Ethical approval was received from the Research Ethics Committee of Taipei Tzu Chi Hospital, Buddhist Tzu Chi Medical Foundation (approval number: 05-X04-007), and Ethics and Governance Council of the Taiwan Biobank (approval number: TWBR10507-02 and TWBR10611-03). All participants signed an informed consent form before participating in the study.

4.2. Clinical Phenotypes and Laboratory Examinations

We measured the following clinical parameters: body height, body weight, and conventional body shape indices such as waist circumference, hip circumference, waist–hip ratio, and BMI, mean heart rate, and systolic, mean, and diastolic BP. We also collected the following biochemistry data: lipid profiles and glucose metabolism parameters such as total HDL and LDL cholesterol and triglyceride levels, fasting plasma glucose levels, and HbA1c. TyG index, TyG-BMI, and TyG-WC were used as IR surrogate markers [60]. BFP was measured using the Body Composition Analyzer BC-420MA (TANITA, TANITA Corporation, Sportlife Tokyo, Japan), a device that sends a weak electric current through the body to measure the impedance (electrical resistance) of the body. ABSI, WHI, and HI were calculated as parameters of allometric body shape indices [35]. The measurement details of BFP and the calculation of conventional and allometric body shape indices and TyG-related parameters are provided in Supplementary Methods. We also performed cross-phenotype analysis for *KLF14* variants and *KLF14* methylation levels and subsequently calculated the standard deviation effect sizes (β) of their effects on the body shape indices and metabolic traits to elucidate whether *KLF14* variations may uncouple excessive adiposity from its comorbidities [26]. The resulting effect sizes are on the same scale, making them comparable across all variants, methylation levels, and traits. By using the criteria of Park et al. [27], we further examined the association of *KLF14* variants and *KLF14* methylation status with metabolically healthy and unhealthy obese phenotypes in TWB participants.

4.3. DNA Isolation and Genotyping of KLF14 Variants

DNA was isolated from blood samples using a PerkinElmer Chemagic 360 instrument (PerkinElmer, Waltham, MA, USA). SNP genotyping was conducted using custom TWB chips and was performed on the Axiom genome-wide array plate system (Affymetrix, Santa Clara, CA, USA).

4.4. Regional Plot Association Studies

To determine the lead SNPs around the *KLF14* region for various studied phenotypes, we performed a QC for regional plot association analysis by including TWB participants enrolled after applying the exclusion criteria (Figure 1). The Axiom genome-wide CHB 1 and 2 array plates, each with 27,720 and 79,774 participants and comprising 611,656 and 640,160 SNPs, respectively, were used for imputation analysis. Genome-wide genotype imputation was performed using SHAPEIT and IMPUTE2, with the data of the East Asian population from the 1000 Genome Project Phase study used as the reference panel. After imputation, indels were removed using VCFtools, and the QC was performed by filtering SNPs with an IMPUTE2 imputation quality score of >0.3. All samples enrolled for the analysis had the identity by descent PI_HAT > 0.1875. For SNP QC, three criteria were used for exclusion from subsequent analyses: (1) an SNP call rate of <3%, (2) a minor allele frequency of <0.01, and (3) the violation of the Hardy–Weinberg equilibrium with $p < 10^{-6}$. After QC, the data of 85,508 participants and 131 SNPs at positions between 130.3 and 130.5 Mb on chromosome 7q32.2 were included in the regional plot association analysis.

4.5. DNA Methylation Analysis

DNA methylation was assessed using sodium-bisulfite-treated DNA from whole blood using the Infinium MethylationEPIC BeadChipEPIC array (Illumina, San Diego, CA, USA). Four criteria were used to perform the QC for DNA methylation analysis: (1) normalization across batches for correction of dye bias, (2) background signal removal, (3) using the median absolute deviation method to eliminate outliers, and (4) elimination of probes with poor detection ($p > 0.05$) and those whose bead counts were <3. Finally, the data of 1702 participants were included in the *KLF14* methylation analysis. Sixty-six participants were further excluded due to fasting for <6 h ($n = 16$) or the absence of any study phenotype ($n = 50$).

4.6. Statistical Analysis

Continuous variables were expressed as median and interquartile ranges when the distribution was strongly skewed. Categorical data distributions were expressed as a percentage. Before analysis, all study parameters with a skew distribution were logarithmically transformed to adhere to the normality assumption. To test for significant differences in the association between different subgroups of sex and obesity, we used a two-sample t test [32]. We assumed the genetic effect to be additive, and a general linear model was used to analyze the studied phenotypes in relation to the predictors of investigated genotypes and confounders. Multivariable logistic regression analysis was used to evaluate the independent effect of genotypes on the risk of hypertension, diabetes mellitus, and metabolic syndrome. The Radar plots were presented with the standardized coefficients (beta) in the linear regression. Genome-wide scans were performed, and the beta coefficient, standard error, and p values by general linear model or logistic regression for genome-wide data were analyzed using the analysis software package PLINK. Genome-wide significance was defined as a p value of $<5 \times 10^{-8}$. For Bonferroni correction of regional plot association studies, the significant value was defined as $p < 1.41 \times 10^{-5}$ calculated as $0.05/(131 \times 27)$ according to a total of 131 variants and 27 traits analyzed. For Bonferroni correction of each genotype–phenotype and methylation–phenotype analysis, the significant value was defined as $p < 2.65 \times 10^{-4}$, calculated as $0.05/(27 \times 7)$ according to 27 traits and 7 subgroups analyzed. Linkage disequilibrium (LD) between each SNP was analyzed using LDmatrix software (https://analysistools.nci.nih.gov/LDlink/?tab=ldmatrix, accessed on 19 July 2021). All statistical analyses were performed using SPSS (version 22; SPSS, IBM Corp., Armonk, NY, USA) or PLINK. Missing data were approached with listwise deletion.

5. Conclusions

Taken together, our data revealed that variations of the *KLF14* gene, including gene variants and methylation levels for body shape indices, IR status, and metabolic traits, are highly dependent on age, sex, or obesity. These results support the critical role of *KLF14* as a key age-, sex-, and obesity-specific transcriptional regulator affecting a large adipose-specific transregulatory network of metabolic traits and adiposity status, which provide further evidence for the differential genetic and epigenetic effects of *KLF14* on the risk of cardiometabolic disorders.

Supplementary Materials: The Supplementary Materials are available online at https://www.mdpi.com/article/10.3390/ijms23084165/s1.

Author Contributions: Conceptualization, Y.-L.K. and S.W.; methodology, Y.-L.K. and M.-S.T.; software and validation, M.-S.T. and S.W.; formal analysis, H.-H.C., Y.-L.K. and L.-A.H.; resources, Y.-L.K. and H.-H.C.; data curation, S.W., M.-S.T. and H.-H.C.; writing—original draft preparation, S.W.; writing—review and editing, Y.-L.K. and S.W.; visualization, M.-S.T. and S.W.; project administration, Y.-L.K. and S.W.; supervision, Y.-L.K.; funding acquisition, M.-S.T., H.-H.C. and Y.-L.K. All authors have read and agreed to the published version of the manuscript.

Funding: This study was supported by grants from Buddhist Tzu Chi Medical Foundation Academic Advancement (TCMF-EP 111-02, CMF-A 107-01-15), grants from the Ministry of Science and Technology (MOST 108-2314-B-303-026-MY3) and Taipei Tzu Chi Hospital, Buddhist Tzu Chi Medical Foundation (TCRD-TPE-MOST-109-05) to Y. L. Ko, and Taipei Tzu Chi Hospital, Buddhist Tzu Chi Medical Foundation (TCRD-TPE-109-RT-1) to H. H. Chou.

Institutional Review Board Statement: The study was conducted according to the guidelines of the Declaration of Helsinki, and approved by the Research Ethics Committee of Taipei Tzu Chi Hospital, Buddhist Tzu Chi Medical Foundation (approval number: 05-X04-007), and Ethics and Governance Council of the Taiwan Biobank (approval number: TWBR10507-02 and TWBR10611-03).

Informed Consent Statement: Informed consent was obtained from all subjects involved in the study.

Data Availability Statement: The data presented in this study are available on request from the corresponding author.

Acknowledgments: We greatly appreciate technical support from the Core Laboratory of Taipei Tzu Chi Hospital, Buddhist Tzu Chi Medical Foundation, and expert statistical analysis assistance from Tsung-Han Hsieh.

Conflicts of Interest: No potential conflict of interest relevant to this article was reported.

Abbreviations

KLFs	Krüpple-like factors
HDL	high-density lipoprotein
GWASs	Genome-wide association studies
IR	insulin resistance
SNPs	single-nucleotide polymorphisms
TWB	Taiwan Biobank
BP	blood pressure
HI	hip index
ABSI	a body shape index
BFP	body fat percentage
TG	triglyceride
BMI	body mass index
WC	waist circumference
WHI	waist–hip index
LDL	low-density lipoprotein
HbA1c	hemoglobin A1c
eQTL	expression quantitative trait locus
MHO	metabolically healthy obesity
MUO	metabolically unhealthy obesity
QC	quality control
IBD	identity by descent
LD	Linkage disequilibrium

References

1. Chen, X.; Shi, W.; Zhang, H. The role of KLF14 in multiple disease processes. *BioFactors* **2020**, *46*, 276–282. [CrossRef] [PubMed]
2. Yang, Q.; Civelek, M. Transcription Factor KLF14 and Metabolic Syndrome. *Front. Cardiovasc. Med.* **2020**, *7*, 91. [CrossRef] [PubMed]
3. Kim, C.K.; He, P.; Bialkowska, A.B.; Yang, V.W. SP and KLF Transcription Factors in Digestive Physiology and Diseases. *Gastroenterology* **2017**, *152*, 1845–1875. [CrossRef] [PubMed]
4. McConnell, B.B.; Yang, V.W. Mammalian Krüppel-like factors in health and diseases. *Physiol. Rev.* **2010**, *90*, 1337–1381. [CrossRef]
5. Pollak, N.M.; Hoffman, M.; Goldberg, I.J.; Drosatos, K. Krüppel-like factors: Crippling and un-crippling metabolic pathways. *JACC Basic Transl. Sci.* **2018**, *3*, 132–156. [CrossRef]
6. Tetreault, M.P.; Yang, Y.; Katz, J.P. Krüppel-like factors in cancer. *Nat. Rev. Cancer* **2013**, *13*, 701–713. [CrossRef]
7. Parker-Katiraee, L.; Carson, A.R.; Yamada, T.; Arnaud, P.; Feil, R.; Abu-Amero, S.N.; Moore, G.E.; Kaneda, M.; Perry, G.H.; Stone, A.C.; et al. Identification of the imprinted KLF14 transcription factor undergoing human-specific accelerated evolution. *PLoS Genet.* **2007**, *3*, e65. [CrossRef]
8. Yang, M.; Ren, Y.; Lin, Z.; Tang, C.; Jia, Y.; Lai, Y.; Zhou, T.; Wu, S.; Liu, H.; Yang, G.; et al. Krüppel-like factor 14 increases insulin sensitivity through activation of PI3K/Akt signal pathway. *Cell. Signal.* **2015**, *27*, 2201–2208. [CrossRef]
9. De Assuncao, T.M.; Lomberk, G.; Cao, S.; Yaqoob, U.; Mathison, A.; Simonetto, D.A.; Huebert, R.C.; Urrutia, R.A.; Shah, V.H. New role for Kruppel-like factor 14 as a transcriptional activator involved in the generation of signaling lipids. *J. Biol. Chem.* **2014**, *289*, 15798–15809. [CrossRef]
10. Wang, J.; Badeanlou, L.; Bielawski, J.; Ciaraldi, T.P.; Samad, F. Sphingosine kinase 1 regulates adipose proinflammatory responses and insulin resistance. *Am. J. Physiol. Endocrinol. Metab.* **2014**, *306*, E756–E768. [CrossRef]
11. Guo, Y.; Fan, Y.; Zhang, J.; Lomberk, G.A.; Zhou, Z.; Sun, L.; Mathison, A.J.; Garcia-Barrio, M.T.; Zhang, J.; Zeng, L.; et al. Perhexiline activates KLF14 and reduces atherosclerosis by modulating ApoA-I production. *J. Clin. Investig.* **2015**, *125*, 3819–3830. [CrossRef] [PubMed]
12. Xie, W.; Li, L.; Zheng, X.L.; Yin, W.D.; Tang, C.K. The role of Krüppel-like factor 14 in the pathogenesis of atherosclerosis. *Atherosclerosis* **2017**, *263*, 352–360. [CrossRef] [PubMed]

13. Scohy, S.; Gabant, P.; van Reeth, T.; Hertveldt, V.; Drèze, P.L.; van Vooren, P.; Rivière, M.; Szpirer, J.; Szpirer, C. Identification of KLF13 and KLF14 (SP6), novel members of the SP/XKLF transcription factor family. *Genomics* **2000**, *70*, 93–101. [CrossRef] [PubMed]
14. Small, K.S.; Todorcevic, M.; Civelek, M.; El-Sayed Moustafa, J.S.; Wang, X.; Simon, M.M.; Fernandez-Tajes, J.; Mahajan, A.; Horikoshi, M.; Hugill, A.; et al. Regulatory variants at KLF14 influence type 2 diabetes risk via a female-specific effect on adipocyte size and body composition. *Nat. Genet.* **2018**, *50*, 572–580. [CrossRef]
15. Chen, X.; Li, S.; Yang, Y.; Yang, X.; Liu, Y.; Liu, Y.; Hu, W.; Jin, L.; Wang, X. Genome-wide association study validation identifies novel loci for atherosclerotic cardiovascular disease. *J. Thromb. Haemost. JTH* **2012**, *10*, 1508–1514. [CrossRef]
16. Teslovich, T.M.; Musunuru, K.; Smith, A.V.; Edmondson, A.C.; Stylianou, I.M.; Koseki, M.; Pirruccello, J.P.; Ripatti, S.; Chasman, D.I.; Willer, C.J.; et al. Biological, clinical and population relevance of 95 loci for blood lipids. *Nature* **2010**, *466*, 707–713. [CrossRef]
17. Voight, B.F.; Scott, L.J.; Steinthorsdottir, V.; Morris, A.P.; Dina, C.; Welch, R.P.; Zeggini, E.; Huth, C.; Aulchenko, Y.S.; Thorleifsson, G.; et al. Twelve type 2 diabetes susceptibility loci identified through large-scale association analysis. *Nat. Genet.* **2010**, *42*, 579–589. [CrossRef]
18. Civelek, M.; Wu, Y.; Pan, C.; Raulerson, C.K.; Ko, A.; He, A.; Tilford, C.; Saleem, N.K.; Stančáková, A.; Scott, L.J.; et al. Genetic Regulation of Adipose Gene Expression and Cardio-Metabolic Traits. *Am. J. Hum. Genet.* **2017**, *100*, 428–443. [CrossRef]
19. Collins, A.R.; Lyon, C.J.; Xia, X.; Liu, J.Z.; Tangirala, R.K.; Yin, F.; Boyadjian, R.; Bikineyeva, A.; Praticò, D.; Harrison, D.G.; et al. Age-accelerated atherosclerosis correlates with failure to upregulate antioxidant genes. *Circ. Res.* **2009**, *104*, e42–e54. [CrossRef]
20. Heida, K.Y.; Franx, A.; van Rijn, B.B.; Eijkemans, M.J.; Boer, J.M.; Verschuren, M.W.; Oudijk, M.A.; Bots, M.L.; van der Schouw, Y.T. Earlier Age of Onset of Chronic Hypertension and Type 2 Diabetes Mellitus After a Hypertensive Disorder of Pregnancy or Gestational Diabetes Mellitus. *Hypertension* **2015**, *66*, 1116–1122. [CrossRef]
21. Okuda, K.; Khan, M.Y.; Skurnick, J.; Kimura, M.; Aviv, H.; Aviv, A. Telomere attrition of the human abdominal aorta: Relationships with age and atherosclerosis. *Atherosclerosis* **2000**, *152*, 391–398. [CrossRef]
22. Steegenga, W.T.; Boekschoten, M.V.; Lute, C.; Hooiveld, G.J.; de Groot, P.J.; Morris, T.J.; Teschendorff, A.E.; Butcher, L.M.; Beck, S.; Müller, M. Genome-wide age-related changes in DNA methylation and gene expression in human PBMCs. *Age* **2014**, *36*, 9648. [CrossRef] [PubMed]
23. Zbieć-Piekarska, R.; Spólnicka, M.; Kupiec, T.; Parys-Proszek, A.; Makowska, Ż.; Pałeczka, A.; Kucharczyk, K.; Płoski, R.; Branicki, W. Development of a forensically useful age prediction method based on DNA methylation analysis. *Forensic Sci. Int. Genet.* **2015**, *17*, 173–179. [CrossRef] [PubMed]
24. Kananen, L.; Marttila, S.; Nevalainen, T.; Jylhävä, J.; Mononen, N.; Kähönen, M.; Raitakari, O.T.; Lehtimäki, T.; Hurme, M. Aging-associated DNA methylation changes in middle-aged individuals: The Young Finns study. *BMC Genom.* **2016**, *17*, 103. [CrossRef]
25. Bacos, K.; Gillberg, L.; Volkov, P.; Olsson, A.H.; Hansen, T.; Pedersen, O.; Gjesing, A.P.; Eiberg, H.; Tuomi, T.; Almgren, P.; et al. Blood-based biomarkers of age-associated epigenetic changes in human islets associate with insulin secretion and diabetes. *Nat. Commun.* **2016**, *7*, 11089. [CrossRef]
26. Huang, L.O.; Rauch, A.; Mazzaferro, E.; Preuss, M.; Carobbio, S.; Bayrak, C.S.; Chami, N.; Wang, Z.; Schick, U.M.; Yang, N.; et al. Genome-wide discovery of genetic loci that uncouple excess adiposity from its comorbidities. *Nat. Metab.* **2021**, *3*, 228–243. [CrossRef]
27. Park, J.M.; Park, D.H.; Song, Y.; Kim, J.O.; Choi, J.E.; Kwon, Y.J.; Kim, S.J.; Lee, J.W.; Hong, K.W. Understanding the genetic architecture of the metabolically unhealthy normal weight and metabolically healthy obese phenotypes in a Korean population. *Sci. Rep.* **2021**, *11*, 2279. [CrossRef]
28. Fan, C.T.; Lin, J.C.; Lee, C.H. Taiwan Biobank: A project aiming to aid Taiwan's transition into a biomedical island. *Pharmacogenomics* **2008**, *9*, 235–246. [CrossRef]
29. Chen, C.H.; Yang, J.H.; Chiang, C.W.K.; Hsiung, C.N.; Wu, P.E.; Chang, L.C.; Chu, H.W.; Chang, J.; Song, I.W.; Yang, S.L.; et al. Population structure of Han Chinese in the modern Taiwanese population based on 10,000 participants in the Taiwan Biobank project. *Hum. Mol. Genet.* **2016**, *25*, 5321–5331. [CrossRef]
30. Civelek, M.; Lusis, A.J. Systems genetics approaches to understand complex traits. *Nat. Rev. Genet.* **2014**, *15*, 34–48. [CrossRef]
31. Consortium, G.T. The Genotype-Tissue Expression (GTEx) project. *Nat. Genet.* **2013**, *45*, 580–585. [CrossRef]
32. Hill, J.H.; Solt, C.; Foster, M.T. Obesity associated disease risk: The role of inherent differences and location of adipose depots. *Horm. Mol. Biol. Clin. Investig.* **2018**, *33*. [CrossRef] [PubMed]
33. Krakauer, N.Y.; Krakauer, J.C. A new body shape index predicts mortality hazard independently of body mass index. *PLoS ONE* **2012**, *7*, e39504. [CrossRef] [PubMed]
34. Krakauer, N.Y.; Krakauer, J.C. An Anthropometric Risk Index Based on Combining Height, Weight, Waist, and Hip Measurements. *J. Obes.* **2016**, *2016*, 8094275. [CrossRef]
35. Christakoudi, S.; Evangelou, E.; Riboli, E.; Tsilidis, K.K. GWAS of allometric body-shape indices in UK Biobank identifies loci suggesting associations with morphogenesis, organogenesis, adrenal cell renewal and cancer. *Sci. Rep.* **2021**, *11*, 10688. [CrossRef]
36. Bertoli, S.; Leone, A.; Krakauer, N.Y.; Bedogni, G.; Vanzulli, A.; Redaelli, V.I.; de Amicis, R.; Vignati, L.; Krakauer, J.C.; Battezzati, A. Association of Body Shape Index (ABSI) with cardio-metabolic risk factors: A cross-sectional study of 6081 Caucasian adults. *PLoS ONE* **2017**, *12*, e0185013. [CrossRef]

37. Christakoudi, S.; Tsilidis, K.K.; Muller, D.C.; Freisling, H.; Weiderpass, E.; Overvad, K.; Söderberg, S.; Häggström, C.; Pischon, T.; Dahm, C.C.; et al. A Body Shape Index (ABSI) achieves better mortality risk stratification than alternative indices of abdominal obesity: Results from a large European cohort. *Sci. Rep.* **2020**, *10*, 14541. [CrossRef]
38. Heyn, H.; Li, N.; Ferreira, H.J.; Moran, S.; Pisano, D.G.; Gomez, A.; Diez, J.; Sanchez-Mut, J.V.; Setien, F.; Carmona, F.J.; et al. Distinct DNA methylomes of newborns and centenarians. *Proc. Natl. Acad. Sci. USA* **2012**, *109*, 10522–10527. [CrossRef]
39. Rönn, T.; Volkov, P.; Gillberg, L.; Kokosar, M.; Perfilyev, A.; Jacobsen, A.L.; Jørgensen, S.W.; Brøns, C.; Jansson, P.A.; Eriksson, K.F.; et al. Impact of age, BMI and HbA1c levels on the genome-wide DNA methylation and mRNA expression patterns in human adipose tissue and identification of epigenetic biomarkers in blood. *Hum. Mol. Genet.* **2015**, *24*, 3792–3813. [CrossRef]
40. Iwaya, C.; Kitajima, H.; Yamamoto, K.; Maeda, Y.; Sonoda, N.; Shibata, H.; Inoguchi, T. DNA methylation of the Klf14 gene region in whole blood cells provides prediction for the chronic inflammation in the adipose tissue. *Biochem. Biophys. Res. Commun.* **2018**, *497*, 908–915. [CrossRef]
41. Zampieri, M.; Ciccarone, F.; Calabrese, R.; Franceschi, C.; Bürkle, A.; Caiafa, P. Reconfiguration of DNA methylation in aging. *Mech. Ageing Dev.* **2015**, *151*, 60–70. [CrossRef] [PubMed]
42. Fabbrini, E.; Sullivan, S.; Klein, S. Obesity and nonalcoholic fatty liver disease: Biochemical, metabolic, and clinical implications. *Hepatology* **2010**, *51*, 679–689. [CrossRef] [PubMed]
43. Klein, S.; Wadden, T.; Sugerman, H.J. AGA technical review on obesity. *Gastroenterology* **2002**, *123*, 882–932. [CrossRef] [PubMed]
44. Neeland, I.J.; Ross, R.; Després, J.P.; Matsuzawa, Y.; Yamashita, S.; Shai, I.; Seidell, J.; Magni, P.; Santos, R.D.; Arsenault, B.; et al. Visceral and ectopic fat, atherosclerosis, and cardiometabolic disease: A position statement. *Lancet Diabetes Endocrinol.* **2019**, *7*, 715–725. [CrossRef]
45. Stefan, N.; Häring, H.U.; Hu, F.B.; Schulze, M.B. Metabolically healthy obesity: Epidemiology, mechanisms, and clinical implications. *Lancet Diabetes Endocrinol.* **2013**, *1*, 152–162. [CrossRef]
46. Loos, R.J. The genetics of adiposity. *Curr. Opin. Genet. Dev.* **2018**, *50*, 86–95. [CrossRef]
47. Appleton, S.L.; Seaborn, C.J.; Visvanathan, R.; Hill, C.L.; Gill, T.K.; Taylor, A.W.; Adams, R.J. Diabetes and cardiovascular disease outcomes in the metabolically healthy obese phenotype: A cohort study. *Diabetes Care* **2013**, *36*, 2388–2394. [CrossRef]
48. Arnlöv, J.; Ingelsson, E.; Sundström, J.; Lind, L. Impact of body mass index and the metabolic syndrome on the risk of cardiovascular disease and death in middle-aged men. *Circulation* **2010**, *121*, 230–236. [CrossRef]
49. Aung, K.; Lorenzo, C.; Hinojosa, M.A.; Haffner, S.M. Risk of developing diabetes and cardiovascular disease in metabolically unhealthy normal-weight and metabolically healthy obese individuals. *J. Clin. Endocrinol. Metab.* **2014**, *99*, 462–468. [CrossRef]
50. Caleyachetty, R.; Thomas, G.N.; Toulis, K.A.; Mohammed, N.; Gokhale, K.M.; Balachandran, K.; Nirantharakumar, K. Metabolically Healthy Obese and Incident Cardiovascular Disease Events Among 3.5 Million Men and Women. *J. Am. Coll. Cardiol.* **2017**, *70*, 1429–1437. [CrossRef]
51. Eckel, N.; Meidtner, K.; Kalle-Uhlmann, T.; Stefan, N.; Schulze, M.B. Metabolically healthy obesity and cardiovascular events: A systematic review and meta-analysis. *Eur. J. Prev. Cardiol.* **2016**, *23*, 956–966. [CrossRef] [PubMed]
52. Hinnouho, G.M.; Czernichow, S.; Dugravot, A.; Nabi, H.; Brunner, E.J.; Kivimaki, M.; Singh-Manoux, A. Metabolically healthy obesity and the risk of cardiovascular disease and type 2 diabetes: The Whitehall II cohort study. *Eur. Heart J.* **2015**, *36*, 551–559. [CrossRef] [PubMed]
53. Lassale, C.; Tzoulaki, I.; Moons, K.G.M.; Sweeting, M.; Boer, J.; Johnson, L.; Huerta, J.M.; Agnoli, C.; Freisling, H.; Weiderpass, E.; et al. Separate and combined associations of obesity and metabolic health with coronary heart disease: A pan-European case-cohort analysis. *Eur. Heart J.* **2018**, *39*, 397–406. [CrossRef] [PubMed]
54. Mørkedal, B.; Vatten, L.J.; Romundstad, P.R.; Laugsand, L.E.; Janszky, I. Risk of myocardial infarction and heart failure among metabolically healthy but obese individuals: HUNT (Nord-Trøndelag Health Study), Norway. *J. Am. Coll. Cardiol.* **2014**, *63*, 1071–1078. [CrossRef] [PubMed]
55. Song, Y.; Manson, J.E.; Meigs, J.B.; Ridker, P.M.; Buring, J.E.; Liu, S. Comparison of usefulness of body mass index versus metabolic risk factors in predicting 10-year risk of cardiovascular events in women. *Am. J. Cardiol.* **2007**, *100*, 1654–1658. [CrossRef] [PubMed]
56. Thomsen, M.; Nordestgaard, B.G. Myocardial infarction and ischemic heart disease in overweight and obesity with and without metabolic syndrome. *JAMA Intern. Med.* **2014**, *174*, 15–22. [CrossRef] [PubMed]
57. Stefan, N. Causes, consequences, and treatment of metabolically unhealthy fat distribution. *Lancet Diabetes Endocrinol.* **2020**, *8*, 616–627. [CrossRef]
58. Lotta, L.A.; Wittemans, L.B.L.; Zuber, V.; Stewart, I.D.; Sharp, S.J.; Luan, J.; Day, F.R.; Li, C.; Bowker, N.; Cai, L.; et al. Association of Genetic Variants Related to Gluteofemoral vs Abdominal Fat Distribution with Type 2 Diabetes, Coronary Disease, and Cardiovascular Risk Factors. *Jama* **2018**, *320*, 2553–2563. [CrossRef]
59. Cannataro, R.; Perri, M.; Luca, G.; Caroleo, M.C.; de Sarro, G.; Cione, E. Ketogenic Diet Acts on Body Remodeling and MicroRNAs Expression Profile. *MicroRNA* **2019**, *8*, 116–126. [CrossRef]
60. Er, L.K.; Wu, S.; Chou, H.H.; Hsu, L.A.; Teng, M.S.; Sun, Y.C.; Ko, Y.L. Triglyceride Glucose-Body Mass Index Is a Simple and Clinically Useful Surrogate Marker for Insulin Resistance in Nondiabetic Individuals. *PLoS ONE* **2016**, *11*, e0149731. [CrossRef]

International Journal of
Molecular Sciences

MDPI

Article

CD33 rs2455069 SNP: Correlation with Alzheimer's Disease and Hypothesis of Functional Role

Fabiana Tortora [1,†], Antonella Rendina [1,†], Antonella Angiolillo [2], Alfonso Di Costanzo [2], Francesco Aniello [3], Aldo Donizetti [1,3,*], Ferdinando Febbraio [1,‡] and Emilia Vitale [1,*,‡]

1 Institute of Biochemistry and Cell Biology, National Research Council (CNR), Via Pietro Castellino 111, 80131 Naples, Italy; fabiana.tortora@ibbc.cnr.it (F.T.); antonella.rendina@ibbc.cnr.it (A.R.); ferdinando.febbraio@cnr.it (F.F.)
2 Centre for Research and Training in Medicine of Aging, Department of Medicine and Health Science "V. Tiberio", University of Molise, 86100 Campobasso, Italy; angiolillo@unimol.it (A.A.); alfonso.dicostanzo@unimol.it (A.D.C.)
3 Department of Biology, University of Naples Federico II, 80126 Naples, Italy; faniello@unina.it
* Correspondence: aldo.donizetti@unina.it (A.D.); emilia.vitale@cnr.it (E.V.)
† These authors contributed equally to this work.
‡ These authors contributed equally to this work.

Abstract: The *CD33* gene encodes for a member of the sialic-acid-binding immunoglobulin-type lectin (Siglec) family, and is one of the top-ranked Alzheimer's disease (AD) risk genes identified by genome-wide association studies (GWAS). Many *CD33* polymorphisms are associated with an increased risk of AD, but the function and potential mechanism of many CD33 single-nucleotide polymorphisms (SNPs) in promoting AD have yet to be elucidated. We recently identified the *CD33* SNP rs2455069-A>G (R69G) in a familial form of dementia. Here, we demonstrate an association between the G allele of the rs2455069 gene variant and the presence of AD in a cohort of 195 patients from southern Italy. We carried out in silico analysis of the 3D structures of *CD33* carrying the identified SNP to provide insights into its functional effect. Structural models of the CD33 variant carrying the R69G amino acid change were compared to the CD33 wild type, and used for the docking analysis using sialic acid as the ligand. Our analysis demonstrated that the CD33-R69G variant may bind sialic acid at additional binding sites compared to the wild type, thus potentially increasing its affinity/specificity for this molecule. Our results led to a new hypothesis of rs2455069-A>G SNP as a risk factor for AD, suggesting that a long-term cumulative effect of the CD33-R69G variant results from the binding of sialic acid, acting as an enhancer of the CD33 inhibitory effects on amyloid plaque degradation.

Keywords: CD33; Alzheimer's disease; sialic acid; phagocytosis

Citation: Tortora, F.; Rendina, A.; Angiolillo, A.; Di Costanzo, A.; Aniello, F.; Donizetti, A.; Febbraio, F.; Vitale, E. CD33 rs2455069 SNP: Correlation with Alzheimer's Disease and Hypothesis of Functional Role. *Int. J. Mol. Sci.* **2022**, *23*, 3629. https://doi.org/10.3390/ijms23073629

Academic Editor: Elixabet Lopez-Lopez

Received: 25 February 2022
Accepted: 22 March 2022
Published: 26 March 2022

1. Introduction

According to recent reports, it is estimated that the number of cases of dementia in the world is expected to rise to 152 million by 2050 as the population ages [1,2], with AD being the most common form of dementia currently [3,4]. Most AD cases are sporadic, with late-onset AD (LOAD) occurring at the age of 65 years or older. Approximately 5% of cases are classified as early-onset AD (EOAD), as reported in a systematic review and meta-analysis [5], where the age of onset is before 65 years and a percentage is considered as familial [6,7]. Except for a small number of early-onset cases who are afflicted due to highly penetrant single-gene mutations, AD, especially in the late-onset form, is genetically heterogeneous, with a polygenic or oligogenic risk inheritance [7]. Genome-wide association studies (GWAS) identified several genetic loci associated with increased susceptibility to LOAD affecting lipid metabolism, tau binding proteins, amyloid precursor protein (APP) metabolism [8], protein catabolism, immune cells, and microglia [9]. Of these, microglial cells are a key participant in LOAD pathogenesis, with many susceptibility loci, including *CD33*, playing a role [10–15]. CD33 belongs to a subfamily of the Ig superfamily comprising a single N-terminal V-set Ig domain that binds sialic-acid-containing glycans through a conserved positively charged arginine that interacts with the negatively charged C-1 carboxyl group on the sialic acid [16]. CD33 may be activated continuously by sialic-acid-containing glycoproteins and glycolipids, which are structural elements of amyloid plaques in the brains of Alzheimer's disease patients, resulting in an inhibition of microglia-mediated immune activation [17,18]. CD33 can affect microglial phagocytosis through crosstalk with the transmembrane receptor TREM2. Knockout of CD33 resulted in improved cognition and decreased amyloid β pathology in 5xFAD mice, whereas these effects were completely abrogated by additional knockout of TREM2 [19]. TREM2, through an interaction with DAP12/TYROBP, stimulates the tyrosine–protein kinase SYK, which in turn triggers the activation of microglial cell phagocytosis [20,21]. This activation could be counteracted by sialic-acid-activated CD33, which stimulates SHP1/SHP2 tyrosine phosphatases, thereby inhibiting microglial phagocytosis [13,21,22].

In accordance with the effect of *CD33* on the inhibition of microglial phagocytosis, primary microglial cells from *CD33* knockout mice [12] or deletion of *CD33* in human macrophages and microglia resulted in an increased phagocytosis of amyloid β1-42 [23]. This was further supported by targeting *CD33* in a mouse model of AD using an adeno-associated virus (AAV) vector encoding a *CD33* microRNA (miRCD33), which demonstrated a reduction in amyloid β40 and β42 levels in brain extracts [24].

Different *CD33* polymorphisms have been positively or negatively correlated with AD susceptibility. These SNPs can affect the expression level, the structure, and the function of CD33, leading to changes in microglia-mediated clearance of amyloid β, likely contributing to the accumulation of senile plaques in the brain [12,25–27]. The CD33 variant rs3865444(A) negatively correlates with AD, has a reduced *CD33* expression [12], and is coinherited with rs12459419(T), which modulates the splicing efficiency of the IgV domain exon involved in sialic acid binding [13,28].

We recently identified the SNP rs2455069 in exon 2 of the human *CD33* gene on chromosome 19, which is involved in an unusual familial form of dementia [29]. This SNP has been previously identified in a block of eight correlated SNPs in *CD33* significantly associated with cognitive decline in univariate analysis in the all-female cohort, with highly significant effects on hyaluronan synthase-1 (HAS1) expression in the temporal cortex [30]. The presence of the minor allele at rs2455069 leads to a change from an arginine to a glycine in the position 69 (R69G). Here, we demonstrate a positive correlation between the *CD33* SNP rs2455069 and Alzheimer's disease in a cohort of patients from a region of southern Italy, and secondly, that the R69G amino acid change affects the CD33 structure in a way that alters its binding affinity for sialic acid residues. We propose that the mechanism of the long-term cumulative effects of the CD33-R69G variant are mediated through an increased binding of sialic acid, which then acts as a more effective enhancer of the CD33 inhibitory effects on amyloid plaque degradation.

2. Results

2.1. Genetic Analysis

Based on the evidence of the possible involvement in an unusual form of dementia of the SNP rs2455069-A>G in exon 2 of *CD33* gene [29], leading to a change from arginine to glycine in position 69, we widened the analysis of this SNP to a cohort of AD patients from southern Italy. We found that the genotype frequency for the rs2455069 polymorphism in the control group was 11.1% (n = 15) for the GG, 53.0% (n = 77) for the AG, and 31.8% (n = 43) for the AA genotypes. In AD patients, the frequency of GG, AG, and AA genotypes were 17.9% (n = 35), 65.1% (n = 127), and 16.9% (n = 33), respectively. While genotypes in the LOAD sample were not in Hardy–Weinberg equilibrium (HWE, p = 0.00013), the genotype frequencies in controls did not deviate significantly from the Hardy–Weinberg equilibrium (HWE) (p = 0.08301). The G allele frequency was 39.6% in controls and 50.5%

in AD patients (p = 0.00582). A significant association between AD and the GG or AG genotype was found (Table 1), particular in the dominant model (AG + GG vs. AA), leading to the hypothesis that the R69G change may have had an impact on the increased risk of LOAD in the analyzed cohort.

Table 1. Association between the genetic polymorphism CD33-rs2455069 and the risk of AD.

CD33 Genotype	OR [a] (95% C.I. [b])	Chi2	p
AA	Ref		
AG	2.149 (1.259–3.668)	8.02	p = 0.00463
GG	3.040 (1.428–6.476)	8.58	p = 0.00341
AG + GG	2.294 (1.363–3.862)	10.03	p = 0.00154
A	Ref		
G	1.555 (1.135–2.130)	7.61	p = 0.00582

[a] OR = odds ratio; [b] C.I. = confidence interval.

2.2. In Silico Structural Analysis

To evaluate the impact of the R69G change on the CD33 protein function, we carried out a series of in silico analyses. Six human CD33 partial structures (Siglec-3) were available in the RCSB Protein Data Bank [31] (Supplementary Table S1, Supplementary Figure S1). We prepared the monomeric form of each CD33 structure, summarized in Supplementary Table S1, using the web-based platform CHARMM-GUI. To reduce possible artifacts due to binding with mimetic substrates, we used a protein model based on the CD33 structure PDB-ID 5IHB, representing the protein region interacting with NAG as a nonspecific ligand. We compared the two structures obtained after CHARMM minimization carrying either arginine or glycine at position 69 of the CD33 protein. We did not observe significant changes in the protein backbone after basic energy minimization, but small rearrangements, both in the α-helix containing the glycine residue (red arrow in Figure 1A) and in a β-sheet of the repeated structural type C2 Ig domain (black arrow in Figure 1A) were evident. Most importantly, few differences were observed in the structures of sidechains, which showed small changes in the connecting region between the two domains (Figure 1B).

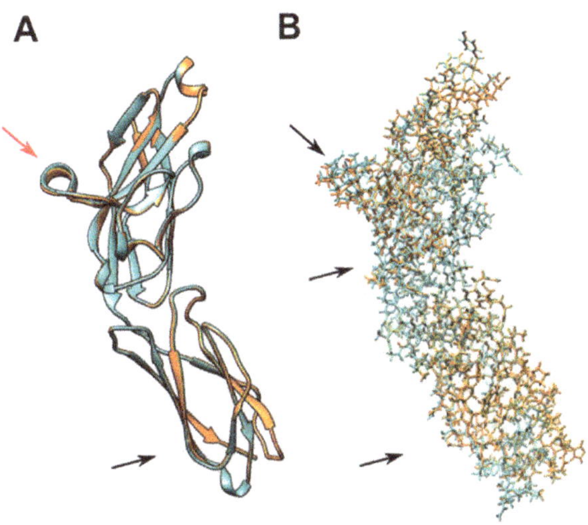

Figure 1. Structural representation of CD33 variants. (**A**) Ribbon representation of the backbone of superimposed CD33 monomers (PDB ID 5IHB) with Arg69 (cyan) and Gly69 (orange) after minimization

by CHARMM. (**B**) Stick representation of backbone and sidechains of structures in (**A**). The arrows indicate the area with major changes in the structure of sidechains after basic energy minimization.

These predictions led us to hypothesize that the effect of the arginine-to-glycine change at position 69 could predominantly affect the reorganization of polar sidechains, and only locally, the rearrangement of the backbone. In agreement with these observations, the analysis carried out on 200 structural models of CD33 (PDB ID 5IHB), 100 with the Arg69 residue and 100 with the Gly69 residue, indicated the presence of few changes in the boundary spatial coordinates (min and max) of the entire CD33 structure (Figure 2). These differences were hypothesized to be due to the increased length of the sidechain consisting of a three-carbon aliphatic straight chain ending in a guanidino group as a consequence of the change from a glycine to an arginine residue in position 69.

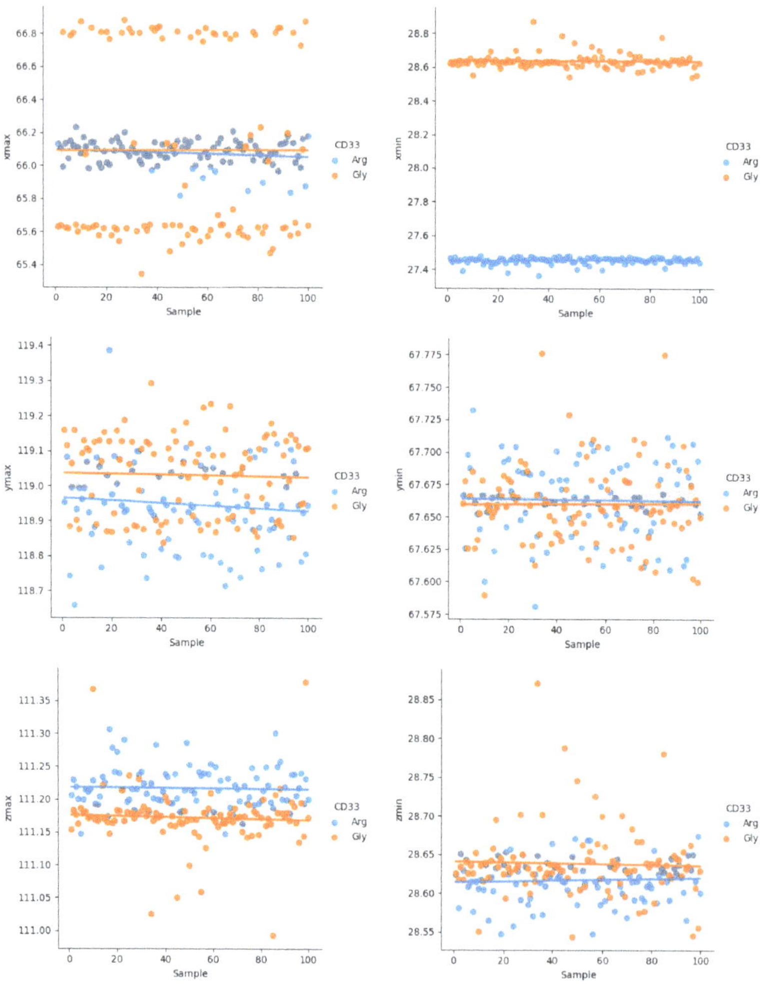

Figure 2. Variations of the maximum and minimum dimensional values in the XYZ axes of the two CD33 variants (PDB ID 5IHB). The variations were obtained after minimization by CHARMM. Blue dots refer to the structures with the arginine residue, whereas the orange dots refer to the structures with a glycine residue in position 69.

2.3. Docking Analysis

The predicted binding sites for sialic acid in the N-terminal domain (aa19-135) of the monomeric form of CD33 (PDB ID 5IHB) carrying either an arginine or a glycine residue in position 69 are reported in Figure 3. The affinity of each binding site toward the ligand is expressed as the binding energy (kcal/mol), with lower values indicating a greater bond affinity. In the Arg69 variant, the values were −4.8 kcal/mol near Arg119 and −5.3 kcal/mol near Arg91, while the values were slightly increased in the form with Gly69, with values of −5.2 kcal/mol near Arg119 and −5.7 kcal/mol near Arg91 (Figure 3).

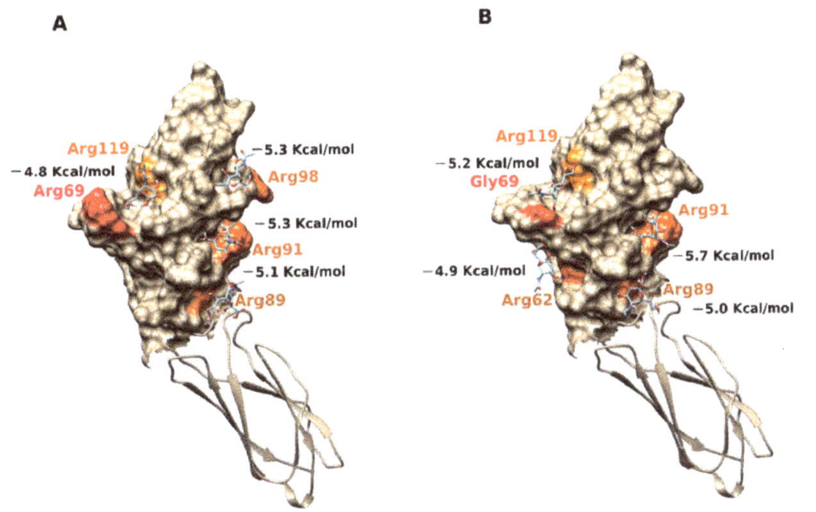

Figure 3. Sialic acid interactions with the two SNPs of CD33 structures. (**A**) Surface representation of functional domain of the CD33 predominant form with Arg in position 69, colored in red (PDB ID 5IHB). (**B**) Surface representation of functional domain of the CD33 variant form with Gly in position 69, colored in red (PDB ID 5IHB). The areas in orange indicate the binding sites of sialic acid, in stick representation, in correspondence with the positively charged arginine residues with which the ligand establishes saline bridges. The values of the binding energy in kcal/mol are reported for each binding site of sialic acid.

We found similar results when we performed docking analyses on the other two structures: the N-terminal domain (aa19-135) (PDB ID 5J06, 5J0B), which carried a glycine residue in position 69; and the repeated structural type C2 Ig domain (aa145-228). In the PDB-ID 5J06 structure, the values were −4.7 kcal/mol in Arg119 proximity and −5.2 kcal/mol in Arg91 proximity; while in the PDB-ID 5J0B structure, the predicted binding affinity values were −4.7 kcal/mol in Arg119 proximity and −5.3 kcal/mol in Arg91 proximity (Supplementary Figure S2). We confirmed these results through the docking analyses carried out on the other three CD33 structures having only the N-terminal domain (PDB ID 6D48, 6D49, and 6D4A). In fact, we always observed a greater binding affinity in the Arg91 residue part than in the Arg119 part of the functional domain (Supplementary Figure S3), suggesting a better chance of binding at this site. This result was in contrast with the evidence from the binding of a sialic acid mimetic ligand FVP in the proximity of Arg119, as found by others [32]. However, by analyzing the asymmetric unit assembly of each resolved structure in that study, we observed the formation of dimers and tetramers involving interfaces near Arg91, which likely prevented the binding of ligands near this residue. Considering that the most reactive side of the D2:IgV domain is represented by the positively charged patch comprising arginine residues 89 and 91, the oligomerization trend involving this surface could be explained easily. Notwithstanding the site of binding, our results suggested a general increase in the binding affinity and specificity of the glycine variant of CD33 toward ligands.

2.4. Virtual Screening Analysis

Plotting the carbon 1 atomic coordinate of the best sialic acid poses against the affinity values (kcal/mol) predicted by the docking analysis, we obtained a conceptual understanding of the ligand distributions on the protein surface with respect to the arginine or glycine variants in position 69 of the CD33 protein. The results highlighted a noteworthy increase in the affinity of several binding sites for sialic acid in the glycine 69 variant of CD33 (Supplementary Figure S4). Indeed, the analysis of the distributions of the best ligands on the D2:IgV domain of CD33, obtained by plotting the XYZ coordinates of the C1 atom of sialic acid structures (Supplementary Figure S5), showed binding in specific sites of CD33 receptors for both variants with arginine and glycine residues in position 69. When comparing the position of arginine residues in the D2:IgV domain of the PDB-ID 5IHB structure with the best ligand poses by overlaying the YZ coordinates of the Cζ atom of arginine sidechains with the coordinates of C1 atom of sialic acid poses (Supplementary Figure S6), we observed a number of sialic-acid-binding molecules close to arginine residues 91, 98, and 119. However, the C1 atom of sialic acid near the approximate Z coordinate 70,000, included in the range of Y coordinates 85,000–95,000 and 110,000–115,000, were instead close to the arginine 62 and 89 residues in the 3D representation, respectively. It was apparent that ligands basically spread out on the surface of the Arg69 structure of CD33, with most of the sialic acid binding near arginine residues 91, 119, 89, 98, and 62, ordered by the number of bonded ligands (Supplementary Figure S6A). In contrast, an increased affinity and number of sialic acid molecules was predicted near the position of arginine 91 in the Gly69 variant of CD33 (Supplementary Figure S6B).

These data suggested the possibility of the presence of more than one binding site for sialic acid derivatives with a similar specificity. In fact, it was not surprising that the more reactive domain near arginine 91 was involved in the binding at the interfaces between the CD33 structures (single and double domains) in the crystallographic unit cell, preventing the recognition of this other sialic-acid-binding site in the crystallographic structures resolved in presence of ligands [32].

The 3D representations in Figure 4 illustrate the best orientations of sialic acid molecules in the CD33 binding sites obtained by virtual screening analysis. In the CD33 structure presenting an arginine residue in position 69, we observed the prevalent distribution of the ligands in the two already identified binding sites involving arginine 91 and 119 (Figure 4A). However, two additional binding sites were predicted near arginine 62 and 89. The number of ligands in these two secondary sites increased in the representation of the second-best poses (Figure 4B), having a lower affinity with respect to the previous ones. In contrast, the best poses of sialic acid molecules were located exclusively near arginine 91 and 119 in the Gly69 variant of CD33 (Figure 4C), and only a few ligand poses were predicted near arginine 62 and 89 (Figure 4D). In addition, when observing the manner the sialic acid binds to the sites on the CD33 surface, we can point out that the orientation of sialic acid molecules was very often preferentially maintained in the binding sites of the Gly69 variant of CD33 (Figure 4). Thus, in addition to the increase in affinity, it was likely that the Gly69 variant of CD33 bound to the sialic acid with a higher specificity than the Arg69 variant. These results supported the idea that the balance of the binding equilibrium constant for CD33–sialic acid interaction changed slightly in a positive direction in the Gly69 variants. This suggested a more efficient binding of sialic acid at a lower concentration by this CD33 variant.

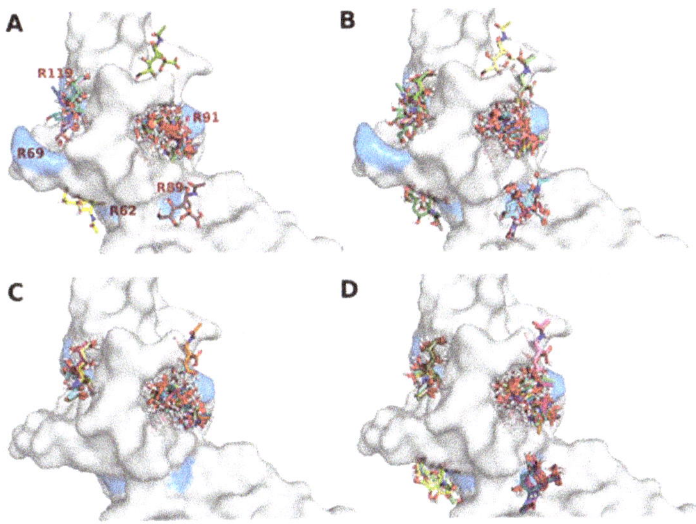

Figure 4. Detail of the binding sites in the D2:IgV domain for sialic acid molecules, as derived from docking analyses. Surface representation, in light gray, of the N-terminal D2:IgV functional domain of arginine 69 models of CD33, in which the docked poses of sialic acid molecules with the best (**A**) and the second best (**B**) ligand mode rmsd are shown in stick figures. Surface representation, in light gray, of the N-terminal D2:IgV functional domain of glycine 69 models of CD33, in which the docked poses of sialic acid molecules with the best (**C**) and the second best (**D**) ligand mode rmsd are shown in stick figures. The surfaces of arginine residues 62, 69, 89, 91, and 119 are presented in light blue.

3. Discussion

Although the association of CD33 with Alzheimer's diseases has not been thoroughly elucidated, literature data supports the hypothesis that through the modulation of immune cell functions (such as phagocytosis, cytokine release, apoptosis, etc.), CD33 could be implicated in the suppression of amyloid fragment uptake [12,18,23,25–27]. Here, we demonstrated a correlation between an SNP variant of the *CD33* gene (SNP rs2455069-A>G) and Alzheimer's disease risk, and proposed a new hypothesis for its functional role. Using different in silico approaches, we demonstrated that the binding site for sialic acid could involve additional regions of the protein to the previously documented Arg119 residue. This residue was demonstrated to be essential for the binding of the sialic acid mimetic [32]. Indeed, it was assumed that the binding site for sialic acid was located at the D2:IgV domain of the CD33 receptor, and contained the arginine 119 residue, which was positively charged at physiological pH. Because sialic acid molecules are typically terminal residues on glycoconjugates, the positively charged arginine forms salt bridges with the sialic acid portion of sugar residues, enabling stable interactions [18,33–35]. Due to the presence of FVP (a subtype-selective sialic acid mimetic ligand) in the 6D49 and 6D4A structures (recently also demonstrated in the 7AW6 structure), the putative binding site was located near the Arg69 and Arg119 positions (Supplementary Figure S7). In agreement with those results, we confirmed the binding of sialic acid to Arg119, but also demonstrated the possibility that the ligand may bind to other regions of the protein with higher affinity that include Arg91. The presence of an additional binding site was recently reported for another Siglec family member [36], pointing out the need to further explore the binding properties of CD33 toward sialic-acid-containing molecules. A mutation of arginine 119 to alanine did not affect phagocytosis of U937 cells [27], raising a question on the role of this residue in sialic-acid-containing glycan binding, and supporting a hypothesis of additional binding sites for sialic acid. Our *in-silico* analyses indeed supported the possibility of additional binding sites for sialic acid. In addition, our studies also supported a hypothesis that the CD33-R69G variant may bind sialic acid with greater specificity and affinity, and results in a more efficient binding of the variant to sialic acid, at lower concentrations than that of the wild type. Interestingly, the dependence of binding on sialic acid concentrations was consistent with observations that sialic acid content increases in old-age females [37], who have an increased risk of LOAD compared to males of the same age [38]. In the light of these observations, new perspectives have opened up in the study of CD33, with novel and interesting understandings of the interactions leading to late-onset Alzheimer's disease. Several polymorphisms in the CD33 gene could affect the structure of this receptor, and in turn, the binding affinity for sialic-acid-containing gangliosides. We hypothesized that the microglial phagocytosis, and in turn, the efficient removal of neurotoxic protein aggregates, might also depend on a variety of factors, including dietary habits, sex-specific accumulation of sialic acid, and aging, among others, affecting ganglioside expression in the brain. This may establish a complex inter-relationship between genetic risk factors and additional predictors in the progression of LOAD. Overall, our results led us to hypothesize that the rs2455069-A>G SNP is a risk factor for LOAD, and that the mechanism of the long-term cumulative effects of the CD33-R69G variant are mediated through an increased binding of sialic acid, which then acts as a more effective enhancer of the CD33 inhibitory effects on amyloid plaque degradation (Figure 5).

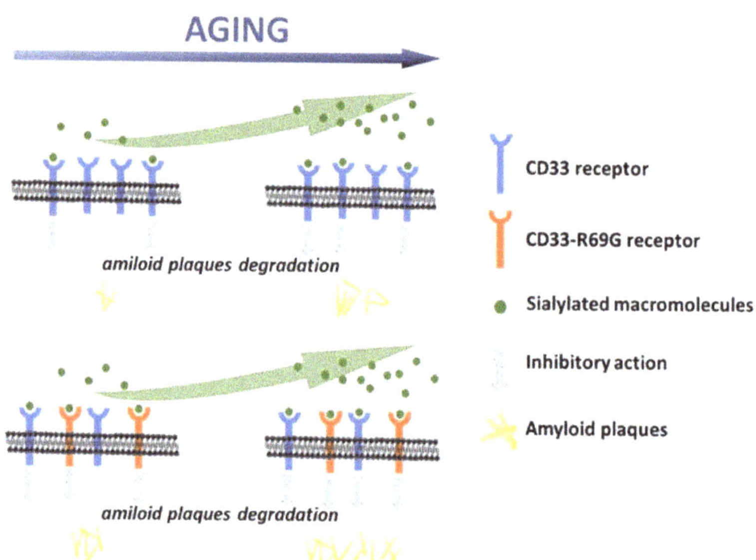

Figure 5. Model describing the plaque accumulation during aging. The increased effect of the CD33-R69G variant in inhibiting the microglial phagocytosis results in a lower extent of clearance of amyloid plaques, supporting late-onset AD.

4. Conclusions

We analyzed the functional role of the SNP rs2455069-A>G, which we found to be associated with Alzheimer's disease (AD) in a cohort of patients from southern Italy. The SNP changed a single amino acid in the sequence of CD33, and resulted in a local change in its protein structure, which we derived by in silico analysis. Our data demonstrated that this small change resulted in a slight increase in the binding affinity and specificity of CD33 toward sialylated molecules, its natural ligands. We propose that this change in binding enhanced the inhibitory effects of CD33 on amyloid plaque degradation, and together with other genetic, epigenetic, and environmental factors, could predispose to the accumulation of amyloid plaques and the onset of AD pathology. These observations support a new approach to the investigation of protein polymorphisms involved in AD, and propose a need for structural analyses to complements genetic studies.

5. Materials and Methods

5.1. Participant Recruitment and Analysis

The participants in the study ($n = 330$) were recruited at the Centre for Research and Training in Medicine of Aging (CeRMA) at the University of Molise (Italy). The AD patients ($n = 195$) fulfilled the National Institute of Aging and Alzheimer's Association (NIA-AA) diagnostic criteria for "probable AD with documented decline" [39]. They scored <24 on the Mini Mental State Examination (MMSE) and >0.5 on the Clinical Dementia Rating (CDR). To rule out other potential causes of cognitive impairment, all patients underwent blood tests (including full blood count, erythrocyte sedimentation rate, blood urea nitrogen and electrolytes, thyroid function, vitamin B12, and folate) and brain imaging. One hundred and thirty-five sex/age-matched cognitively healthy subjects (HS) were recruited as a control group. DNA from all participants was analyzed for the presence of the rs2455069 SNP by high-resolution melting analysis (HRM). Genomic DNA was extracted from whole blood with a QIAamp DNA Blood Mini Kit (QIAGEN). PCR amplifications were performed on a SureCycler 8800 Thermal Cycler (Agilent, Santa Clara, CA, USA). The study was conducted in accordance with ethical principles stated in the Declaration of Helsinki, and with approved national and international guidelines for human research. The Institutional Review Board (IRB) of the University of Molise approved the study (IRB Prot. n. 16/2020). Written informed consent was obtained from each participant or caregiver.

5.2. High-Resolution Melting Analysis (HRM)

The high-resolution melting analysis (HRM) was performed as previously described [29] on a LightCycler 480 Instrument II (Roche, Basel, Switzerland) and analyzed using LightCycler® 480 Gene Scanning Software 1.5.1. High-resolution melting Mastermix (Roche) was used. Specific primers used for amplification and sequencing by the Sanger method and the electropherograms were analyzed with Chromas 2.22 software and aligned according to reference sequences present in GenBank (http://www.ncbi.nlm.nih.gov/GenBank/index.html; accessed on 1 September 2019). Templates were amplified in 96-well plates and ~30 ng analyzed in a 20 µL final volume reaction as previously specified [29].

5.3. CD33 Structure Editing and Minimization

Computer simulations were carried out on the 3D crystallographic structures available in the Protein Data Bank [31] with ID numbers 5IHB, 5J06, 5J0B, 6D48, 6D49, 6D4A, and 7AW6. The first four included the N-terminal domain (aa19-135) and the repeated structural domain Ig of type C2 (aa145-228), while the last three structures included only the N-terminal domain. All PDB structures of CD33 (excluding 6D48) incorporated ligands. Characteristics of the seven human CD33 structures (Siglec-3) are summarized in Supplementary Table S1. The PDB files were edited by using Pymol [40]. Any molecules bound in the protein catalytic sites and water molecules bound to the protein during the crystallization process were removed. The CHARMM-GUI web-based platform (http://www.charmm-gui.org; accessed on 15 February 2022) was used to extract a single monomer from each of the CD33 structures contained in the PDB files, and to generate the two single-nucleotide polymorphisms (SNPs) at position 69 for each monomer. All generated structures were optimized offline using the CHARMM program [41] by performing a basic energy minimization with the steepest descent (SD) algorithm of the initial 25–50 steps to remove bad van der Waals contact. A more precise minimization was carried out subsequently with the Newton–Raphson method (ABNR) of 1,000,000 steps to remove potential problems, such as incorrect contacts, collisions, and nonphysical contacts/interactions. When the step-to-step energy change was less than 0.001 kcal/mol, the structure was considered sufficiently minimized.

5.4. Ligand Structure Editing and Minimization

The sialic acid and similar ligand (3′-sialyllactose, 6′-sialyllactose) 3D structures were generated using Avogadro software (http://avogadro.cc; accessed on 15 February 2022), and the structures were optimized through the MMFF94 force field with the steepest descent algorithm until the ΔE reach a value approaching 0 (<10–10).

5.5. Docking Analysis

The minimized monomeric structures of CD33 were used for the docking analysis of sialic acid and similar ligand structures. The docking analysis was carried out using Autodock Vina [42]. The ADT software package was used to determine the grid sizes, visualize the ligand poses, and add hydrogen atoms to the templates [43]. We superimposed the different CD33 structures and generated a single grid box that included all the CD33 structures. The coordinates of the grid box were put in a configuration file (config file in Supplementary Materials) used to perform both the docking and the virtual screening analysis. During the docking procedure, both the proteins and ligands were considered as rigid, and the only torsion that was allowed was for the carboxylic and alcoholic groups of the ligand. A bash shell command was used to pass receptors, ligands, and grid box parameters to Autodock Vina, producing an output file containing the predicted models (ligand poses) and a log file for each analysis. The structures were analyzed and the images were produced by using the PyMOL [40] and the UCSF Chimera 1.13.1 (http://www.cgl.ucsf.edu/chimera; accessed on 15 February 2022) [44] molecular graphics software programs.

5.6. Virtual Screening Analysis

We prepared two datasets of 100 models each of the Arg69 and Gly69 variants of the CD33 structure (PDB ID 5IHB). We performed a basic energy minimization using the CHARMM program [41], as previously described, to make them more relaxed for subsequent analysis. Using an ad hoc script (see Supplementary Materials) to transfer receptors and grid box parameters for automating the docking process, we submitted the two structure datasets (200 structures total) for analysis by Autodock Vina against the sialic acid as ligand, retrieving at least eight poses for each CD33 structure.

5.7. Statistical Analysis

Data were analyzed using the SPSS (v. 17.0) statistical software package (SPSS Inc., Chicago, IL, USA). The frequencies of genotypes and allelotypes of the CD33 gene polymorphism were calculated and determined for deviation from the Hardy–Weinberg equilibrium (HWE). The Chi-squared test and odds ratio (OR) were used to evaluate the association of AD risk with different genotypes and alleles. A p-value of <0.05 was considered statistically significant.

Supplementary Materials: The following supporting information can be downloaded at: https://www.mdpi.com/article/10.3390/ijms23073629/s1.

Author Contributions: F.T. and F.F. performed the molecular modeling and simulation; A.R. performed the in vitro experiments; A.A. and A.D.C. performed the collection of data from patients; A.R., A.D. and E.V. performed the analysis and comparison of experimental data; F.F. performed the analysis of in silico simulations; F.F. and E.V. conceived and designed the research; A.D., F.A., A.D.C., F.F. and E.V. interpreted the results; F.T., A.D., F.F. and E.V. wrote the manuscript; all authors edited the manuscript. All authors have read and agreed to the published version of the manuscript.

Funding: A.R. was supported by the EMBO Short-Term Fellowship 7532-2018-2019. F.T. was supported by the FESR Campania 2014-2020 (Asse 1—Obiettivo Specifico 1.2—Azione 1.2.2) (ex PON03PE_00060_4). A.D.C. and A.A. were supported by the MIUR/PRIN grant N. 2017T9JNLT.

Institutional Review Board Statement: The study was conducted in accordance with ethical principles stated in the Declaration of Helsinki, and with approved national and international guidelines for human research. The Institutional Review Board (IRB) of the University of Molise approved the study (IRB Prot. n. 16/2020).

Informed Consent Statement: Written informed consent was obtained from each participant or caregiver.

Data Availability Statement: All data are available in the main text or the Supplementary Materials.

Acknowledgments: Figure 5 contains graphs provided by SMART (Servier Medical Art) (https://smart.servier.com; accessed on 15 February 2022).

Conflicts of Interest: The authors declare no conflict of interest.

References

1. Alzheimer's Disease International. World Alzheimer Report 2019: Attitudes to Dementia. Available online: https://www.alzint.org/u/WorldAlzheimerReport2019.pdf (accessed on 1 September 2019).
2. GBD 2019 Dementia Forecasting Collaborators. Estimation of the global prevalence of dementia in 2019 and forecasted prevalence in 2050: An analysis for the Global Burden of Disease Study 2019. *Lancet Public Health* **2022**, *7*, e105–e125. [CrossRef]
3. Alzheimer's Association. 2019 Alzheimer's disease facts and figures. *Alzheimer's Dement.* **2019**, *15*, 321–387. [CrossRef]
4. Winblad, B.; Amouyel, P.; Andrieu, S.; Ballard, C.; Brayne, C.; Brodaty, H.; Cedazo-Minguez, A.; Dubois, B.; Edvardsson, D.; Feldman, H.; et al. Defeating Alzheimer's disease and other dementias: A priority for European science and society. *Lancet Neurol.* **2016**, *15*, 455–532. [CrossRef]
5. Zhu, X.-C.; Tan, L.; Wang, H.-F.; Jiang, T.; Cao, L.; Wang, C.; Wang, J.; Tan, C.-C.; Meng, X.-F.; Yu, J.-T. Rate of early onset Alzheimer's disease: A systematic review and meta-analysis. *Ann. Transl. Med.* **2015**, *3*, 38, Erratum in *Ann. Transl. Med.* **2016**, *4*, E4. [CrossRef]
6. Cacace, R.; Sleegers, K.; Van Broeckhoven, C. Molecular genetics of early-onset Alzheimer's disease revisited. *Alzheimer's Dement.* **2016**, *12*, 733–748. [CrossRef]
7. Kamboh, M.I. Genomics and Functional Genomics of Alzheimer's Disease. *Neurotherapeutics* **2021**, 1–21. [CrossRef]
8. Kunkle, B.W.; Grenier-Boley, B.; Sims, R.; Bis, J.C.; Damotte, V.; Naj, A.C.; Boland, A.; Vronskaya, M.; Van Der Lee, S.J.; Amlie-Wolf, A.; et al. Genetic meta-analysis of diagnosed Alzheimer's disease identifies new risk loci and implicates Aβ, tau, immunity and lipid processing. *Nat. Genet.* **2019**, *51*, 414–430, Erratum in *Nat. Genet.* **2019**, *51*, 1423–1424. [CrossRef]
9. Wightman, D.P.; Jansen, I.E.; Savage, J.E.; Shadrin, A.A.; Bahrami, S.; Holland, D.; Rongve, A.; Børte, S.; Winsvold, B.S.; Drange, O.K.; et al. A genome-wide association study with 1,126,563 individuals identifies new risk loci for Alzheimer's disease. *Nat. Genet.* **2021**, *53*, 1276–1282, Erratum in *Nat. Genet.* **2021**, *53*, 1722. [CrossRef]
10. Bertram, L.; Lange, C.; Mullin, K.; Parkinson, M.; Hsiao, M.; Hogan, M.F.; Schjeide, B.M.; Hooli, B.; DiVito, J.; Ionita, I.; et al. Genome-wide Association Analysis Reveals Putative Alzheimer's Disease Susceptibility Loci in Addition to APOE. *Am. J. Hum. Genet.* **2008**, *83*, 623–632. [CrossRef]
11. Hollingworth, P.; Harold, D.; Sims, R.; Gerrish, A.; Lambert, J.-C.; Carrasquillo, M.M.; Abraham, R.; Hamshere, M.L.; Pahwa, J.S.; Moskvina, V.; et al. Common variants at ABCA7, MS4A6A/MS4A4E, EPHA1, CD33 and CD2AP are associated with Alzheimer's disease. *Nat. Genet.* **2011**, *43*, 429–435. [CrossRef]
12. Griciuc, A.; Serrano-Pozo, A.; Parrado, A.R.; Lesinski, A.N.; Asselin, C.N.; Mullin, K.; Hooli, B.; Choi, S.H.; Hyman, B.T.; Tanzi, R.E. Alzheimer's Disease Risk Gene CD33 Inhibits Microglial Uptake of Amyloid Beta. *Neuron* **2013**, *78*, 631–643. [CrossRef]
13. Malik, M.; Simpson, J.F.; Parikh, I.; Wilfred, B.R.; Fardo, D.W.; Nelson, P.T.; Estus, S. CD33 Alzheimer's Risk-Altering Polymorphism, CD33 Expression, and Exon 2 Splicing. *J. Neurosci.* **2013**, *33*, 13320–13325. [CrossRef]
14. Naj, A.C.; Jun, G.; Beecham, G.W.; Wang, L.-S.; Vardarajan, B.N.; Buros, J.; Gallins, P.J.; Buxbaum, J.D.; Jarvik, G.P.; Crane, P.K.; et al. Common variants at MS4A4/MS4A6E, CD2AP, CD33 and EPHA1 are associated with late-onset Alzheimer's disease. *Nat. Genet.* **2011**, *43*, 436–441. [CrossRef]
15. Bao, J.; Wang, X.-J.; Mao, Z.-F. Associations Between Genetic Variants in 19p13 and 19q13 Regions and Susceptibility to Alzheimer Disease: A Meta-Analysis. *Med. Sci. Monit.* **2016**, *22*, 234–243. [CrossRef]
16. Duan, S.; Paulson, J.C. Siglecs as Immune Cell Checkpoints in Disease. *Annu. Rev. Immunol.* **2020**, *38*, 365–395. [CrossRef]
17. Jiang, T.; Yu, J.-T.; Hu, N.; Tan, M.-S.; Zhu, X.-C.; Tan, L. CD33 in Alzheimer's Disease. *Mol. Neurobiol.* **2014**, *49*, 529–535. [CrossRef]
18. Zhao, L. CD33 in Alzheimer's Disease—Biology, Pathogenesis, and Therapeutics: A Mini-Review. *Gerontology* **2019**, *65*, 323–331. [CrossRef]
19. Griciuc, A.; Patel, S.; Federico, A.N.; Choi, S.H.; Innes, B.J.; Oram, M.K.; Cereghetti, G.; McGinty, D.; Anselmo, A.; Sadreyev, R.I.; et al. TREM2 Acts Downstream of CD33 in Modulating Microglial Pathology in Alzheimer's Disease. *Neuron* **2019**, *103*, 820–835.e7. [CrossRef]
20. Ulland, T.K.; Colonna, M. TREM2—A key player in microglial biology and Alzheimer disease. *Nat. Rev. Neurol.* **2018**, *14*, 667–675. [CrossRef]
21. Neumann, H.; Daly, M.J. Variant TREM2 as Risk Factor for Alzheimer's Disease. *N. Engl. J. Med.* **2013**, *368*, 182–184. [CrossRef]

22. Linnartz, B.; Wang, Y.; Neumann, H. Microglial Immunoreceptor Tyrosine-Based Activation and Inhibition Motif Signaling in Neuroinflammation. *Int. J. Alzheimer's Dis.* **2010**, *2010*, 587463. [CrossRef]
23. Wißfeld, J.; Nozaki, I.; Mathews, M.; Raschka, T.; Ebeling, C.; Hornung, V.; Brüstle, O.; Neumann, H. Deletion of Alzheimer's disease-associated CD33 results in an inflammatory human microglia phenotype. *Glia* **2021**, *69*, 1393–1412. [CrossRef]
24. Griciuc, A.; Federico, A.N.; Natasan, J.; Forte, A.M.; McGinty, D.; Nguyen, H.; Volak, A.; LeRoy, S.; Gandhi, S.; Lerner, E.P.; et al. Gene therapy for Alzheimer's disease targeting CD33 reduces amyloid beta accumulation and neuroinflammation. *Hum. Mol. Genet.* **2020**, *29*, 2920–2935. [CrossRef]
25. Efthymiou, A.G.; Goate, A.M. Late onset Alzheimer's disease genetics implicates microglial pathways in disease risk. *Mol. Neurodegener.* **2017**, *12*, 43. [CrossRef]
26. Estus, S.; Shaw, B.C.; Devanney, N.; Katsumata, Y.; Press, E.E.; Fardo, D.W. Evaluation of CD33 as a genetic risk factor for Alzheimer's disease. *Acta Neuropathol.* **2019**, *138*, 187–199. [CrossRef]
27. Bhattacherjee, A.; Jung, J.; Zia, S.; Ho, M.; Eskandari-Sedighi, G.; St. Laurent, C.D.; McCord, K.A.; Bains, A.; Sidhu, G.; Sarkar, S.; et al. The CD33 short isoform is a gain-of-function variant that enhances Aβ1–42 phagocytosis in microglia. *Mol. Neurodegener.* **2021**, *16*, 19. [CrossRef]
28. Raj, T.; Ryan, K.J.; Replogle, J.M.; Chibnik, L.B.; Rosenkrantz, L.; Tang, A.; Rothamel, K.; Stranger, B.E.; Bennett, D.A.; Evans, D.A.; et al. CD33: Increased inclusion of exon 2 implicates the Ig V-set domain in Alzheimer's disease susceptibility. *Hum. Mol. Genet.* **2014**, *23*, 2729–2736. [CrossRef]
29. Rendina, A.; Drongitis, D.; Donizetti, A.; Fucci, L.; Milan, G.; Tripodi, F.; Giustezza, F.; Postiglione, A.; Pappatà, S.; Ferrari, R.; et al. CD33 and SIGLECL1 Immunoglobulin Superfamily Involved in Dementia. *J. Neuropathol. Exp. Neurol.* **2020**, *79*, 891–901. [CrossRef]
30. Nettiksimmons, J.; Tranah, G.; Evans, D.S.; Yokoyama, J.S.; Yaffe, K. Gene-based aggregate SNP associations between candidate AD genes and cognitive decline. *Age* **2016**, *38*, 41. [CrossRef]
31. Burley, S.K.; Berman, H.M.; Bhikadiya, C.; Bi, C.; Chen, L.; Di Costanzo, L.; Christie, C.; Dalenberg, K.; Duarte, J.M.; Dutta, S.; et al. RCSB Protein Data Bank: Biological macromolecular structures enabling research and education in fundamental biology, biomedicine, biotechnology and energy. *Nucleic Acids Res.* **2019**, *47*, D464–D474. [CrossRef]
32. Miles, L.A.; Hermans, S.J.; Crespi, G.A.; Gooi, J.; Doughty, L.; Nero, T.L.; Markulić, J.; Ebneth, A.; Wroblowski, B.; Oehlrich, D.; et al. Small Molecule Binding to Alzheimer Risk Factor CD33 Promotes Aβ Phagocytosis. *iScience* **2019**, *19*, 110–118. [CrossRef] [PubMed]
33. Crocker, P.R.; Paulson, J.C.; Varki, A. Siglecs and their roles in the immune system. *Nat. Rev. Immunol.* **2007**, *7*, 255–266. [CrossRef] [PubMed]
34. Avril, T.; Floyd, H.; Lopez, F.; Vivier, E.; Crocker, P. The Membrane-Proximal Immunoreceptor Tyrosine-Based Inhibitory Motif Is Critical for the Inhibitory Signaling Mediated by Siglecs-7 and -9, CD33-Related Siglecs Expressed on Human Monocytes and NK Cells. *J. Immunol.* **2004**, *173*, 6841–6849. [CrossRef] [PubMed]
35. Varki, A.; Angata, T. Siglecs—The major subfamily of I-type lectins. *Glycobiology* **2006**, *16*, 1R–27R. [CrossRef]
36. Yamakawa, N.; Yasuda, Y.; Yoshimura, A.; Goshima, A.; Crocker, P.R.; Vergoten, G.; Nishiura, Y.; Takahashi, T.; Hanashima, S.; Matsumoto, K.; et al. Discovery of a new sialic acid binding region that regulates Siglec-7. *Sci. Rep.* **2020**, *10*, 8647. [CrossRef]
37. Olaru, O.G.; Constantin, G.I.; Pena, C.M. Variation of total serum sialic acid concentration in postmenopausal women. *Exp. Ther. Med.* **2020**, *20*, 2455–2459. [CrossRef]
38. Mosconi, L.; Rahman, A.; Diaz, I.; Wu, X.; Scheyer, O.; Hristov, H.W.; Vallabhajosula, S.; Isaacson, R.S.; De Leon, M.J.; Brinton, R.D. Increased Alzheimer's risk during the menopause transition: A 3-year longitudinal brain imaging study. *PLoS ONE* **2018**, *13*, e0207885. [CrossRef]
39. McKhann, G.M.; Knopman, D.S.; Chertkow, H.; Hyman, B.T.; Jack, C.R., Jr.; Kawas, C.H.; Klunk, W.E.; Koroshetz, W.J.; Manly, J.J.; Mayeux, R.; et al. The diagnosis of dementia due to Alzheimer's disease: Recommendations from the National Institute on Aging-Alzheimer's association workgroups on diagnostic guidelines for Alzheimer's disease. *Alzheimers Dement.* **2011**, *7*, 263–269. [CrossRef]
40. The PyMOL Molecular Graphics System, version 1.8; New York (USA) Schrödinger, LLC: 2015. Available online: Pymol.org/2/support.html (accessed on 15 February 2022).
41. Brooks, B.R.; Brooks, C.L., III; MacKerell, A.D., Jr.; Nilsson, L.; Petrella, R.J.; Roux, B.; Won, Y.; Archontis, G.; Bartels, C.; Boresch, S.; et al. CHARMM: The biomolecular simulation program. *J. Comput. Chem.* **2009**, *30*, 1545–1614. [CrossRef]
42. Trott, O.; Olson, A.J. AutoDock Vina: Improving the speed and accuracy of docking with a new scoring function, efficient optimization, and multithreading. *J. Comput. Chem.* **2010**, *31*, 455–461. [CrossRef]
43. Morris, G.M.; Huey, R.; Lindstrom, W.; Sanner, M.F.; Belew, R.K.; Goodsell, D.S.; Olson, A.J. AutoDock4 and AutoDockTools4: Automated docking with selective receptor flexibility. *J. Comput. Chem.* **2009**, *30*, 2785–2791. [CrossRef]
44. Pettersen, E.F.; Goddard, T.D.; Huang, C.C.; Couch, G.S.; Greenblatt, D.M.; Meng, E.C.; Ferrin, T.E. UCSF Chimera—A visualization system for exploratory research and analysis. *J. Comput. Chem.* **2004**, *25*, 1605–1612. [CrossRef]

International Journal of *Molecular Sciences*

MDPI

Review

rs7041 and rs4588 Polymorphisms in Vitamin D Binding Protein Gene (VDBP) and the Risk of Diseases

Dominika Rozmus [1,*], Janusz Płomiński [2,3], Klaudia Augustyn [4] and Anna Cieślińska [1]

1 Faculty of Biology and Biotechnology, University of Warmia and Mazury, 10-719 Olsztyn, Poland; anna.cieslinska@uwm.edu.pl
2 Clinical Department of Trauma-Orthopedic Surgery and Spine Surgery of the Provincial Specialist Hospital in Olsztyn, 10-561 Olsztyn, Poland; plominsky@poczta.onet.pl
3 Department and Clinic of Orthopaedics and Traumatology, Collegium Medicum, University of Warmia and Mazury, 10-719 Olsztyn, Poland
4 Faculty of Medicine, Collegium Medicum, University of Warmia and Mazury, 10-082 Olsztyn, Poland; klaudia.augustyn@student.uwm.edu.pl
* Correspondence: dominika.rozmus@uwm.edu.pl

Abstract: The purpose of the study was to investigate the role of vitamin D binding protein (VDBP, DBP) and its polymorphism in the vitamin D pathway and human health. This narrative review shows the latest literature on the most popular diseases that have previously been linked to VDBP. Vitamin D plays a crucial role in human metabolism, controlling phosphorus and calcium homeostasis. Vitamin D binding protein bonds vitamin D and its metabolites and transports them to target tissues. The most common polymorphisms in the VDBP gene are rs4588 and rs7041, which are located in exon 11 in domain III of the VDBP gene. rs4588 and rs7041 may be correlated with differences not only in vitamin D status in serum but also with vitamin D metabolites. This review supports the role of single nucleotide polymorphisms (SNPs) in the VDBP gene and presents the latest data showing correlations between VDBP variants with important human diseases such as obesity, diabetes mellitus, tuberculosis, chronic obstructive pulmonary disease, and others. In this review, we aim to systematize the knowledge regarding the occurrence of diseases and their relationship with vitamin D deficiencies, which may be caused by polymorphisms in the VDBP gene. Further research is required on the possible influence of SNPs, modifications in the structure of the binding protein, and their influence on the organism. It is also important to mention that most studies do not have a specific time of year to measure accurate vitamin D metabolite levels, which can be misleading in conclusions due to the seasonal nature of vitamin D.

Keywords: VDBP; vitamin D binding protein; rs7041; rs4588; bone density; diabetes; obesity; COPD; pulmonary tuberculosis; SNP; MD; PD

Citation: Rozmus, D.; Płomiński, J.; Augustyn, K.; Cieślińska, A. rs7041 and rs4588 Polymorphisms in Vitamin D Binding Protein Gene (VDBP) and the Risk of Diseases. *Int. J. Mol. Sci.* **2022**, *23*, 933. https://doi.org/10.3390/ijms23020933

Academic Editor: Elixabet Lopez-Lopez

Received: 29 November 2021
Accepted: 13 January 2022
Published: 15 January 2022

1. Introduction

Recent data suggest that vitamin D deficiency is widespread across Europe. Analysis of 14 population studies revealed that 13% of the 55,844 European individuals had average yearly serum 25(OH)D concentrations <30 nmol/L, regardless of age group, ethnic mix, and the latitude of study populations [1]. According to the US Endocrine Society definition of vitamin D deficiency (<50 nmol/L), the prevalence was 40.4% and dark-skinned ethnic subgroups were more likely to be vitamin D deficient [2]. Due to the darker skin color, melanin blocks the UVB radiation, which is necessary for vitamin D synthesis. Hypovitaminosis D was highly prevalent among pregnant Bangladeshi women, and parity and gestational age were common risk factors of vitamin D deficiency [3].

$1,25(OH)_2D$ is considered to be the most powerful physiological agent. It stimulates the active transport of calcium, phosphorus, and magnesium. Disorders in vitamin D action can lead to a decrease in the net flux of minerals to the extracellular compartment, which can lead to hypocalcemia and secondary hyperparathyroidism [4]. In addition, low concentrations of calcium and phosphorus will lead to defective mineralization of the bone matrix and rickets [5,6]. Vitamin D is also a regulator of the immune system, where the expression of CYP27B1 in macrophages leads to local production of 1,25-dihydroxyvitamin D ((1,25[OH]$_2$D)), which induces the expression of genes encoding antimicrobial peptides [7]. (1,25[OH]$_2$D) induces and stimulates autophagy resulting in enhanced bacterial killing, suppresses production of pro-inflammatory cytokines, and prevents overstimulated immune response [8].

The purpose of the study is to investigate and systematize the current knowledge regarding the impact of VDBP polymorphisms on the risk of incidence of various diseases and human health. The "Vitamin D" (calciferol) term refers to two secosteroids: vitamin D2 (ergocalciferol) and vitamin D3 (cholecalciferol). They are both produced from sterol precursors with light in the UVB spectrum of 280 to 320 nm. In fungi and plants, ergosterol is the vitamins' D2 precursor, while the vitamins' D3 precursor is 7-dehydrocholesterol (7-DHC), and its high concentration was found in the skin [9]. Vitamin D2 and D3 differ in side chains, but both are converted to 25-hydroxyvitamin D 25(OH)D and 1,25-dihydroxyvitamin D [9,10]. 25OHD is considered to be the best reflection of the vitamin D level in serum [10]. Vitamin D3 can be synthesized endogenously under ultraviolet (UV) light [11].

Vitamin D photoproduction starts with 7-dehydrocholesterol (7-DHC), which is synthesized and built into the membranes of the epidermis and dermis [12]. During sunlight exposure, the epidermal 7-DHC is converted into pre-vitamin D3 [12,13]. Pre-D3 is thermo-isomerized to the vitamin D3 in the cell membrane. The produced cholecalciferol is removed into extracellular space and reaches the skin's capillary by diffusion [12]. Prolonged exposure to solar ultraviolet radiation while pre-vitamin D3 synthesis reaches a plateau of 10 to 15 percent of the original 7-DHC has an increasing effect only on lumisterol and tachysterol, two biologically inactive photoisomers [13]. Lumisterol can revert back to pre-D3 in the dark, but maximum levels of pre-D3 lead to the accumulation of inactive luminsterol with continued UV exposure. The production of lumisterol and tachysterol has a protective effect against the production of toxic amounts of D3 [9]. During activation and inactivation processes, cytochrome P450 (CYP) enzymes are involved throughout the vitamin D3 pathway [14,15]. The pathway is presented in Figure 1 (based on [12,15–18].

The first step is the conversion of vitamin D to 25OHD in the liver by 25-hydroxylase enzymes. Enzymes with this activity are identified as: CYP3A4, CYP2R1, CYP27A1, CYP2J1, CYP2C11, CYP2D25 [19]. The production of the hormonally active form: 1α,24-dixydroxyvitamin D (1α,25(OH)$_2$D3) is catalyzed by CYP27B1: 25-hydroxyvitamin D-1 α-hydroxylase. An active form acts with the vitamin D receptor (VDR) [19]. There is also an alternative pathway that starts with the action of CYP11A1 on D3 and produces 20(OH)D3, 22(OH)D3, 20,23(OH)2D3, 20,22(OH)$_2$D3, and 17,20,23(OH)$_3$D3. Hydroxylation of some of these metabolites can occur through the activity of CYP27B1 at C1α, by CYP24A1 at C24 and C25, and by CYP27A1 at C25 and C26 [16,20]. Possible pathways are shown in Figure 2 based on SNPedia base [21]. Mitochondrial CYP24A1 catalyzes the first step of 25(OH)D3 and 1,25(OH)$_2$D3 degradation by 24- or 23-hydroxylation [22]. CYP11A1 was also found to be expressed in extrarenal and extragonadal tissues [23] and also in the immune system [17]. CYP11A1 vitamin D metabolites are also detectable in the serum [16].

Vitamin D receptor is a member of the nuclear receptor superfamily and plays a crucial role in the actions of vitamin D [24]. VDR mediates many genomic and non-genomic effects of vitamin D. Many biological pathways and networks are influenced by VDR, for example metabolism [25], including bone metabolism and remodeling [26,27], immunity and immune response [24,28], cell proliferation and differentiation [24], and cell health [29]. VDR regulates the expression of numerous genes and is involved in calcium/phosphate homeostasis [24]. Many genes are up-regulated (CYP24A1, osteocalcin or *Rankl*) or down-regulated (parathormone—PTH, CYP27B1) due to VDR activation [24]. Slominski et al. reported the alternatives to VDR nuclear receptors that are involved in vitamin D metabolite signaling pathways: retinoic acid-related orphan receptors (RORα–γ; NR1F1–3) [30], AhR [31], and LXR [32].

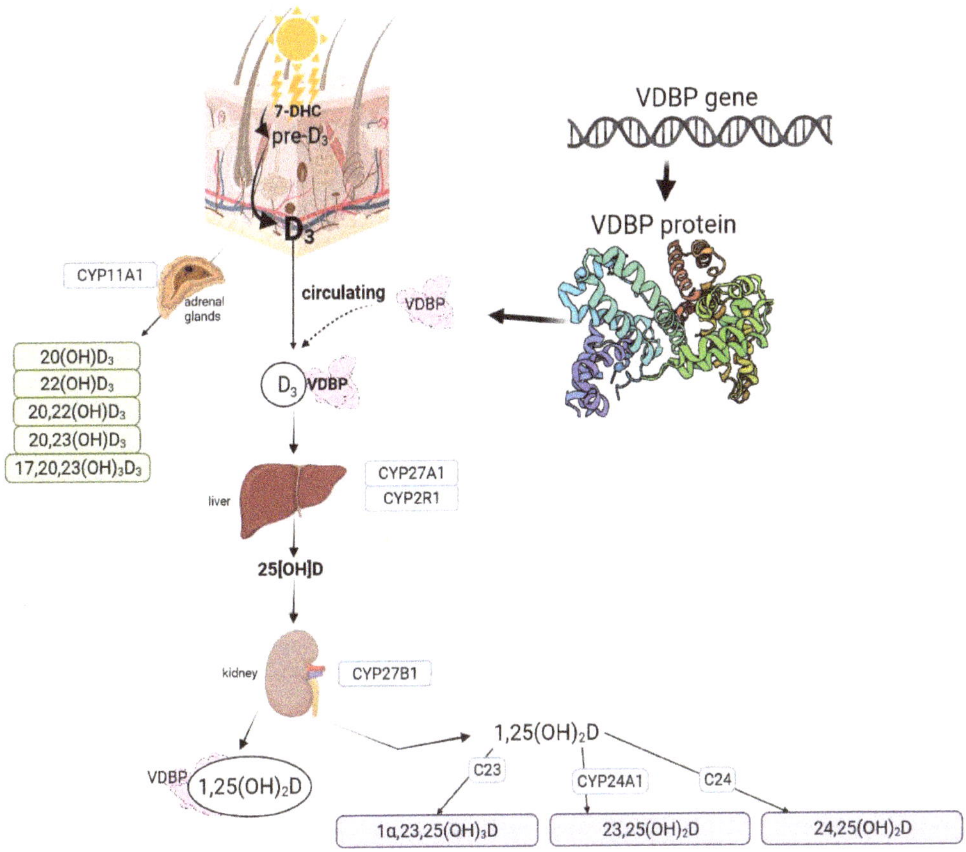

Figure 1. Vitamin D metabolic pathways based on Bikle (2014), Slominski et al. (2012, 2015, 2020), Rozmus et al. (2020) [12,15–18]. Figure fully created with biorender.com (accessed on 1 November 2021).

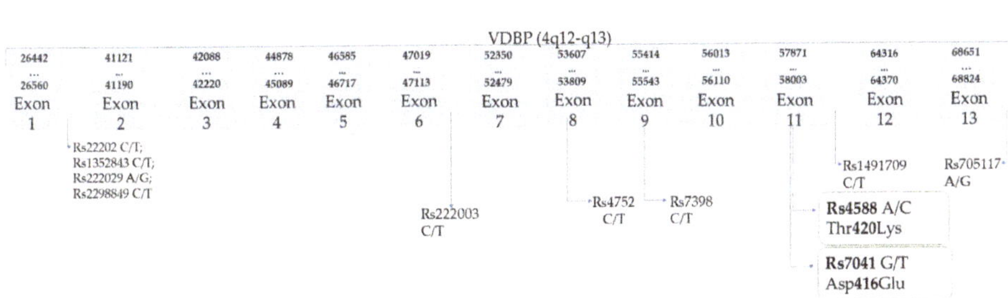

Figure 2. VDBP gene and its polymorphisms (based on the SNPedia database [21]).

Vitamin D binding protein was initially named the 'group-specific component' (Gc) by Hirschfield in 1959 after isolation from the α2-globulin portion of plasma [33]. As a result of the binding and transport of vitamin D analogs, the DBP name was adopted. After discovering macrophage-stimulating activities, VDBP was renamed as the macrophage-activating factor (GcMAF/DBP-MAF) [34]. The name has been changed several times, as many different biological functions of VDBP have been discovered [34,35]. VDBP binds to fatty acids and actin monomers and also has immune functions independent of vitamin D transport [35], such as binding to leukocyte membrane proteoglycans and the activation of the complement C5 system [18]. VDBP is well known for its single nucleotide polymorphisms (SNP), and the most common are rs7041 and rs4588, located in exon 11 of the VDBP gene [18,36]. SNPs are the most abundant genetic variants in genomes [37]. SNPs may affect protein stability, folding, flexibility, and aggregation; functional sites, reaction kinetics, and dependence on environmental parameters, such as pH, salt concentration, and temperature; protein expression and subcellular localization; and protein–small molecule, protein–protein, protein–DNA, and protein–membrane interactions [38]. Many studies have shown associations between SNPs and the concentration of protein as well as substance protein transport via VDBP in this particular case [18,39,40]. The effects of vitamin D supplementation according to the most common polymorphisms of the vitamin D binding protein was studied by Al-Daghri et al., and it was shown that 25[OH]D concentrations were significantly higher among people with the major homozygous rs7041 genotype. Post supplementation 25[OH]D was higher in participants carrying homozygous major genotypes in rs4588 and rs7041 compared to other genotypes [41].

The vitamin D binding protein has a single binding site for all vitamin D metabolites and has a high affinity for 25OHD and 1,25(OH)2D [42], but has no affinity for lumisterol and minimum affinity to tachysterol [13].

Possible haplotypes are as follows: Gc1f(1f): rs7041(T) + rs4588(C); Gc1s(1s): rs7041(G) + rs4588(C); Gc2(2): rs7041(T) + rs4588(A). According to some publications, in which rs2282679 is used as a proxy for rs4588, rs2282679(A) is typically coinherited with rs4588(C) and vice versa. Since human have two copies of each gene, it leads to six possible VDBP phenotypes [21] presented in Table 1. Table 2 present SNPs of VDBP, and chromosome location. Table 3 present the frequency of rs7041 and rs4588 among populations and geographic regions. Figure 2 show the location of the most common SNPs of the VDBP gene, and their loci in exons and introns are pointed out with arrows.

Table 1. Characteristics of vitamin D binding protein polymorphisms in most common variants (based on Rozmus et al. 2020 with modifications [18,21]).

Variant	Version
GC1S/1S	rs7041(G;G)
GC1S/1F	rs7041(G;T) AND rs4588(C;C), or, rs7041(G;T) AND rs2282679(A;A)
GC1S/2	rs7041(G;T) AND rs4588(A;C), or, rs7041(G;T) AND rs2282679(A;C)
GC1F/1F	rs7041(T;T) AND rs4588(C;C), or, rs7041(T;T) AND rs2282679(A;A)
GC1F/2	rs7041(T;T) AND rs4588(A;C), or, rs7041(T;T) AND rs2282679(A;C)
GC2/2	rs4588(A;A) or rs2282679(C;C)

Table 2. SNPs alleles and chromosome location (based on SNPedia [21]).

SNP	Location	Major Allele	Minor Allele
rs7041	exon 11	G	T
rs4588	exon 11	C	A
rs1155563	intron 1	T	C
rs1352844	intron 1	C	T
rs1352845	intron 1	A	G
rs222016	intron 2	A	G
rs2282679	intron 11	A	C
rs705119	intron 11	C	A
rs12512631	3′ downstream	T	C
rs222049	3′ downstream	C	G
rs3733359	5′ UTR	G	A

Table 3. Frequencies of alleles in rs4588 and rs7041 among different populations and geographic regions based on [43,44].

Geographic Region/Population	Sample Size (*n*)		Allele Frequencies	Allele Frequencies	References
	rs4588 *	rs7041 **	rs4588-T	rs7041-A	
Estonian	4480	4480	0.3036	0.4125	[41,42]
Korean	2930	nd. ***	0.2843	nd. ***	
Northern Sweden	600	600	0.242	0.375	
Daghestan	1136	1134	0.2764	0.4462	
Vietnamese	614	nd.	0.22	nd. ***	
Finland	304	304	0.188	0.355	
Quatari	216	216	0.199	0.486	
Siberian	nd. ***	34	nd. ***	0.26	
European	263394	285118	0.281206	0.433	
African	10488	11716	0.09392	0.8182	
African American	10118	11306	0.09567	0.81523	
Asian	6536	6908	0.2852	0.7351	
East Asian	4624	4946	0.2885	0.7351	
Other Asian	1912	1962	0.2772	0.7243	
Latin American individuals with Afro-Caribbean ancestry	1252	1488	0.2236	0.541	
Latin American individuals with mostly European and Native American Ancestry	2188	7238	0.1846	0.4823	
South Asian	314	5226	0.226	0.4351	
Other	21820	18956	0.26801	0.49625	

* [42]. ** [43]. *** nd means no data.

Our review focuses on the most important correlations between VDBP polymorphisms and selected diseases described in the latest scientific reports garding obesity, polycystic ovary syndrome, metabolic syndrome, diabetes mellitus, asthma, pulmonary tuberculosis, chronic obstructive pulmonary disease, coronary artery disease, multiple sclerosis, and Parkinson's Disease. This study is supplementation of our previous review concerning the role of VDBP on malignant tumors [18].

2. Diseases

2.1. Bone Density

Osteoporosis is a skeletal disease that affects women older than 50 years of age. During the past few years, it has been a serious public health problem because of the high socioeconomic burden. Patients suffer from deterioration of bone microarchitecture, low bone mineral density (BMD), and increased risk of fragility fractures [45]. The first studies on the effect of vitamin D supplementation on bone density showed that vitamin D (with calcium) reduced bone loss measured in the femoral neck, spine, and total body during the 3-year study and reduced the incidence of non-vertebral fractures [46]. The study by Martinez-Aguilar et al. supports the correlation of low serum VDBP levels with low BMD (osteopenic and osteoporotic). VDBP could be considered a novel, potential, and non-invasive biomarker for the early detection of osteoporosis [45]. The study by Rivera-Paredez et al. supports the association of VDBP and bone health. The article showed that the rs7041 G allele is associated with a higher level of VDBP and BMD compared to homozygous TT. The A allele of rs4588 was associated with a lower VDBP and BMD compared to homozygous CC. Among men, no association was found between these polymorphisms and VDBP, but GC variants were associated with VDBP levels. In both the women and men subgroup, no association was observed between free and bioavailable 25(OH)D and BMD [47]. Among women and adolescents, the GC genotype was associated with susceptibility to low 25(OH)D levels. The study included 198 healthy girls aged 10–18 years. The AA genotype of rs4588, TT genotype of rs7041, and CT-AT/AT-AT (GC 1f-2/2-2) genotypes were significantly associated with lower 25(OH)D levels, even after adjustment for age and season at the time of blood collection [48]. Studies of Lauridsen et al. also support the VDBP role in premenopausal bone fracture risk among white women aged 45–58, as Gc2-2 is considered to increase the risk of bone fracture compared to Gc1-1 [49]. The results of Ezura et al. indicated a complex combined effect of VDBP SNPs that underlie susceptibility to low BMD and osteoporosis. The genotyping of 13 SNPs among 384 participants and the analysis of results showed that not only a single SNP, but also a combination of them could act as a risk factor of osteoporosis. Five SNPs (39C > T, IVS1 + 827C > T, IVS1 + 1916C > T, IVS1 − 1154A > G, and IVS11 + 1097G > C) had a significant correlation with radial BMD, and IVS11 + 1097G > C located in intron 11 was the most correlated [50].

2.2. Obesity

Obesity contributes to reduced life expectancy, poor quality of life, cardiovascular diseases, type 2 diabetes, osteoarthritis, and cancer [51]. Serum vitamin D was found to be lower in obese people [52]. Obesity increases the risk of vitamin D deficiency among different population groups. Higher body mass index (BMI), waist circumference, and the sum of skinfolds were statistically significantly associated with lower 25[OH]D levels and with higher levels of PTH [53]. According to genetic studies, higher adiposity also causes an increased concentration of 25-hydroxyvitamin D, which is used as a vitamin D status indicator [54]. Another study showed that among the haplotypes rs7041 and rs4588, GC2-2 (rs7041 AA and rs4588 TT) has the lowest 25[OH]D levels compared to other haplotypes that contained at least one copy of the Gc1 allele ($p < 0.0001$) [55]. Interestingly, it was also observed that VDBP gene rs7041 polymorphism might be associated with the risk of obesity. In obese patients, a difference was found in the gene type TG + GG and TT frequency of rs7041 between obesity and control groups ($p = 0.020$). The G allele frequency was higher compared to the control group ($p = 0.023$). The TG and TG + GG of VDBP gene rs7041 polymorphism increased the risk of obesity after including age and gender [39]. The study of Almesri et al. showed that rs7041-G and the rare GG genotype were associated with an increase in BMI ($p = 0.007$ and $p = 0.012$, respectively) and had no influence on 25OHD3 levels. On the other hand, rs2282679 (A) and rs4588 (C) were associated with low 25[OH]D3 plasma levels ($p = 0.039$ and $p = 0.021$, respectively). There was no association between rs2282679 (A), rs4588 (C), and BMI in general, but after categorizing patients into subgroups based on their sex, it was shown that rs7041 GG was associated with high BMI in females ($p = 0.003$), and rs4588 CC was associated with high BMI in females ($p = 0.034$) and low levels of 25OHD3 in males ($p = 0.009$). Furthermore, rs12721377 AA was associated with low 25[OH]D3 levels in females ($p = 0.039$) [56].

2.3. Polycystic Ovarian Syndrome (PCOS) and Metabolic Syndrome (MetS)

All participants from Santos et al. were genotyped for polymorphisms rs2282679, rs4588, and rs7041, and serum 25(OH)D levels were determined. Women with PCOS were at a younger age and had significantly higher body mass index, blood pressure, and insulin resistance than the control group ($p < 0.05$). The 25(OH)D levels were lower among PCOS women with MetS, but no association was observed between PCOS and polymorphisms of VDBP. Above that, PCOS participants with MetS had a higher frequency of the TT genotype in rs7041 [57]. The study by Haldar et al. did not show significant differences in the frequency of rs7041, rs4588, and rs2060793 genotypes in PCOS and control women. The GT allele of rs7041, as well as the allelic combination of Gc1F/1F (T allele of rs4588 and C allele of rs7041; p value = 0.03), were associated with an increased risk of developing PCOS in vitamin D deficient women [36].

2.4. Postmenopausal Women

Postmenopausal women can exhibit biochemical signs of vitamin D insufficiency. Vitamin D is related to bone integrity, and 25-hydroxyvitamin D is a reliable clinical indicator of vitamin D status. Low levels of vitamin D have been linked to secondary hyperparathyroidism, increased bone turnover, reduced BMD, and increased risk of osteoporotic fractures [58]. As VDBP plays a crucial role in vitamin D transport, the study of Pop et al. showed that lower estradiol levels are associated with lowering VDBP levels [59]. Studies by Sinotte et al. showed that 25(OH)D concentrations in premenopausal women are strongly associated with higher VDBP polymorphisms. Rs7041 and rs4588 were associated with lower 25[OH]D concentrations. Rare alleles of rs7041 (TT genotype) and rs4588 (AA genotype) were associated with the lowest levels of vitamin D3 in a period of low (November to April) and high (May to October) vitamin D load [60]. The study by Alharazy et al. shows that rs7041 among postmenopausal women in Saudi Arabia was associated with total 25(OH)D, and rs4588 did not show an association with total or free levels of 25(OH)D [40]. The results presented by Santos et al. suggest that rs2282679 and the DBP GC2 isoform are related to lower serum levels of DBP and with susceptibility to 25(OH)D deficiency in adults and postmenopausal women [57]. Lauridsen et al. showed that the DBP-phenotype is linked with premenopausal bone fracture risk in perimenopausal white women (595 subjects, age 45–58). There was a significant difference in bone fracture risk among women with different DBP-phenotypes (relative risk of 0.32 in Gc2-2, compared with Gc1-1) [49].

2.5. Diabetes Mellitus (DB)

Diabetes mellitus is a group of dysfunctions as a result of hyperglycemia characterized by insulin resistance (IR) and secretion or excessive secretion of glucagon. Of the two types of diabetes, type 2 (T2D) is more common and connects a problem of impaired glucose regulation with a combination of dysfunctional pancreatic beta cells and insulin resistance. However, the main risk is obesity (where abdominal is the highest risk of all types). Type 1 (T1D) is an autoimmune disorder that leads to the destruction of pancreatic beta-cells [61].

2.5.1. Diabetes Type 1 (T1D)

The meta-analysis of the five studies presented showed no association between rs7041 and rs4588 polymorphisms with the risk of T1D [62].

2.5.2. Diabetes Type 2 (T2DM)

The study by Zhao et al. showed that there was a significant multiplicative interaction between rs7041 and body mass index (BMI) associated with elevated blood glucose levels and a higher BMI (>28.47), and the carrying allele G was given a stronger effect than the genotype of TT. In conclusion, the interactions between GC rs7041–CYP2R1 (enzyme in the vitamin D metabolic pathway), rs1993116, and GC rs7041-BMI may explain the mechanisms by which these may increase the risk of developing T2DM [63]. Another study by Fawzy et al. examined the frequency distribution of GC-rs7041 and showed no difference between patients and healthy controls, while GC-rs4588 showed an association with T2DM in all genetic models. The rs4588 AA variant was correlated with higher serum GC globulin, albuminuria, and poor glycemic control. On the other hand, a higher frequency of rs7041*TT and rs4588*AA was noticed in the macroalbuminuria vs. normoalbuminuric group. Patients with the GC-2 haplotype were approximately 2.5 times more likely to develop diabetes and had higher levels of albuminuria [64]. Other results were obtained in the experiment by Klahold et al. during a case-control cohort study that was conducted to investigate an association of SNPs in the vitamin D metabolic pathway with T2D. Up to 464 T2D patients and 292 healthy controls were genotyped. Patients with genotypes CYP27B1 rs10877012 "CC" (pc = 4×10^{-5}), VDBP rs7041 "GG" (pc = 0.003), rs4588 "CC" (pc = 3×10^{-4}), CYP24A1 rs2585426 "CG" (pc = 0.006), and rs2248137 "CG" (pc = 0.001) showed lower 25(OH)D3 and VDBP rs4588 "CC" lower 1,25(OH)2D3 levels (pc = 0.005). This study supports that vitamin D deficiency is highly prevalent in type 2 diabetes and most patients are also functionally affected by low levels of the active metabolite 1,25(OH)2D3. Furthermore, vitamin D system genes can affect the risk of type 2 diabetes and 25(OH)D3 concentration when compared to the healthy group. Despite this, the underlying mechanism has not been clarified, and trials, as well as functional studies, appear to be necessary to identify mechanisms by which the vitamin D system affects the pathophysiology of T2DM [65].

2.6. *Asthma*

Asthma is a chronic disease affecting inflammation in the lungs and airways. Common symptoms of the disease are a cough, chest tightness, and shortness of breath. Inflammation causes an overabundance of eosinophils, mast cells, activated T helper lymphocytes, and aids in identifying inflammatory mediators [66,67]. Vitamin D has been studied in asthma progression. Levels of vitamin D were significantly decreased in asthmatic patients in comparison to control patients. Results of genotyping the rs7041 among Kurdish patients showed that the GG genotype, as well as VDBP levels, was increased among the asthmatic group compared to the healthy controls ($p = 0.003$). Asthma progression was increased among patients carrying the rs7041 GG genotype [68]. A study by Fawzy et al. on s group of Egyptian patients showed that the rs7041 GG genotype is correlated significantly with asthma disease, while rs4588CA and AA genotypes were found as protective [69]. Another study supported the importance of rs7041 with the GC1S haplotype, especially among children diagnosed with asthma. The GC1S haplotype was considered to increase the risk of respiratory syncytial virus bronchiolitis in childhood and subsequent asthma development. The GC1s haplotype is associated with higher VDBP levels, which results in less free available vitamin D [70]. A study by Paraskakis et al. on a group of children with asthma showed a higher frequency of the rs7041 G allele and the A allele in rs4588 as a lower frequency allele in children with controlled asthma [71].

2.7. *Pulmonary Tuberculosis (PTB, TB)*

Tuberculosis is most often acquired through mycobacterial inhalation. It starts as an infection focus in the lung parenchyma (primary tuberculosis). It begins with necrotizing bronchopneumonia and progresses rapidly to necrotizing granuloma. Mycobacteria from the lungs spread only through the lymphatic system to the lymph nodes of the hilum drainage, where they cause necrotizing granulomatous inflammation [72]. As innate immunity plays an important role in the pathophysiology of tuberculosis, vitamin D with its transporter protein VDBP and its nuclear receptor vitamin D receptor can play a potential role in altering host defense against Mycobacterium tuberculosis. Decreased serum levels of vitamin D were observed in active TB patients as compared to healthy controls ($p < 0.001$) 209 [73]. The study by Zhang et al. showed [74] two less common VDBP polymorphisms, rs3733359 GA and rs16847024 CT, that were significantly associated with a reduced risk of PTB, as well as alleles rs3733359 A and rs16847024 T that were associated with a decreased susceptibility to PTB. In one of the most common polymorphisms, rs4588, the GT genotype was significantly higher in patients with PTB when compared to controls. The findings of Harishankar et al. [75] support those mentioned above, as the CA genotype rs4588 was associated with susceptibility to TB [OR: 1.47 (0.85–2.55); $p = 0.049$] and associated with 47.4% deficiency of 25(OH)D in patients with PTB, but the AA genotype was significantly associated with protection from TB [OR: 0.14 (0.02–1.29); $p = 0.042$]. No association was found with rs7041 polymorphism. Gene variants with 25(OH)D deficiency did not reveal a significant association due to the limited sample size, but the results showed a tendency towards 25(OH)D deficiency in rs7041 TG and rs4588 CA [75].

2.8. Chronic Obstructive Pulmonary Disease

Vitamin D deficiency was associated with increased risks of chronic obstructive pulmonary disease (COPD). However, the mechanism remains unknown. The vitamin D metabolite 1,25(OH)2D3 reinforced physical interactions between the vitamin D receptor with NF-κB p65 and c-Jun. The results of Fu et al. show that vitamin D is inversely correlated with inflammatory signaling in patients with COPD, and vitamin D may be a vital mediator of the progress of COPD in patients with low vitamin D levels [76]. The study of Li et al. also shows that COPD patients are at high risk of vitamin D deficiency, and the severity of COPD is inversely correlated with vitamin D levels. Furthermore, the homozygous carrier of the rs7041 T allele influences serum levels of 25OHD and is related to the susceptibility of COPD, which could be a potential candidate gene for screening COPD [77]. Among COPD smokers, high frequencies of rs7041/rs4588 haplotypes were homozygous GC1S/1S (42.5%), and higher levels of VDBP in the sputum were observed in stage I and II of COPD, only in the genotype GC1S/1S compared to non-smokers ($p = 0.034$ and $p = 0.002$, respectively) [78]. The studies of Horita et al. included 1712 patients and 1181 non COPD controls among Asians and Caucasians. The prevalence of each allele among the control group was: GC-1F 14.0%, GC-1S 53.8%, and GC-2 31.9%. Compared to GC-1S, the GC-1F allele and the GC-2 allele were associated with the risk of COPD with pooled odds ratios of 1.44 (95% CI 1.14–1.83, $p = 0.002$) and 0.83 (95% CI 0.69–0.996, $p = 0.045$), respectively. In comparison to the 1S-1S genotype, the 1F-1F genotype was a risk factor of COPD with a pooled odds ratio of 2.64 (95% CI 1.29–5.39, $p = 0.008$). The VDBP GC-1F allele was a risk for COPD in the recessive model [79]. Another study showed that patients carrying C allele at rs4588 exhibited a higher frequency of exacerbations ($p = 0.0048$), and a greater susceptibility to chronic obstructive pulmonary disease ($p = 0.0003$), as well as emphysema ($p = 0.0029$), and a tendency for rapid decline of airflow obstruction ($p = 0.0927$) [80]. The meta-analysis of Khanna et al. proves that VDBP is a major determinant of vitamin-D metabolism and transport, showing that alleles GC1F and GC1F/1F posed a risk of COPD. GC1S-1S was found to be a risk only among European participants in these studies [81].

2.9. Coronary Artery Disease

Cardiovascular diseases (CVDs) are the leading cause of death worldwide, and among CVD coronary artery disease (CAD), they are almost half of all cardiovascular deaths and the most common cause of death [82]. VDBP and its genetic polymorphisms have been linked as susceptible components for CAD [83]. The study by Peri et al. showed evidence of the association of rs4588 and rs7051 with CAD cases among patients after acute myocardial infarction and correlations of these polymorphisms with serum levels of 25-hydroxyvitamin D 25(OH)D. Rs4588 T/T was determined as a susceptibility factor for anteroseptal myocardial infarction, where the same genotype was generally more prevalent in smokers [82]. A study by Tarighi et al. among the Iranian population showed a significant association between the GG genotype (rs7041) and CAD ($p = 0.02$, OR = 0.5737 95% CI = 0.304–0.944) [83]. Daffara et al., in a group of 1080 patients, proved that 57% carried the mutated G allele of rs7041, while 22% carried the A allele of rs4588. In addition, higher C-reactive protein levels were observed in the carriers of the G allele of rs7041 ($p = 0.02$), and 25-hydroxyvitamin D levels were similar between the groups. The rs4588 A allele was associated with a higher prevalence of lesions in the left anterior descending artery and a longer lesion length ($p = 0.04$ and $p = 0.03$, respectively). Rs7041 and rs4588 did not affect the prevalence of CAD [84].

2.10. Multiple Sclerosis (MS)

Numerous studies suggest that vitamin D levels affect the risk of multiple sclerosis development and modify disease activity [85–89]. Munger et al. found an inverse relationship between serum 25(OH)D level and risk of MS among Caucasians. No similar association was found among those with darker skin or Hispanics [88]. According to the research, these groups have lower 25-hydroxyvitamin D levels with no significant health consequences [90,91]. There are racial/ethnical differences in the polymorphism of the vitamin D-binding protein-SNPs at rs7041 and rs4588. The dominant allele in rs7041 in Caucasians is the minor allele in those with African heritage. The VDBP isoform most commonly found in individuals with darker skin is the most efficient transporter of 25-hydroxyvitamin D and its metabolites. The MS Sunshine study found that there is a strong association between higher lifetime sun exposure and MS risk across racial/ethnic groups. Low ultraviolet radiation from the sun should lead to low vitamin D status and can explain the geographic distribution of the disease. There is a lack of association between 25-hydroxyvitamin D and MS risk among those with African heritage and Hispanics [92]. Langer-Gould et al. suggested that these differences cannot be explained by racial/ethnic variations in bioavailable vitamin D [92]. Xin Zhang et al. provide evidence that VDBP rs7041 and 4588 polymorphism may not be associated with an increased risk of multiple sclerosis in the meta-analysis [62].

2.11. *Parkinson's Disease*

Low vitamin D status is suggested to be associated with Parkinson's Disease [93–95]. Knekt et al. found that individuals with a serum vitamin D concentration of at least 50 nmol/L had a 65% lower risk than those with values below 25 nmol/L, after the adjustment of several potential cofounders, in the follow-up for Parkinson's disease occurrence of the Mini-Finland Health survey [96]. Zheng Lv et al. suggested that patients with 25(OH)D level <50 nmol/L experienced a twofold increased risk of PD in the meta-analysis [97]. In the prospective cohort study of 137 patients with Parkinson's disease, circulating 25-hydroxivitamin D levels were deficient in one-half of the patients [98]. GC polymorphism was associated with 25(OH)D levels-TT carriers for GC1, and AA carriers for GC2 had lower vitamin D status. There was no significant association between GC polymorphism and 1,25(OH)D levels. SNPs of VDBP showed no significant association with the severity of PD. In another study of 382 patients and 242 controls in a Turkish cohort, only rs7041 was associated with PD risk [94]. Significantly higher levels of serum 25-hydroxyvitamin D was observed in the group of homozygous major allele carriers for rs2282679, rs3755967, and rs2298850 with slower progression of the disease. In the proteomic studies, a decreased level of VDBP in the CSF has been suggested to be a biomarker of disease [99].

All the information above in the Section 2 is gathered into summary table—Table 4.

Table 4. Summary of mentioned diseases in the Section 2. Most common VDBP polymorphisms with effects.

	Polymorphisms	Effects	Group Characteristics	References
Bone density	rs7041 "G"	Low VDBP = low BMD;	women, aged ≥ 45 years old, 446 participants	[45]
		Higher VDBP and higher BMDlevels;	1853 adults, aged ≥ 18	[47]
	rs7041 "T"	Lower 25(OH)D levels;	198 girls, aged 10–18 years old	[48]
	rs4588 "A"	Lower VDBP and lower BMD levels;	1853 adults, aged ≥18	[47]
		Lower 25(OH)D levels;	198 girls, aged 10–18 years old	[48]
		Increasing bone fracture risk;	595 women	[49]

Table 4. *Cont.*

	Polymorphisms	Effects	Group Characteristics	References
Obesity	rs7041 "G"	G and GG associated with higher BMI in females; low 25OHD in males	406 adults	[56]
		Increasing BMI; no effect on 25(OH)D levels;		
	rs2282679 "A"	Lower 25[OH]D3		
	rs4588 "C"			
		High BMI in females		
	rs12721377 "A"	Low 25[OH]D3 levels in females		
PCOS	rs7041 "T"	PCOS + metabolic syndrome: significantly higher body mass index, blood pressure, and insulin resistance	443 healthy women aged 20–62 years, 359 of them were postmenopausal	[57]
	rs4588 "T" and rs7041 "C"	Increased risk of developing PCOS in vitamin D deficient women	100 women, 50 healthy and 50 with PCOS	[36]
Diabetes mellitus T2	rs7041 "G"	Elevated blood glucose levels; higher BMI	2271 adults	[63]
	rs7041 "G"	lower 25(OH)D3 and VDBP levels	553 patients, 916 controls	[65]
	rs4588 "C"			
	rs4588 "CC"	lower 1,25(OH)2D3 levels		
	rs4588 "A"	Higher serum GC globulin, albuminuria, and poor glycemic control (Patients more likely to develop diabetes)	200 participants. 120 with DMT2, 80 controls	[64]
Asthma	rs7041 "G"	Increasing VDBP levels; increasing asthma progression	110 patients with asthma, 110 healthy controls	[68]
		Correlated significantly with asthma	192 children and adolescents (96 with asthma and 96 healthy controls)	[69]
		Increasing the risk of respiratory syncytial virus bronchiolitis in infancy and subsequent asthma development	198 healthy children with families	[70]
	rs4588 CA and AA	Protective effect	192 children and adolescents (96 with asthma and 96 healthy controls)	[69]
Tuberculosis	rs3733359 "A"	Decreased susceptibility to PTB	490 PTB cases and 489 healthy controls)	[74]
	rs16847024 "T"			
	rs4588 CA	Associated with susceptibility to TB	125 PTB cases and 125 healthy controls	[75]
		Associated with 47.4% deficiency of 25(OH)D in patients with PTB		
	rs4588 CA	Protective effect		

Table 4. *Cont.*

	Polymorphisms	Effects	Group Characteristics	References
COPD	rs7041 "T"	Related to susceptibility of COPD	250 participants: 116 COPD patients with smoking history and 134 healthy smokers	[77]
		Associated with the risk of COPD	1712 subjects: 531 COPD cases and 1181 controls.	[79]
	Rs4588 "C"	Susceptibility to COPD, emphysema	361 COPD patients and 219 control	[80]
CAD	rs7041 "G"	Significant association with CAD	143 men with CAD and 145 healthy	[83]
	Rs4588 "A"	Higher prevalence of lesions in the left anterior descending artery and a longer lesion length	1080 patients	[84]
PD	rs7041, rs4588	No significant association with the severity of disease	137 patients	[98]
	rs7041	Rs7041 associated with PD risk ($p < 0.05$)	N = 382 PD patients and 242 healthy controls in a Turkish cohort	[94]
	rs2282679	higher levels of serum 25-hydroxyvitamin D in slower progression of disease		
	rs3755967			
	rs2298850			
MS	Rs7041 Rs4588	No significant association of polymorphism with the risk of MS	Meta-analysis of six studies	[62]

3. Research Limitations on Vitamin D

Vitamin D and its metabolites have some measurement limitations. Even if both serum and plasma can be used for vitamin D metabolite measurements, serum is preferred due to the fact that it is free from anticoagulants (heparin, EDTA, citrate). EDTA, heparin, and citrate may interfere with measurements [100]. The second problem in measuring is the stability of vitamin D metabolites because metabolites are only stable due to the binding to VDBP and proper storing: room temperature, 4 °C, or frozen. In case of separation from VDBP, storing at −70 °C is required [101]. Seasonal variation of vitamin D also needs to be considered as vitamin D biosynthesis is sun-dependent and depends on geographical locations and seasons (highest levels of vitamin D during the summer, lowest during the winter) [102]. This information may provide many significant differences in studies concerning vitamin D and its metabolites and its connection to polymorphisms occurring in the vitamin D pathway. Genetic variants of VDBP [103], DHCR7 [103], CYP2R1 [104], CYP24A1 [104], VDR [104], CYP3A4 [105], CYP2R1 [105], CYP27B [105], and LRP2 [105] were found to be associated with 25(OH)D levels. There are still many polymorphisms that have not yet been included in any studies.

Other important factors are age, sex, BMI, and lifestyle. Age can play a role, mostly in the group of elderly people >75 years old, due to reduced calcium absorption, intestinal resistance of calcium absorption to circulating $1,25(OH)_2D$, decreased ability of skin in vitamin D producing, deficiency of vitamin D substrate [106], and less sunshine exposure [107]. BMI is also considered to increase with age. Vitamin D deficiency is prevalent in a group of obese people which suggests the correlation of higher levels of adipose tissue and lower levels of vitamin D status [108]. Skin color also plays a role in vitamin D levels, as darker skin tones protect from UVB irradiation and, consequently, increase the risk of vitamin D deficiency. In addition, darker-skinned individuals have slower vitamin D synthesis [100].

The liver and kidneys are the two most important organs in the metabolism of vitamin D. Decreased kidney and liver functions may provide calcitriol deficiency and disruptions in overall vitamin D catabolism [100]. Diseases associated with these organs and their relationship with vitamin D and its metabolic pathway still require more research.

We suggest that all experiments that include vitamin D and its metabolism should contain more specific information concerning the studied group, age, sex, and geographical location. Articles without such information may lead to misleading conclusions.

4. Conclusions

Vitamin D and vitamin D binding protein have an undeniable impact on human health. Polymorphisms occurring in the VDBP gene can be a significant risk factor or prevalence factor in many diseases associated with obesity, diabetes, PCOS, MD, or PD. Many studies have shown VDBP and vitamin D level as a biomarker in many diseases, and if that is so, knowing the role of SNPs in proteins may contribute to finding new approaches for many syndromes and diseases associated with the vitamin D metabolic pathway.

In bone density, low levels of VDBP were associated with lowering BMD in the rs7041 "G" allele., while rs4588 "A" was linked to lower 25(OH)D levels and increased bone fracture risk. In obesity, the rs4588 "C" allele was linked to lower 25(OH)D levels and higher BMI scores among females. Rs4588 "T" and rs7041 "C" alleles increased risk in developing PCOS among women who had vitamin D deficiency. In COPD, the rs7041 "T" allele was associated with a higher risk of disease.

We believe that VDBP polymorphisms may affect the levels of vitamin D metabolites and thus contribute to the development of certain diseases. However, we are aware that most studies do not have a specific time of the year to measure vitamin D accurate metabolite levels, which can be misleading due to its seasonal nature. Research to date, although linking some diseases with vitamin D deficiency or VDBP polymorphisms, is not sufficient if the underlying mechanism is not elucidated. For this purpose, further research is required regarding the possible influence of SNP polymorphisms, their modifications in the structure of the binding protein, and their influence on the organism.

Author Contributions: Conceptualization, A.C. and D.R.; investigation, J.P., D.R. and K.A.; writing D.R., K.A. and J.P.; writing—review and editing, D.R. and A.C.; visualization, D.R.; supervision, A.C. All authors have read and agreed to the published version of the manuscript.

Funding: This research received no external funding.

Acknowledgments: Figure 1 was created thanks to Natalia Kordulewska and biorender.com.

Conflicts of Interest: The authors declare no conflict of interest.

References

1. Cashman, K.D.; Dowling, K.G.; Škrabáková, Z.; Gonzalez-Gross, M.; Valtueña, J.; De Henauw, S.; Moreno, L.; Damsgaard, C.T.; Michaelsen, K.F.; Mølgaard, C.; et al. Vitamin D deficiency in Europe: Pandemic? *Am. J. Clin. Nutr.* **2016**, *103*, 1033–1044. [CrossRef]
2. Amrein, K.; Scherkl, M.; Hoffmann, M.; Neuwersch-Sommeregger, S.; Köstenberger, M.; Berisha, A.T.; Martucci, G.; Pilz, S.; Malle, O. Vitamin D deficiency 2.0: An update on the current status worldwide. *Eur. J. Clin. Nutr.* **2020**, *74*, 1498–1513. [CrossRef]
3. Ames, B.N.; Grant, W.B.; Willett, W.C. Does the High Prevalence of Vitamin D Deficiency in African Americans Contribute to Health Disparities? *Nutrients* **2021**, *13*, 499. [CrossRef]
4. Liberman, U.A. Disorders in Vitamin D Action. In *Endotext*; Feingold, K.R., Anawalt, B., Boyce, A., Chrousos, G., de Herder, W.W., Dhatariya, K., Dungan, K., Hershman, J.M., Hofland, J., Kalra, S., et al., Eds.; MDText.com, Inc.: South Dartmouth, MA, USA, 2000.
5. Van Driel, M.; van Leeuwen, J.P. Vitamin D endocrinology of bone mineralization. *Mol. Cell. Endocrinol.* **2017**, *453*, 46–51. [CrossRef] [PubMed]
6. Mays, S.; Brickley, M.B. Vitamin D deficiency in bioarchaeology and beyond: The study of rickets and osteomalacia in the past. *Int. J. Paleopathol.* **2018**, *23*, 1–5. [CrossRef]
7. Ismailova, A.; White, J.H. Vitamin D, infections and immunity. *Rev. Endocr. Metab. Disord.* **2021**, 1–13. [CrossRef] [PubMed]
8. Dimitrov, V.; Barbier, C.; Ismailova, A.; Wang, Y.; Dmowski, K.; Salehi-Tabar, R.; Memari, B.; Groulx-Boivin, E.; White, J.H. Vitamin D-regulated Gene Expression Profiles: Species-specificity and Cell-specific Effects on Metabolism and Immunity. *Endocrinology* **2021**, *162*, bqaa218. [CrossRef] [PubMed]
9. Bikle, D.D. Vitamin D: An ancient hormone. *Exp. Dermatol.* **2010**, *20*, 7–13. [CrossRef]

10. Balachandar, R.; Pullakhandam, R.; Kulkarni, B.; Sachdev, H.S. Relative Efficacy of Vitamin D_2 and Vitamin D_3 in Improving Vitamin D Status: Systematic Review and Meta-Analysis. *Nutrients* **2021**, *13*, 3328. [CrossRef]
11. Kennel, K.A.; Drake, M.T.; Hurley, D.L. Vitamin D Deficiency in Adults: When to Test and How to Treat. *Mayo Clin. Proc.* **2010**, *85*, 752–758. [CrossRef]
12. Bikle, D.D. Vitamin D Metabolism, Mechanism of Action, and Clinical Applications. *Chem. Biol.* **2014**, *21*, 319–329. [CrossRef]
13. Holick, M.F.; MacLaughlin, J.A.; Doppelt, S.H. Regulation of Cutaneous Previtamin D 3 Photosynthesis in Man: Skin Pigment Is Not an Essential Regulator. *Science* **1981**, *211*, 590–593. [CrossRef]
14. Prosser, D.E.; Jones, G. Enzymes involved in the activation and inactivation of vitamin D. *Trends Biochem. Sci.* **2004**, *29*, 664–673. [CrossRef]
15. Slominski, A.T.; Kim, T.; Shehabi, H.Z.; Semak, I.; Tang, E.K.Y.; Nguyen, M.N.; Benson, H.A.E.; Korik, E.; Janjetovic, Z.; Chen, J.; et al. In vivo evidence for a novel pathway of vitamin D 3 metabolism initiated by P450scc and modified by CYP27B1. *FASEB J.* **2012**, *26*, 3901–3915. [CrossRef]
16. Slominski, A.T.; Kim, T.-K.; Li, W.; Postlethwaite, A.; Tieu, E.W.; Tang, E.K.Y.; Tuckey, R.C. Detection of novel CYP11A1-derived secosteroids in the human epidermis and serum and pig adrenal gland. *Sci. Rep.* **2015**, *5*, 14875. [CrossRef] [PubMed]
17. Slominski, R.M.; Tuckey, R.; Manna, P.R.; Jetten, A.M.; Postlethwaite, A.; Raman, C.; Slominski, A.T. Extra-adrenal glucocorticoid biosynthesis: Implications for autoimmune and inflammatory disorders. *Genes Immun.* **2020**, *21*, 150–168. [CrossRef]
18. Rozmus, D.; Ciesielska, A.; Płomiński, J.; Grzybowski, R.; Fiedorowicz, E.; Kordulewska, N.; Savelkoul, H.; Kostyra, E.; Cieślińska, A. Vitamin D Binding Protein (VDBP) and Its Gene Polymorphisms—The Risk of Malignant Tumors and Other Diseases. *Int. J. Mol. Sci.* **2020**, *21*, 7822. [CrossRef] [PubMed]
19. Jenkinson, C. The vitamin D metabolome: An update on analysis and function. *Cell Biochem. Funct.* **2019**, *37*, 408–423. [CrossRef] [PubMed]
20. Slominski, A.T.; Kim, T.-K.; Hobrath, J.V.; Oak, A.S.; Tang, E.K.; Tieu, E.W.; Li, W.; Tuckey, R.C.; Jetten, A.M. Endogenously produced nonclassical vitamin D hydroxy-metabolites act as "biased" agonists on VDR and inverse agonists on RORα and RORγ. *J. Steroid Biochem. Mol. Biol.* **2017**, *173*, 42–56. [CrossRef] [PubMed]
21. GC—SNPedia. Available online: https://www.snpedia.com/index.php/GC (accessed on 10 November 2021).
22. Kägi, L.; Bettoni, C.; Pastor-Arroyo, E.M.; Schnitzbauer, U.; Hernando, N.; Wagner, C.A. Regulation of vitamin D metabolizing enzymes in murine renal and extrarenal tissues by dietary phosphate, FGF23, and 1,25(OH)2D3. *PLoS ONE* **2018**, *13*, e0195427. [CrossRef]
23. Slominski, R.; Raman, C.; Elmets, C.; Jetten, A.; Slominski, A.; Tuckey, R. The significance of CYP11A1 expression in skin physiology and pathology. *Mol. Cell. Endocrinol.* **2021**, *530*, 111238. [CrossRef] [PubMed]
24. Wang, Y.; Zhu, J.; DeLuca, H.F. Where is the vitamin D receptor? *Arch. Biochem. Biophys.* **2012**, *523*, 123–133. [CrossRef] [PubMed]
25. Bozic, M.; Guzmán, C.; Benet, M.; Sánchez-Campos, S.; García-Monzón, C.; Gari, E.; Gatius, S.; Valdivielso, J.M.; Jover, R. Hepatocyte vitamin D receptor regulates lipid metabolism and mediates experimental diet-induced steatosis. *J. Hepatol.* **2016**, *65*, 748–757. [CrossRef] [PubMed]
26. Ye, C.F.; Pan, Y.M.; Zhou, H. Regulation of vitamin D receptor and Genistein on bone metabolism in mouse osteoblasts and the molecular mechanism of osteoporosis. *J. Boil. Regul. Homeost. agents* **2018**, *32*, 497–505.
27. Nakamichi, Y.; Udagawa, N.; Suda, T.; Takahashi, N. Mechanisms involved in bone resorption regulated by vitamin D. *J. Steroid Biochem. Mol. Biol.* **2018**, *177*, 70–76. [CrossRef] [PubMed]
28. Trochoutsou, A.I.; Kloukina, V.; Samitas, K.; Xanthou, G. Vitamin-D in the Immune System: Genomic and Non-Genomic Actions. *Mini-Reviews Med. Chem.* **2015**, *15*, 953–963. [CrossRef]
29. Ricca, C.; Aillon, A.; Bergandi, L.; Alotto, D.; Castagnoli, C.; Silvagno, F. Vitamin D Receptor Is Necessary for Mitochondrial Function and Cell Health. *Int. J. Mol. Sci.* **2018**, *19*, 1672. [CrossRef]
30. Slominski, A.T.; Kim, T.; Takeda, Y.; Janjetovic, Z.; Brożyna, A.A.; Skobowiat, C.; Wang, J.; Postlethwaite, A.; Li, W.; Tuckey, R.C.; et al. RORα and ROR γ are expressed in human skin and serve as receptors for endogenously produced noncalcemic 20-hydroxy- and 20,23-dihydroxyvitamin D. *FASEB J.* **2014**, *28*, 2775–2789. [CrossRef]
31. Slominski, A.T.; Kim, T.-K.; Janjetovic, Z.; Brozyna, A.; Żmijewski, M.A.; Xu, H.; Sutter, T.R.; Tuckey, R.C.; Jetten, A.M.; Crossman, D.K. Differential and Overlapping Effects of 20,23(OH)2D3 and 1,25(OH)2D3 on Gene Expression in Human Epidermal Keratinocytes: Identification of AhR as an Alternative Receptor for 20,23(OH)2D3. *Int. J. Mol. Sci.* **2018**, *19*, 3072. [CrossRef]
32. Slominski, A.T.; Kim, T.-K.; Qayyum, S.; Song, Y.; Janjetovic, Z.; Oak, A.S.W.; Slominski, R.M.; Raman, C.; Stefan, J.; Mier-Aguilar, C.A.; et al. Vitamin D and lumisterol derivatives can act on liver X receptors (LXRs). *Sci. Rep.* **2021**, *11*, 8002. [CrossRef]
33. Svasti, J.; Kurosky, A.; Bennett, A.; Bowman, B.H. Molecular basis for the three major forms of human serum vitamin D binding protein (group-specific component). *Biochemistry* **1979**, *18*, 1611–1617. Available online: https://pubs.acs.org/doi/pdf/10.1021/bi00575a036 (accessed on 10 November 2021). [CrossRef]
34. Gomme, P.T.; Bertolini, J. Therapeutic potential of vitamin D-binding protein. *Trends Biotechnol.* **2004**, *22*, 340–345. [CrossRef] [PubMed]
35. Chun, R.F. New perspectives on the vitamin D binding protein. *Cell Biochem. Funct.* **2012**, *30*, 445–456. [CrossRef] [PubMed]
36. Haldar, D.; Agrawal, N.; Patel, S.; Kambale, P.R.; Arora, K.; Sharma, A.; Tripathi, M.; Batra, A.; Kabi, B.C. Association of VDBP and CYP2R1 gene polymorphisms with vitamin D status in women with polycystic ovarian syndrome: A north Indian study. *Eur. J. Nutr.* **2016**, *57*, 703–711. [CrossRef] [PubMed]

37. Zhang, F.; Zhao, Z. The influence of neighboring-nucleotide composition on single nucleotide polymorphisms (SNPs) in the mouse genome and its comparison with human SNPs. *Genomics* **2004**, *84*, 785–795. [CrossRef] [PubMed]
38. Teng, S.; Madej, T.; Panchenko, A.; Alexov, E. Modeling Effects of Human Single Nucleotide Polymorphisms on Protein-Protein Interactions. *Biophys. J.* **2009**, *96*, 2178–2188. [CrossRef] [PubMed]
39. McGrath, J.J.; Saha, S.; Burne, T.; Eyles, D. A systematic review of the association between common single nucleotide polymorphisms and 25-hydroxyvitamin D concentrations. *J. Steroid Biochem. Mol. Biol.* **2010**, *121*, 471–477. [CrossRef] [PubMed]
40. Alharazy, S.; Naseer, M.I.; Alissa, E.; Robertson, M.D.; Lanham-New, S.; Alqahtani, M.H.; Chaudhary, A.G. Association of SNPs in GC and CYP2R1 with total and directly measured free 25-hydroxyvitamin D in multi-ethnic postmenopausal women in Saudi Arabia. *Saudi J. Biol. Sci.* **2021**, *28*, 4626–4632. [CrossRef]
41. Al-Daghri, N.M.; Mohammed, A.K.; Bukhari, I.; Rikli, M.; Abdi, S.; Ansari, M.G.A.; Sabico, S.; Hussain, S.D.; Alenad, A.; Al-Saleh, Y.; et al. Efficacy of vitamin D supplementation according to vitamin D-binding protein polymorphisms. *Nutrition* **2019**, *63-64*, 148–154. [CrossRef]
42. Bouillon, R.; Schuit, F.; Antonio, L.; Rastinejad, F. Vitamin D Binding Protein: A Historic Overview. *Front. Endocrinol.* **2020**, *10*, 910. [CrossRef]
43. Rs7041 RefSNP Report—DbSNP—NCBI. Available online: https://www.ncbi.nlm.nih.gov/snp/rs7041 (accessed on 7 January 2022).
44. Rs4588 RefSNP Report—DbSNP—NCBI. Available online: https://www.ncbi.nlm.nih.gov/snp/rs4588 (accessed on 7 January 2022).
45. Martínez-Aguilar, M.M.; Aparicio-Bautista, D.I.; Ramírez-Salazar, E.G.; Reyes-Grajeda, J.P.; De La Cruz-Montoya, A.H.; Antuna-Puente, B.; Hidalgo-Bravo, A.; Rivera-Paredez, B.; Ramírez-Palacios, P.; Quiterio, M.; et al. Serum Proteomic Analysis Reveals Vitamin D-Binding Protein (VDBP) as a Potential Biomarker for Low Bone Mineral Density in Mexican Postmenopausal Women. *Nutrients* **2019**, *11*, 2853. [CrossRef]
46. Dawson-Hughes, B.; Harris, S.S.; Krall, E.A.; Dallal, G.E. Effect of Calcium and Vitamin D Supplementation on Bone Density in Men and Women 65 Years of Age or Older. *N. Engl. J. Med.* **1997**, *337*, 670–676. [CrossRef]
47. Rivera-Paredez, B.; Hidalgo-Bravo, A.; León-Reyes, G.; Antuna-Puente, B.; Flores, Y.; Salmerón, J.; Velázquez-Cruz, R. Association of GC Variants with Bone Mineral Density and Serum VDBP Concentrations in Mexican Population. *Genes* **2021**, *12*, 1176. [CrossRef]
48. Santos, B.R.; Mascarenhas, L.P.G.; Boguszewski, M.C.; Spritzer, P.M. Variations in the Vitamin D-Binding Protein (DBP) Gene Are Related to Lower 25-Hydroxyvitamin D Levels in Healthy Girls: A Cross-Sectional Study. *Horm. Res. Paediatr.* **2013**, *79*, 162–168. [CrossRef] [PubMed]
49. Lauridsen, A.L.; Vestergaard, P.; Hermann, A.P.; Møller, H.J.; Mosekilde, L.; Nexo, E. Female Premenopausal Fracture Risk Is Associated With Gc Phenotype. *J. Bone Miner. Res.* **2004**, *19*, 875–881. [CrossRef] [PubMed]
50. Ezura, Y.; Nakajima, T.; Kajita, M.; Ishida, R.; Inoue, S.; Yoshida, H.; Suzuki, T.; Shiraki, M.; Hosoi, T.; Orimo, H.; et al. Association of Molecular Variants, Haplotypes, and Linkage Disequilibrium Within the Human Vitamin D-Binding Protein (DBP) Gene With Postmenopausal Bone Mineral Density. *J. Bone Miner. Res.* **2003**, *18*, 1642–1649. [CrossRef] [PubMed]
51. Blüher, M. Metabolically Healthy Obesity. *Endocr. Rev.* **2020**, *41*, 405–420. [CrossRef]
52. Walsh, J.; Bowles, S.; Evans, A.L. Vitamin D in obesity. *Curr. Opin. Endocrinol. Diabetes Obes.* **2017**, *24*, 389–394. [CrossRef]
53. Snijder, M.B.; Van Dam, R.M.; Visser, M.; Deeg, D.J.H.; Dekker, J.M.; Bouter, L.M.; Seidell, J.; Lips, P. Adiposity in Relation to Vitamin D Status and Parathyroid Hormone Levels: A Population-Based Study in Older Men and Women. *J. Clin. Endocrinol. Metab.* **2005**, *90*, 4119–4123. [CrossRef]
54. Hyppönen, E.; Boucher, B.J. Adiposity, vitamin D requirements, and clinical implications for obesity-related metabolic abnormalities. *Nutr. Rev.* **2018**, *76*, 678–692. [CrossRef]
55. Li, L.-H.; Yin, X.-Y.; Wu, X.-H.; Zhang, L.; Pan, S.-Y.; Zheng, Z.-J.; Wang, J.-G. Serum 25(OH)D and vitamin D status in relation to VDR, GC and CYP2R1 variants in Chinese. *Endocr. J.* **2014**, *61*, 133–141. [CrossRef]
56. Almesri, N.; Das, N.S.; Ali, M.E.; Gumaa, K.; Giha, H.A. Independent associations of polymorphisms in vitamin D binding protein (GC) and vitamin D receptor (VDR) genes with obesity and plasma 25OHD3 levels demonstrate sex dimorphism. *Appl. Physiol. Nutr. Metab.* **2016**, *41*, 345–353. [CrossRef]
57. Santos, B.; Costa, N.; Silva, T.; Casanova, G.; Oppermann, K.; Spritzer, P. SAT-234 DBP Gene Polymorphisms in Adult and Postmenopausal Women: Association with DBP and Vitamin D Serum Levels. *J. Endocr. Soc.* **2019**, *3*. [CrossRef]
58. Mezquita-Raya, P.; Muñoz-Torres, M.; Luna, J.D.D.; Luna, V.; Lopez-Rodriguez, F.; Torres-Vela, E.; Escobar-Jiménez, F. Relation Between Vitamin D Insufficiency, Bone Density, and Bone Metabolism in Healthy Postmenopausal Women. *J. Bone Miner. Res.* **2001**, *16*, 1408–1415. [CrossRef] [PubMed]
59. Pop, L.C.; Shapses, S.A.; Chang, B.; Sun, W.; Wang, X. Vitamin D-Binding Protein in Healthy Pre- and Postmenopausal Women: Relationship with Estradiol Concentrations. *Endocr. Pr.* **2015**, *21*, 936–942. [CrossRef]
60. Sinotte, M.; Diorio, C.; Bérubé, S.; Pollak, M.; Brisson, J. Genetic polymorphisms of the vitamin D binding protein and plasma concentrations of 25-hydroxyvitamin D in premenopausal women. *Am. J. Clin. Nutr.* **2008**, *89*, 634–640. [CrossRef]
61. Blair, M. Diabetes Mellitus Review. *Urol. Nurs.* **2016**, *36*, 27–36. [CrossRef] [PubMed]
62. Zhang, X.; Gao, B.; Xu, B. No association between the vitamin D-binding protein (DBP) gene polymorphisms (rs7041 and rs4588) and multiple sclerosis and type 1 diabetes mellitus: A meta-analysis. *PLoS ONE* **2020**, *15*, e0242256. [CrossRef]

63. Zhao, Y.; Wang, G.; Li, Y.; Liu, X.; Liu, L.; Yang, K.; Wang, C.; Wei, S. Evaluation of the Associations of GC and CYP2R1 Genes and Gene-Obesity Interactions with Type 2 Diabetes Risk in a Chinese Rural Population. *Ann. Nutr. Metab.* **2020**, *76*, 175–182. [CrossRef] [PubMed]
64. Fawzy, M.S.; Toraih, E.A.; Al Ageeli, E.; Mohamed, A.M.; Abu AlSel, B.T.; Kattan, S.W.; Alelwani, W. Group-specific component exon 11 haplotypes (D432E and T436K) and risk of albuminuria in type 2 diabetes mellitus patients. *Arch. Physiol. Biochem.* **2019**, 1–10. [CrossRef]
65. Penna-Martinez, M.; Badenhoop, K.; Klahold, E.; Bruns, F.; Seidl, C.; Wicker, S. Vitamin D in Type 2 Diabetes: Genetic Susceptibility and the Response to Supplementation. *Horm. Metab. Res.* **2020**, *52*, 492–499. [CrossRef]
66. Barnes, P.J. Immunology of asthma and chronic obstructive pulmonary disease. *Nat. Rev. Immunol.* **2008**, *8*, 183–192. [CrossRef]
67. Kikly, K.K.; Bochner, B.S.; Freeman, S.D.; Tan, K.; Gallagher, K.T.; D'Alessio, K.J.; Holmes, S.D.; Abrahamson, J.A.; Erickson-Miller, C.L.; Murdock, P.R.; et al. Identification of SAF-2, a novel siglec expressed on eosinophils, mast cells, and basophils. *J. Allergy Clin. Immunol.* **2000**, *105*, 1093–1100. [CrossRef]
68. Nasiri-Kalmarzi, R.; Abdi, M.; Hosseini, J.; Tavana, S.; MokariZadeh, A.; Rahbari, R. Association of vitamin D genetic pathway with asthma susceptibility in the Kurdish population. *J. Clin. Lab. Anal.* **2020**, *34*, e23039. [CrossRef]
69. Fawzy, M.S.; Elgazzaz, M.G.; Ibrahim, A.; Hussein, M.H.; Khashana, M.S.; Toraih, E.A. Association of Group-specific Component Exon 11 Polymorphisms with Bronchial Asthma in Children and Adolescents. *Scand. J. Immunol.* **2018**, *89*, e12740. [CrossRef]
70. Randolph, A.G.; Yip, W.-K.; Falkenstein-Hagander, K.; Weiss, S.T.; Janssen, R.; Keisling, S.; Bont, L. Vitamin D-binding protein haplotype is associated with hospitalization for RSV bronchiolitis. *Clin. Exp. Allergy* **2014**, *44*, 231–237. [CrossRef]
71. Paraskakis, E.; Iordanidou, M.; Tavridou, A.; Chatzimichael, A.; Manolopoulos, V.G. Vitamin D Receptor and Vitamin D Binding Protein Polymorphisms Are Associated with Asthma Control in Children. *Eur. Respir. J.* **2012**, *40*, 4569.
72. Jain, D.; Ghosh, S.; Teixeira, L.; Mukhopadhyay, S. Pathology of pulmonary tuberculosis and non-tuberculous mycobacterial lung disease: Facts, misconceptions, and practical tips for pathologists. *Semin. Diagn. Pathol.* **2017**, *34*, 518–529. [CrossRef] [PubMed]
73. Panda, S.; Tiwari, A.; Luthra, K.; Sharma, S.; Singh, A. Status of vitamin D and the associated host factors in pulmonary tuberculosis patients and their household contacts: A cross sectional study. *J. Steroid Biochem. Mol. Biol.* **2019**, *193*, 105419. [CrossRef] [PubMed]
74. Zhang, T.-P.; Chen, S.-S.; Zhang, G.-Y.; Shi, S.-J.; Wei, L.; Li, H.-M. Association of vitamin D pathway genes polymorphisms with pulmonary tuberculosis susceptibility in a Chinese population. *Genes Nutr.* **2021**, *16*, 6. [CrossRef] [PubMed]
75. Harishankar, M.; Sampath, P.; Athikesavan, V.; Chinnaiyan, P.; Velayutham, B.; Putcha, U.K.; Tripathy, S.P.; Ranganathan, U.D.; Selvaraj, P.; Bethunaickan, R. Association of rs7041 and rs4588 polymorphisms of vitamin D binding protein gene in pulmonary tuberculosis. *Meta Gene* **2020**, *26*, 100822. [CrossRef]
76. Fu, L.; Fei, J.; Tan, Z.-X.; Chen, Y.-H.; Hu, B.; Xiang, H.-X.; Zhao, H.; Xu, D.-X. Low Vitamin D Status Is Associated with Inflammation in Patients with Chronic Obstructive Pulmonary Disease. *J. Immunol.* **2021**, *206*, 515–523. [CrossRef]
77. Li, X.; Liu, X.; Xu, Y.; Xiong, W.; Zhao, J.; Ni, W.; Chen, S. The correlation of vitamin D level and vitamin D-binding protein gene polymorphism in chronic obstructive pulmonary disease. *Zhonghua Nei Ke Za Zhi* **2014**, *53*, 303–307. [PubMed]
78. Gao, J.; Törölä, T.; Li, C.-X.; Ohlmeier, S.; Toljamo, T.; Nieminen, P.; Hattori, N.; Pulkkinen, V.; Iwamoto, H.; Mazur, W. Sputum Vitamin D Binding Protein (VDBP) GC1S/1S Genotype Predicts Airway Obstruction: A Prospective Study in Smokers with COPD. *Int. J. Chronic Obstr. Pulm. Dis.* **2020**, *ume 15*, 1049–1059. [CrossRef]
79. Horita, N.; Miyazawa, N.; Tomaru, K.; Inoue, M.; Ishigatsubo, Y.; Kaneko, T. Vitamin D binding protein genotype variants and risk of chronic obstructive pulmonary disease: A meta-analysis. *Respirology* **2015**, *20*, 219–225. [CrossRef] [PubMed]
80. Ishii, T.; Motegi, T.; Kamio, K.; Gemma, A.; Kida, K. Association of group component genetic variations in COPD and COPD exacerbation in a Japanese population. *Respirology* **2014**, *19*, 590–595. [CrossRef]
81. Khanna, R.; Nandy, D.; Senapati, S. Systematic Review and Meta-Analysis to Establish the Association of Common Genetic Variations in Vitamin D Binding Protein With Chronic Obstructive Pulmonary Disease. *Front. Genet.* **2019**, *10*, 413. [CrossRef] [PubMed]
82. Peršić, V.; Raljević, D.; Markova-Car, E.; Cindrić, L.; Miškulin, R.; Žuvić, M.; Pavelić, S.K. Vitamin D-binding protein (rs4588) T/T genotype is associated with anteroseptal myocardial infarction in coronary artery disease patients. *Ann. Transl. Med.* **2019**, *7*, 374. [CrossRef]
83. Tarighi, S.; Najafi, M.; Hossein-Nezhad, A.; Ghaedi, H.; Meshkani, R.; Moradi, N.; Fadaei, R.; Kazerouni, F.; Shanaki, M. Association Between Two Common Polymorphisms of Vitamin D Binding Protein and the Risk of Coronary Artery Disease: A Case-Control Study. *J. Med Biochem.* **2017**, *36*, 349–357. [CrossRef]
84. Daffara, V.; Verdoia, M.; Rolla, R.; Nardin, M.; Marino, P.; Bellomo, G.; Carriero, A.; De Luca, G. Impact of polymorphism rs7041 and rs4588 of Vitamin D Binding Protein on the extent of coronary artery disease. *Nutr. Metab. Cardiovasc. Dis.* **2017**, *27*, 775–783. [CrossRef]
85. Cortese, M.; Munger, K.L.; Martínez-Lapiscina, E.H.; Barro, C.; Edan, G.; Freedman, M.S.; Hartung, H.-P.; Montalbán, X.; Foley, F.W.; Penner, I.K.; et al. Vitamin D, smoking, EBV, and long-term cognitive performance in MS. *Neurology* **2020**, *94*, e1950–e1960. [CrossRef]
86. Kampman, M.T.; Wilsgaard, T.; Mellgren, S.I. Outdoor activities and diet in childhood and adolescence relate to MS risk above the Arctic Circle. *J. Neurol.* **2007**, *254*, 471–477. [CrossRef] [PubMed]

87. Lucas, R.M.; Ponsonby, A.-L.; Dear, K.; Valery, P.C.; Pender, M.; Taylor, B.V.; Kilpatrick, T.; Dwyer, T.; Coulthard, A.; Chapman, C.; et al. Sun exposure and vitamin D are independent risk factors for CNS demyelination. *Neurology* **2011**, *76*, 540–548. [CrossRef]
88. Munger, K.L.; Levin, L.I.; Hollis, B.W.; Howard, N.S.; Ascherio, A. Serum 25-Hydroxyvitamin D Levels and Risk of Multiple Sclerosis. *JAMA: J. Am. Med Assoc.* **2006**, *296*, 2832–2838. [CrossRef] [PubMed]
89. Salzer, J.; Hallmans, G.; Nyström, M.; Stenlund, H.; Wadell, G.; Sundström, P. Vitamin D as a protective factor in multiple sclerosis. *Neurology* **2012**, *79*, 2140–2145. [CrossRef]
90. Powe, C.E.; Evans, M.K.; Wenger, J.; Zonderman, A.B.; Berg, A.H.; Nalls, M.; Tamez, H.; Zhang, D.; Bhan, I.; Karumanchi, S.A.; et al. Vitamin D–Binding Protein and Vitamin D Status of Black Americans and White Americans. *N. Engl. J. Med.* **2013**, *369*, 1991–2000. [CrossRef]
91. Yetley, E.A. Assessing the vitamin D status of the US population. *Am. J. Clin. Nutr.* **2008**, *88*, 558S–564S. [CrossRef]
92. Langer-Gould, A.; Lucas, R.; Xiang, A.H.; Chen, L.H.; Wu, J.; Gonzalez, E.; Haraszti, S.; Smith, J.B.; Quach, H.; Barcellos, L.F. MS Sunshine Study: Sun Exposure But Not Vitamin D Is Associated with Multiple Sclerosis Risk in Blacks and Hispanics. *Nutrients* **2018**, *10*, 268. [CrossRef]
93. Evatt, M.L.; Delong, M.R.; Khazai, N.; Rosen, A.; Triche, S.; Tangpricha, V. Prevalence of Vitamin D Insufficiency in Patients With Parkinson Disease and Alzheimer Disease. *Arch. Neurol.* **2008**, *65*, 1348–1352. [CrossRef] [PubMed]
94. Gezen-Ak, D.; Alaylıoğlu, M.; Genç, G.; Gündüz, A.; Candaş, E.; Bilgiç, B.; Atasoy, I.L.; Apaydın, H.; Kızıltan, G.; Gürvit, H.; et al. GC and VDR SNPs and Vitamin D Levels in Parkinson's Disease: The Relevance to Clinical Features. *NeuroMolecular Med.* **2017**, *19*, 24–40. [CrossRef]
95. Newmark, H.L.; Newmark, J. Vitamin D and Parkinson's disease—A hypothesis. *Mov. Disord.* **2007**, *22*, 461–468. [CrossRef]
96. Knekt, P.; Kilkkinen, A.; Rissanen, H.; Marniemi, J.; Sääksjärvi, K.; Heliövaara, M. Serum Vitamin D and the Risk of Parkinson Disease. *Arch. Neurol.* **2010**, *67*, 808–811. [CrossRef]
97. Lv, Z.; Qi, H.; Wang, L.; Fan, X.; Han, F.; Wang, H.; Bi, S. Vitamin D status and Parkinson's disease: A systematic review and meta-analysis. *Neurol. Sci.* **2014**, *35*, 1723–1730. [CrossRef]
98. Suzuki, M.; Yoshioka, M.; Hashimoto, M.; Murakami, M.; Bs, K.K.; Noya, M.; Ms, D.T.; Urashima, M. 25-hydroxyvitamin D, vitamin D receptor gene polymorphisms, and severity of Parkinson's disease. *Mov. Disord.* **2012**, *27*, 264–271. [CrossRef]
99. Zhang, J.; Sokal, I.; Peskind, E.R.; Quinn, J.F.; Jankovic, J.; Kenney, C.; Chung, K.A.; Millard, S.P.; Nutt, J.G.; Montine, T.J. CSF Multianalyte Profile Distinguishes Alzheimer and Parkinson Diseases. *Am. J. Clin. Pathol.* **2008**, *129*, 526–529. [CrossRef]
100. Makris, K.; Sempos, C.; Cavalier, E. The measurement of vitamin D metabolites: Part I—metabolism of vitamin D and the measurement of 25-hydroxyvitamin D. *Hormones* **2020**, *19*, 81–96. [CrossRef]
101. Hayden, Y.; Pillay, T.; Marx, G.; De Lange, W.; Kuyl, J.M. Pre-analytical stability of 25(OH)-vitamin D in primary collection tubes. *Clin. Chem. Lab. Med.* **2015**, *53*, 55–57. [CrossRef] [PubMed]
102. Webb, A.R.; Kline, L.; Holick, M.F. Influence of Season and Latitude on the Cutaneous Synthesis of Vitamin D_3: Exposure to Winter Sunlight in Boston and Edmonton Will Not Promote Vitamin D3 Synthesis in Human Skin. *J. Clin. Endocrinol. Metab.* **1988**, *67*, 373–378. [CrossRef]
103. Lu, L.; Sheng, H.; Li, H.; Gan, W.; Liu, C.; Zhu, J.; Loos, R.; Lin, X. Associations between common variants in GC and DHCR7/NADSYN1 and vitamin D concentration in Chinese Hans. *Qual. Life Res.* **2011**, *131*, 505–512. [CrossRef] [PubMed]
104. Barry, E.L.; Rees, J.R.; Peacock, J.L.; Mott, L.A.; Amos, C.I.; Bostick, R.M.; Figueiredo, J.C.; Ahnen, D.J.; Bresalier, R.; Burke, C.A.; et al. Genetic Variants inCYP2R1,CYP24A1, andVDRModify the Efficacy of Vitamin D3Supplementation for Increasing Serum 25-Hydroxyvitamin D Levels in a Randomized Controlled Trial. *J. Clin. Endocrinol. Metab.* **2014**, *99*, E2133–E2137. [CrossRef] [PubMed]
105. Shao, B.; Jiang, S.; Muyiduli, X.; Wang, S.; Mo, M.; Li, M.; Wang, Z.; Yu, Y. Vitamin D pathway gene polymorphisms influenced vitamin D level among pregnant women. *Clin. Nutr.* **2018**, *37*, 2230–2237. [CrossRef] [PubMed]
106. Gallagher, J.C. Vitamin D and Aging. *Endocrinol. Metab. Clin. North Am.* **2013**, *42*, 319–332. [CrossRef] [PubMed]
107. Gloth, F.M.; Gundberg, C.M.; Hollis, B.W.; Haddad, J.G.; Tobin, J.D. Vitamin D Deficiency in Homebound Elderly Persons. *JAMA: J. Am. Med Assoc.* **1995**, *274*, 1683–1686. [CrossRef]
108. Migliaccio, S.; Di Nisio, A.; Mele, C.; Scappaticcio, L.; Savastano, S.; Colao, A.; Obesity Programs of nutrition, Education, Research and Assessment (OPERA) Group. Obesity and hypovitaminosis D: Causality or casualty? *Int. J. Obes. Suppl.* **2019**, *9*, 20–31. [CrossRef] [PubMed]

International Journal of
Molecular Sciences

MDPI

Review

Genetic Variants of the NF-κB Pathway: Unraveling the Genetic Architecture of Psoriatic Disease

Rubén Queiro [1,2,*], Pablo Coto [3], Leire González-Lara [4] and Eliecer Coto [2,5]

1 Rheumatology & ISPA Translational Immunology Division, Hospital Universitario Central de Asturias, 33011 Oviedo, Spain
2 Department of Medicine, Oviedo University School of Medicine, 33011 Oviedo, Spain; eliecer.coto@sespa.princast.es
3 Dermatology Division, Hospital Vital Alvarez Buylla, 33611 Mieres, Spain; pablocotosegura@gmail.com
4 Dermatology Division, Hôpital Ambroise-Paré, 92100 Boulogne-Billancourt, France; leiregonzalezlara@gmail.com
5 Molecular Genetics Unit, Hospital Universitario Central Asturias, 33011 Oviedo, Spain
* Correspondence: rubenque7@yahoo.es

Abstract: Psoriasis is a multifactorial genetic disease for which the genetic factors explain about 70% of disease susceptibility. Up to 30–40% of psoriasis patients develop psoriatic arthritis (PsA). However, PsA can be considered as a "disease within a disease", since in most cases psoriasis is already present when joint complaints begin. This has made studies that attempt to unravel the genetic basis for both components of psoriatic disease enormously difficult. Psoriatic disease is also accompanied by a high burden of comorbid conditions, mainly of the cardiometabolic type. It is currently unclear whether these comorbidities and psoriatic disease have a shared genetic basis or not. The nuclear factor of kappa light chain enhancer of activated B cells (NF-κB) is a transcription factor that regulates a plethora of genes in response to infection, inflammation, and a wide variety of stimuli on several cell types. This mini-review is focused on recent findings that highlight the importance of this pathway both in the susceptibility and in the determinism of some features of psoriatic disease. We also briefly review the importance of genetic variants of this pathway as biomarkers of pharmacological response. All the above may help to better understand the etiopathogenesis of this complex entity.

Keywords: psoriasis; psoriatic arthritis; NF-κB; comorbidities; genetic architecture

Citation: Queiro, R.; Coto, P.; González-Lara, L.; Coto, E. Genetic Variants of the NF-κB Pathway: Unraveling the Genetic Architecture of Psoriatic Disease. *Int. J. Mol. Sci.* **2021**, *22*, 13004. https://doi.org/10.3390/ijms222313004

Academic Editor: Elixabet Lopez-Lopez

Received: 14 November 2021
Accepted: 29 November 2021
Published: 30 November 2021

1. Introduction

Psoriasis is a chronic immune-mediated inflammatory dermopathy that affects 2–3% of the general population. Its most common companion is psoriatic arthritis (PsA), a chronic arthritis included under the spondyloarthritis concept that affects one-third of psoriasis patients [1]. According to recent estimates, PsA can affect almost 0.6% of the adult population [2]. Both psoriasis and PsA are the main poles of what is now regarded as psoriatic disease, a systemic nature entity, where apart from the skin and musculoskeletal condition, there are a wide variety of comorbidities, the most relevant being those of cardiometabolic type [1,3]. In fact, we know of the tight connections between the inflammatory burden of psoriatic disease and higher cardiovascular risk [1,4].

Psoriatic disease itself, as well as its pleomorphic clinical manifestations, result from complex stochastic interactions between elements of genetic predisposition, immunopathological alterations, and environmental factors [1]. In the case of PsA, it is currently thought that the genetic substrate of the disease favors certain alterations of the gut microbiome of these patients, which gives rise to an IL17-type immune response, which have the potential capacity to migrate to distant sites such as entheses and joints, where the activation of certain lineages of innate cellularity (ILC-3 and γδ T cells), responsible for the beginning of the manifestations of the disease, would prevail (gut–joint axis theory) [1,5]. Adaptive immunity would act, in this case, as positive feedback for the primal responses of innate immunity [1,5].

Psoriasis is a multifactorial genetic disease for which the genetic factors explain about 70% of disease susceptibility [1,6,7]. However, PsA can be considered as a "disease within a disease", since in most cases psoriasis is already present when arthritis begins [1]. This has made studies that attempt to unravel the genetic basis for both components of psoriatic disease enormously difficult. Moreover, this information is essential as it would help predict which psoriasis patients are at increased risk of developing arthritis over time. Genetic factors are evident from the high prevalence of PsA among first-degree relatives of PsA probands, and a recurrence risk ratio of 30–35 [1,6,7]. Also, the psoriatic disease concordance rate for monozygotic twins is higher compared to that of dizygotic twins [1]. Like psoriasis, both dominant and recessive inheritance, as well as an excessive paternal transmission pattern, has been proposed for PsA but neither apply, thus PsA has also been considered a multifactorial polygenetic disease [1].

Genome wide association studies (GWAS) have revealed more than 80 risk loci associated with psoriatic disease that are basically incorporated into three fundamental networks: (i) genes involved in the maintenance of the cutaneous barrier function, (ii) genes that control innate immune responses mediated by NF-κB and interferon signaling, (iii) genes that control adaptive immune responses that involves CD8 lymphocytes and Th17 signaling [1,7]. However, most of them are related to psoriasis risk, while only a few seem specific to PsA. Genes with genome-wide significance for PsA include *HLA-B/C, HLA-B, IL12B, IL23R, TNP1, TRAF3IP3*, and *REL* [1,7]. Although recent GWAS studies have identified numerous risk loci of psoriatic disease, almost 50% of the heritability of the disease remains unknown, and most of the genes identified so far have a small weight in the genetic etiology of the disease. This "lost" heritability can be attributed, among others, to common variants that have a very small weight in risk, copy number variants, epigenetic or epistatic interactions, or lack of power of current detection tools [1,7].

As we have commented so far, the concept of psoriatic disease not only refers to the skin or musculoskeletal domains of the disease, but also to the wide variety of comorbidities that accompany it [1,3]. Therefore, it would be interesting to know if this complex network of manifestations presents common or different genetic restriction elements. This knowledge could be key in the future to lay the foundations of truly personalized medicine for these patients.

This mini-review will focus on the importance of the NF-κB pathway, both in the predisposition to psoriatic disease, as well as in the differentiation of the genetic architecture of its different components, including the determinism of some of its comorbidities.

2. The NF-κB Signaling Pathway

The nuclear factor of kappa light chain enhancer of activated B cells (NF-κB) is a transcription factor that regulates a plethora of genes in response to infection, inflammation, and a wide variety of stimuli on immune cells [8,9]. This transcription factor represents a family of structurally related proteins (p100 or NFκB2, p105 or NFκB1, p65 or RelA, RelB, or c-Rel), which exist as homo- or heterodimers [8,9]. Effects of NF-κB are mediated through three pathways. In the canonical pathway, (i) phosphorylation of inhibitory IκB proteins (IkBα) leads to release of NF-κB and its nuclear translocation to promote inflammation and cell survival. The p105 pathway (ii) is dependent on phosphorylation of p105 proteins, leading to nuclear translocation of p52 heterodimer complexes to promote inflammation. Different to the above two pathways, the alternate p100 pathway (iii) does not depend on the NF-κB essential modulator (NEMO)-IKKa-IKKb (NEMO-IKK) complex for phosphorylation, but rather on NF-κB inducing kinase (NIK), and IKKa heterodimers phosphorylate p100 and allows nuclear translocation of p52/RelB heterodimers [8,9].

The pathogenetic role of NF-κB has been demonstrated in many diseases including cancer, immune-mediated inflammatory diseases (IMIDs), as well as cardiovascular diseases [9]. Genome-wide association studies have revealed several psoriatic disease susceptibility genes associated with the NF-κB pathway [10]. In addition, the role of NF-κB in psoriasis is supported by studies that reported its differential expression in normal vs. affected skin, but also by studies that showed that the inhibition of this transcriptional factor by several compounds ameliorated skin inflammation [11,12].

Due to its relevance to this review, we will refer in detail to a specific pathway, which involves TNF-α and the NF-κB transcription complex (Figure 1) with special focus on genetic variations in *CARD14*, *TNFSRF1B*, *NFKB1*, *NFKBIA*, and *NFKBIZ*.

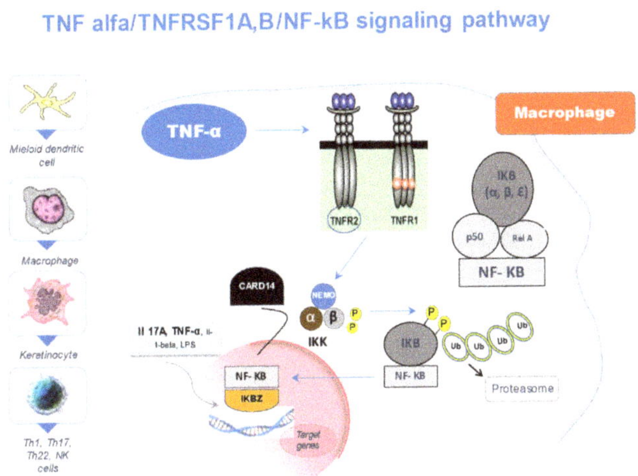

Figure 1. The TNF-α/TNFRSF1A,B/NF-KB signaling pathway. TNF-α is a pro-inflammatory cytokine synthesized by various cell lineages, mainly macrophages, dendritic cells, keratinocytes as well as various T cell types. It exerts its functions through binding to type 1 (TNFR1) and 2 (TNFR2) cell membrane receptors. TNFR2 (TNFRSF1B) shows higher affinity for TNF-α. NF-κB is a dimer made up of Rel family proteins (p50, p52, Rel A/p65, cRel, and Rel B), the p50/p65 heterodimer being the most common. In the absence of TNF-α stimulation, the activity of NF-κB is regulated at cytoplasmic level by an inhibitory complex of proteins called IKB or Kappa Beta inhibitors (IkB alpha, beta, and epsilon) that regulate its function by blocking its translocation to the nucleus. The binding of TNF-α to its receptors triggers the activation of the inhibitor kinases of NF-κB or "IKK complex" consisting of IKK-alpha, IKK-beta and IKK-gamma or NEMO. The characteristic event is the phosphorylation of IKB-alpha by the IKKa/b complex, inducing its ubiquitination and degradation by the proteasome; and thus, allowing the release of NF-κB. Once in the nucleus, NF-κB meets another inhibitory protein called IKBZ. Likewise, IKBZ can act as a transcription factor regulated by IL-17A, IL-1beta, and to a lesser extent, by TNF-α. In turn, IKBZ plays an important role in the development and expansion of Th17 lineages. Finally, CARD14 activates NF-κB by partially known mechanisms (see text for more details), although some authors suggest that CARD14 would activate the IKK complex, consequently regulating the activity of NF-κB, leading to an increase in its transcriptional activity. TNF-α: tumor necrosis factor-alpha. TNFR1 (TNFRSF1A): TNF-α type 1 receptor. TNFR2 (TNFRSF1B): TNF-α type 2 receptor. NF-κB: nuclear factor of kappa light chain enhancer of activated B cells. IKB: kappa-beta inhibitors. IKK: inhibitor kinases of NF-κB. IKBZ: z-inhibitor protein of NF-κB. IL: interleukin. Th: T-helper. CARD14: caspase recruitment domain family member 14.

Activation of the NF-κB pathway leads to increased transcription of numerous genes including proinflammatory cytokines, chemokines, and growth factors, all of which are involved in the onset and perpetuation of the inflammatory response in psoriatic disease [11,12]. Interestingly, the activation of NF-κB induces the production of TNF-α, which in turn can stimulate this pathway, thus resulting in positive proinflammatory feedback, and playing a key role in the chronicity of the skin/joint inflammatory response [11,12]. Moreover, an association between high levels of TNF-α and the activation of NF-κB has been found in the skin of patients with psoriasis; and therefore, a possible mechanism of action of anti-TNFα drugs would be, at least in part, the inhibition of the transcriptional activity of NF-κB [13], as we will see later.

3. Role in the Etiology of Psoriatic Disease

The caspase recruitment domain family member 14 (CARD14) is an intracellular scaffold protein that is prominently expressed in keratinocytes and mediates NF-κB activation through the formation of a CBM (CARD14-BCL10-MALT1) signaling complex [14]. It has been shown that gain-of-function mutations of *CARD14* enhance CBM complex formation in keratinocytes driving hyperactivation of the NF-κB pathway [15]. This leads to transcription of several chemokines (CCL20, CXCL1, and CXCL2), cytokines (IL-36 and IL-19), and antimicrobial peptides, followed by recruitment and activation of neutrophils, dendritic cells, and T cells [15]. Activated dendritic cells in turn release IL-23, which drives the downstream expression of IL-17 and IL-22, two central cytokines in the pathogenesis of the disease [15]. Variations in the *CARD14* gene have been associated with psoriatic disease. Jordan et al. identified 15 nucleotide variants in the *CARD14* gene significantly associated with the risk of psoriasis [16]. In a meta-analysis, they confirmed the association between rsll652075 (p. Arg820Trp) and psoriasis [17]. This association was later confirmed by two large meta-analysis [18,19]. We also found a statistically significant association between rsll652075 (CC genotype) and the risk of developing psoriasis (OR 1.59), regardless of the Cw6 genotype. However, we did not find genotypic differences related to psoriasis severity, age at disease onset (above or below 40 years), PsA, or family history of psoriasis [20].

The *TNFSRF1B* gene encodes the type 2 membrane receptor for TNF-α. The SNP TNFRSF1B rs1061622 (p. Met196Arg) has been implicated in the risk of several IMIDs [21,22]. This variant has hardly been studied in psoriatic disease. In our population, we did not find differences in allelic and genotypic frequencies between patients and controls for the SNP rs1061622. Neither were differences found in psoriasis severity, nor between the presence or absence of PsA. However, we did find an increased frequency of the G allele in the Cw6 + patients (OR 1.69), which suggests an increased risk of developing psoriasis conditioned by the HLA-Cw6 allele [23].

The *NFKB1* gene encodes the p50 protein, responsible for binding to the consensus sequence of the promotor region of many genes [8,9]. The most studied and best characterized polymorphism of this gene is a 4 nucleotide indel in the promoter region (rs28362491; -94 of the ATTG). The copies of the gene with the deletion would have less promoter activity and, therefore, translate into lower protein levels, which could explain the genetic associations described for this variant [24]. Other variants located at the chromosomal region where *NFKB1* is found have been investigated and related to the risk of psoriasis, specifically rs1020760 and rs1609798 [25]. A study in a Han Chinese population has recently established the association between the SNP rs1020760, psoriasis, and a positive family history of psoriasis [26]. We studied the SNP rs230526 A/G in the *NFKB1* gene, which is in complete linkage disequilibrium (LD) with rs28362491 (-94 of ATTG). Therefore, the genotypes of rs230526 would allow us to infer those of rs28362491. Despite this, we did not find any difference between allelic or genotypic frequencies in patients and controls. We also did not see a significant effect on its severity, PsA, or the age of onset of psoriasis; nor differences in relation to the presence of HLA-Cw6 [27].

NFKBIA gene encodes IκBα, a cytoplasmic inhibitor of the NF-κB transcriptional complex. That is, its binding to the complex keeps it inactive in the cytoplasm by blocking its ability to move to the cell nucleus. In response to certain stimuli, IκBα is phosphorylated, and this releases it from the complex, which can now migrate to the nucleus and act as a multigenic transcription factor [8,9]. Variants in this gene have been associated with the risk of myocardial infarction and several IMIDs [28,29]. In 2010, two GWAS established the association of two *NFKBIA* SNPs and the risk of psoriasis, while in 2015 a third GWAS confirmed the association of an additional SNP with psoriasis [30–32]. We have studied the rs7152376 SNP, which is in complete LD with rs12883343, a SNP related in the literature with both psoriasis and PsA. We did not find any association between rs7152376 and the risk of psoriasis, psoriasis severity, or the age at disease onset. However, we did find a statistically significant association between the rare G allele and the risk of PsA, with a dominant effect (AG + GG vs. AA). The rare NFKBIA rs7152376 C was significantly more frequent in PsA vs. controls (OR 2.03). The difference was even higher between PsA vs. "pure" psoriasis (OR 3.2) [33]. These findings are in line with the results of a meta-analysis that described the association of rs12883343, in complete LD with rs7152376, and the risk of PsA [34]. In this meta-analysis, the rare allele was also more frequent in PsA than in controls (meta-OR 1.22). Therefore, this variation in *NFKBIA* could be a true risk marker for PsA within the psoriatic complex.

The *NFKBIZ* gene encodes the IKBζ protein. The activity of NF-κB is regulated by nuclear inhibitors, such as IKBζ, which block its binding to DNA [8,9]. On the other hand, these peptides have a dual effect since by themselves they can act as transcription factors activating the expression of genes regulated by IL-17 or TNFα [35]. In 2015, Tsoi et al. described the genetic association between the *NFKBIZ* SNP rs7637230 and psoriasis [36]. This SNP is located outside the gene at the 3 'end, so it could be an indirect marker of some variant in *NFKBIZ* associated with its expression and/or function. The best candidate described so far would be an indel-type polymorphism in intron 10, rs3217713. Given its proximity to the 5 'end, it could influence the processing of total RNA into messenger RNA, an aspect that should be further investigated in vitro. We have investigated this indel SNP in psoriatic disease. When studying rs3217713, we did not find significant differences between patients and controls, and between the different clinical groups of patients, except for Cw6. Interestingly, the risk of developing psoriasis conferred by the combination of Cw6 + and rs3217713 ins/ins was statistically higher (OR 3.61) than the risk conferred by each genotype separately, which suggests some mechanism of genetic interaction or epistasis between both loci [37].

4. NF-κB Pathway Variants and Drug Response

Due to the various levels of regulation, the NF-κB signaling pathway can be potentially targeted at various levels including kinases, phosphatases, ubiquitination, nuclear translocation, DNA binding, protein acetyl transferases, and methyl transferases [38]. Currently available drugs for the treatment of psoriasis and PsA interact in some way with these targets within the NF-kB signaling pathway. For example, NSAIDs are commonly used in the treatment of PsA and have been shown to inhibit the activity of IKKβ, preventing the degradation of IκB and blocking NF-κB nuclear translocation. Glucocorticoids are highly effective in the topical treatment of active psoriatic lesions but also as systemic agents in PsA and have been shown to inhibit NF-κB through both indirect and direct mechanisms. Indirectly, glucocorticoids induce the transcription and synthesis of IκBα, enhancing the retention of NF-κB in the cytoplasm and effectively inhibiting its activation. However, under certain conditions, glucocorticoids can directly inhibit activated NF-κB via competition between p65 and the glucocorticoid receptor for limited nuclear supplies of coactivator proteins. Both receptors require these coactivators for maximal activity, and by sequestering the available stores, glucocorticoids interfere with p65-dependent gene activation. Stimulation of human synovial cells with TNFα results in NF-κB activation and subsequent cell proliferation, with NF-κB blockade able to activity inhibit this TNFα-induced proliferation. Anti-TNFα biologics effectively remove these upstream activator stimuli, and as such, the inhibition of NF-κB activation can be considered part of their mechanism of action [38]. Secukinumab, an anti–IL-17A antibody, mediates some of its antipsoriatic effects by rapidly inhibiting IκBζ and subsequently IκBζ signature genes, which highly suggests that IL-17A/IκBζ signaling is a key driver of the complex psoriatic phenotype. These data strongly indicate that an important and very early mechanism of action of anti–IL-17A therapy in patients with psoriasis is a reduction in IκBζ expression and a concomitant reduction in expression of IκBζ signature genes. It is possible that the IL-17A/IκBζ signaling axis also plays a role in the pathogenesis of psoriatic arthritis [39].

The search for genetic markers of good/bad response to drugs is a booming field, but still in its infancy. In this case, we will focus on the associations between NF-κB pathway variants and the response to anti-TNFα drugs. Anti-TNFα blocking agents could negatively modulate the signaling of the NF-κB pathway [13]. Genetic variations in the components of this pathway, particularly those that are related to the risk of psoriatic disease, could serve as pharmacogenetic markers. With respect to *CARD14* gene, we did find a significant association between rsll652075 (CC genotype) and the response to anti-TNFα drugs measured as PASI 75 at 24 weeks (OR 3.71) [40]. In our study cohorts, carriers of the G allele (GT + GG genotypes) of rs1061622 *TNFRSF1B* were associated with a worse response to anti-TNFα drugs (OR 2.34) [23]. However, the literature shows disparate results both in psoriatic disease and in other IMIDs, in this regard [41]. In our experience with anti-TNFα drugs, we did not observe any association between the genetic variants in *NFKB1*, *NFKBIA*, and drug response. However, in the case of *NFKBIZ*, the insertion allele was found more frequently in the group of non-responders, with a recessive effect (II vs. ID + DD $p = 0.02$), and OR for the negative drug response of 3.01 (95%CI: 1.15–7.88) [42]. However, the results of the literature in this regard are also disparate [43]. Anyway, it must be considered that this insertion allele is associated with the risk of psoriasis conditioned by the presence of HLA-Cw6, and the latter has also been linked to drug response in this disease [44].

5. NF-κB Genetic Variants and Cardiometabolic Comorbidity

Psoriatic disease is accompanied by extracutaneous and extra-articular manifestations, as well as by a wide range of comorbidities. Compared with controls, patients with psoriatic disease have a higher incidence rate of other autoimmune diseases, cardiovascular disease, obesity/overweight, depression, anxiety, smoking, cancer, diabetes, alcohol abuse, osteoporosis, uveitis, and fatty liver disease [45]. One of the common connectors between the disease and its comorbidities is inflammation. For example, chronic inflammation in both psoriatic disease and atherosclerosis promotes increased production of adipokines and pro-inflammatory cytokines (e.g., TNF) with consequent insulin resistance and endothelial dysfunction. Also, data from animal models and human studies highlight the proatherogenic role of IL-17 on vascular inflammation, acting in synergy with TNF and IL-6. IL-17 is involved in endothelial dysfunction, hypertension, plaque progression and destabilization, stroke, and myocardial infarction. IL-17 also links inflammation with insulin resistance and adipocytes dysfunction (two common aspects of the so-called psoriatic metainflammation). In mice models with systemic inflammation and insulin resistance, there is accumulation of IL-6/IL-17 co-expressing T cells in the adipose tissue; these cells enhance leptin production, which eventually acts in synergy with IL-6 and IL-17 to promote Th17 differentiation. This complex network drives local and systemic insulin resistance, and, in fact, IL-17 neutralization improves glucose uptake. Another example of these relationships is seen in osteoporosis (OP), another common comorbidity among psoriatic patients. Cytokines involved in PsA pathogenesis have an effect on bone cell activity with possible inhibitory or stimulatory stimuli on osteoclasts and osteoblasts. TNF, IL-6, and IL-1, for example, exert stimulatory activity toward osteoclasts and inhibitory activity toward osteoblasts. IL-12 and IL-23 are also critical to inflammation-induced bone resorption. Specifically, IL-23 upregulates the receptor activator of nuclear factor kappa- B (RANK) on preosteoclasts and induces Th17 cells to produce IL-17. IL-17, one of PsA signature cytokines, promotes bone resorption via RANK ligand upregulation. Indeed, Th17 cells have been found to be highly increased in blood and tissues of patients with OP [45].

In any case, for the purposes of this review, we will focus on the most important comorbidity in these patients, such as cardiovascular disease. Although psoriatic disease is frequently accompanied by cardiovascular comorbidity, the potential genetic connections between this comorbidity and psoriatic disease have received little attention. Furthermore, we do not know whether the genetic pathways to both cardiovascular comorbidity and psoriatic disease are common or differ. Although some studies suggest shared genes between psoriasis and cardiometabolic comorbidity, others suggest that the genetic architecture of psoriatic disease and cardiometabolic traits are largely distinct [46,47]. The NF-κB pathway might play a role in the pathogenesis of renal disease and type 2 diabetes mellitus (T2DM). In that sense, we studied 487 individuals, all Caucasian and aged 65–85 years. A total of 104 (21%) had impaired renal function and 30% were diabetics. The NFKB1 variants were significantly associated with T2DM: rs7667496 (OR 1.68) and rs28362491 (OR 1.67). There was a trend toward the association of these variants with impaired renal function. The two NFKB1 variants were in LD, and homozygous for the two non-risk alleles (rs7667496 CC + rs28362491 II), were significantly more common in the non-diabetics [48]. Therefore, *NFKB1* gene variations may be related to T2DM and an impaired renal function, both aspects clearly present in psoriatic patients [1,3].

Several reports show that the transcription factor NF-κB also activates genes involved in various cardiovascular diseases, in the pathogenesis of cardiac remodeling, and heart failure [49]. NF-κB is activated in the heart in many conditions: during acute ischemia and reperfusion, during unstable angina or in response to preconditioning [49]. An association between inflammation and cardiovascular risk has been suggested by the evidence that inflammatory cytokines, including IL-1β, IL-6, IL-18, and TNF-α, are increased in patients with heart failure, and increased inflammatory markers, such as C-reactive protein, predict a worse survival during acute coronary syndromes [49]. The association between psoriatic disease and cardiovascular risk has been clearly established over the last decade [1,3,4]. We determined whether common variants in three NF-κB genes were associated with early-onset coronary artery disease (CAD). Allele and genotype frequencies for the NFKB1 rs28362491 (-94 del ATTG) and NFKBIA rs8904 were not significantly different between patients and controls. For the NFKBIZ rs3217713, the deletion allele was significantly more frequent in early-onset CAD patients. Deletion-carriers were more frequent in CAD (OR 1.48). Therefore, NFKBIZ variant was an independent risk factor for developing early-onset CAD [50]. However, in a recent meta-analysis of 13 case-control studies with 17 individual cohorts containing 9378 cases and 10,738 controls, the mutant D allele in NFKB1 rs28362491 locus increased the risk of CAD [51].

Table 1 summarizes the main results of this review.

Table 1. Genetic variants within the TNF-α/NF-κB pathway and psoriatic disease features.

Genetic Variant	Disease Feature
CARD14 rsll652075 (p. Arg820Trp)	- Increased risk of developing psoriasis [16–20]. Good anti-TNFα drug response among CC genotype carriers [40].
TNFSRF1B rs1061622 (p. Met196Arg)	- Increased risk of developing psoriasis conditioned by HLA-Cw6 positivity [23]. Worse response to anti-TNFα drugs among G genotype carriers [23].
NFKB1 * rs28362491 (−94 of ATTG)	- No apparent association with disease risk and/or disease phenotypes [27]. Increased risk for coronary artery disease [51].
NFKB1 rs7667496/rs28362491	- Type 2 diabetes. Impaired renal function [48].

Table 1. *Cont.*

Genetic Variant	Disease Feature
NFKBIA rs7152376 C (complete LD with rs12883343)	- Significant association with PsA [33,34] - Risk of myocardial infarction and several IMIDs [28,29]. - Increased psoriasis risk [30–32]. Increased psoriasis risk [30–32].
NFKBIZ rs7637230 *NFKBIZ* rs3217713 ins/ins	- Increased risk of developing psoriasis [36]. - Increased risk of developing psoriasis conditioned by HLA-Cw6 positivity [37]. Insertion allele more frequent among anti-TNFα non-responders, with a recessive effect [42].
NFKBIZ rs3217713 del/del	- Early-onset coronary artery disease [50].

* Other variants located at the chromosomal region where *NFKB1* is found have been investigated and related to psoriasis risk, specifically rs1020760 (also associated with a positive family history) and rs1609798 [25,26]. LD: linkage disequilibrium. IMIDs: immune-mediated inflammatory diseases. CARD14: caspase recruitment domain family member 14 gene. TNFSRF1B: Type 2-TNFα receptor gene. NFKB1: NFκB subunit 1 gene. NFKBIA: NFκB alpha-inhibitor gene. NFKBIZ: NFκB zeta-inhibitor gene.

Obesity, the core element of metabolic syndrome (MetS), is common among patients with psoriatic disease [1,3]. The NF-κB pathway is involved in the low-grade chronic inflammation associated with obesity. The noncanonical IkB kinase ε and TBK1 are up-regulated by overnutrition and may therefore be suitable potential therapeutic targets for MetS [52]. Also, NF-κB modulates apical components of metabolic processes including metabolic hormones such as insulin and glucagon, and therefore this pathway is also involved in both insulin resistance and sensitivity, another key component of psoriatic disease-associated MetS [3,49]. The importance between metabolic derangements and NFκB-mediated inflammation associated with psoriatic disease is already moving to the practical realm. Thus, it has recently been set that obese, diabetic, and hypertensive patients with PsA tend to present a better response and persistence to secukinumab, an anti-IL17A therapy [53].

6. Conclusions

The NF-κB is an essential regulator of inflammatory and metabolic processes associated with psoriatic disease. Genetic variants of this pathway are associated not only with the risk of suffering the condition but may also be at the base of the comorbidities that frequently accompany it. These variants can also help improve treatment selection. In recent years, genetic variant analysis of the NF-κB pathway has contributed to a finer dissection of the genetic architecture of psoriatic disease. However, we still need further analysis of these genetic variants, especially with an emphasis on non-canonical regulators of this pathway, which currently have a direct implication in the appearance of important comorbidities of psoriatic disease such as diabetes and obesity. Thus, for example, an IKK/TBK1 inhibitor has been tested in obese individuals with type 2 diabetes showing encouraging results [52]. These findings may have future implications for a disease approach focused on the needs of each psoriatic patient.

Funding: This research received no external funding.

Conflicts of Interest: The authors declare no conflict of interest.

References

1. Queiro, R.; Lorenzo, A.; Pardo, E.; Cañete, J.D. Psoriatic arthritis: A current vision. In *Encyclopedia of Biomedical Gerontology*; Rattan, S.I.S., Ed.; Elsevier, Academic Press: Cambridge, MA, USA, 2020; Volume 3, pp. 105–119.
2. Pérez, A.R.; Queiro, R.; Seoane-Mato, D.; Graell, E.; Chamizo, E.; Chaparro, L.C.; Herrera, S.R.; Dolset, J.P.; Ostáriz, M.A.P.; Garrido, S.R.-A.; et al. Higher prevalence of Psoriatic arthritis in the adult population in Spain? A population-based cross-sectional study. *PLoS ONE* **2020**, *15*, e0234556.
3. Perez-Chada, L.M.; Merola, J.F. Comorbidities associated with Psoriatic arthritis: Review and update. *Clin. Immunol.* **2020**, *214*, 108397. [CrossRef]
4. Polachek, A.; Touma, Z.; Anderson, M.; Eder, L. Risk of cardiovascular morbidity in patients with psoriatic arthritis: A meta-analysis of observational studies. *Arthritis Care Res.* **2017**, *69*, 67–74. [CrossRef]
5. Gracey, E.; Vereecke, L.; McGovern, D.; Fröhling, M.; Schett, G.; Danese, S.; De Vos, M.; Bosch, F.V.D.; Elewaut, D. Revisiting the gut-joint axis: Links between gut inflammation and spondyloarthritis. *Nat. Rev. Rheumatol.* **2020**, *16*, 415–433. [CrossRef] [PubMed]
6. Lonnberg, A.S.; Skov, L.; Skytthe, A.; Kyvik, K.O.; Pedersen, O.B.; Thomsen, S.F. Heritability of Psoriasis in a large twin sample. *Br. J. Dermatol.* **2013**, *169*, 412–416. [CrossRef] [PubMed]
7. Rahmati, S.; Li, Q.; Rahman, P.; Chandran, V. Insights into the pathogenesis of Psoriatic arthritis from genetic studies. *Semin. Immunopathol.* **2021**, *43*, 221–234. [CrossRef] [PubMed]
8. Hayden, M.S.; Ghosh, S. Shared principles in NF-κB signaling. *Cell* **2008**, *132*, 344–362. [CrossRef]
9. Hoesel, B.; Schmid, J.A. The complexity of NF-κB signaling in inflammation and cancer. *Mol. Cancer* **2013**, *12*, 86. [CrossRef]
10. Ogawa, K.; Okada, Y. The current landscape of Psoriasis genetics in 2020. *J. Dermatol. Sci.* **2020**, *99*, 2–8. [CrossRef] [PubMed]
11. Goldminz, A.M.; Au, S.C.; Kim, N.; Gottlieb, A.B.; Lizzul, P.F. NF-κB: An essential transcription factor in Psoriasis. *J. Dermatol. Sci.* **2013**, *69*, 89–94. [CrossRef] [PubMed]
12. Rebholz, B.; Haase, I.; Eckelt, B.; Paxian, S.; Flaig, M.J.; Ghoreschi, K.; Nedospasov, S.A.; Mailhammer, R.; Debey-Pascher, S.; Schultze, J.; et al. Crosstalk between keratinocytes and adaptive immune cells in an IκBα protein-mediated inflammatory disease of the skin. *Immunity* **2007**, *27*, 296–307. [CrossRef] [PubMed]
13. Lizzul, P.F.; Aphale, A.; Malaviya, R.; Sun, Y.; Masud, S.; Dombrovskiy, V.; Gottlieb, A.B. Differential expression of phosphorylated NF-κB/RelA in normal and psoriatic epidermis and downregulation of NF-κB in response to treatment with etanercept. *J. Invest. Dermatol.* **2005**, *124*, 1275–1283. [CrossRef]
14. Howes, A.; O'Sullivan, P.A.; Breyer, F.; Ghose, A.; Cao, L.; Krappmann, D.; Bowcock, A.M.; Ley, S.C. Psoriasis mutations disrupt CARD14 autoinhibition promoting BCL10-MALT1-dependent NF-kappaB activation. *Biochem. J.* **2016**, *473*, 1759–1768. [CrossRef]
15. Mellett, M.; Meier-Schiesser, B.; Mohanan, D.; Schairer, R.; Cheng, P.; Satoh, T.; Kiefer, B.; Ospelt, C.; Nobbe, S.; Thome, M.; et al. CARD14 gain-of-function mutation alone is sufficient to drive IL-23/IL-17—mediated psoriasiform skin inflammation in vivo. *J. Invest. Dermatol.* **2018**, *138*, 2010–2023. [CrossRef] [PubMed]
16. Jordan, C.T.; Cao, L.; Roberson, E.; Pierson, K.C.; Yang, C.-F.; Joyce, C.E.; Ryan, C.; Duan, S.; Helms, C.A.; Liu, Y.; et al. PSORS2 is due to mutations in CARD14. *Am. J. Hum. Genet.* **2012**, *90*, 784–795. [CrossRef]
17. Jordan, C.T.; Cao, L.; Roberson, E.; Duan, S.; Helms, C.A.; Nair, R.P.; Duffin, K.C.; Stuart, P.E.; Goldgar, D.; Hayashi, G.; et al. Rare and common variants in CARD14, encoding an epidermal regulator of NF-kappaB, in Psoriasis. *Am. J. Hum. Genet.* **2012**, *90*, 796–808. [PubMed]
18. Tsoi, L.C.; Spain, S.L.; Knight, J.; Ellinghaus, E.; Stuart, P.E.; Capon, F.; Ding, J.; Li, Y.; Tejasvi, T.; Gudjonsson, J.E.; et al. Identification of 15 new Psoriasis susceptibility loci highlights the role of innate immunity. *Nat. Genet.* **2012**, *44*, 1341–1348. [PubMed]
19. Shi, G.; Li, S.J.; Wang, T.T.; Cheng, C.M.; Fan, Y.M.; Zhu, K.J. The common CARD14 gene missense polymorphism rs11652075 (c.C2458T/p.Arg820Trp) is associated with Psoriasis: A meta-analysis. *Genet. Mol. Res.* **2016**, *15*. [CrossRef]
20. González-Lara, L.; Coto-Segura, P.; Penedo, A.; Eiris, N.; Díaz, M.; Santos-Juanes, J.; Queiro, R.; Coto, E. SNP rs11652075 in the CARD14 gene as a risk factor for Psoriasis (PSORS2) in a Spanish cohort. *DNA Cell Biol.* **2013**, *32*, 601–604. [CrossRef] [PubMed]
21. Morita, C.; Horiuchi, T.; Tsukamoto, H.; Hatta, N.; Kikuchi, Y.; Arinobu, Y.; Otsuka, T.; Sawabe, T.; Harashima, S.I.; Nagasawa, K.; et al. Association of tumor necrosis factor receptor type II polymorphism 196R with systemic lupus erythematosus in the Japanese: Molecular and functional analysis. *Arthritis Rheum.* **2001**, *44*, 2819–2827. [CrossRef]
22. Tolusso, B.; Sacco, S.; Gremese, E.; La Torre, G.; Tomietto, P.; Ferraccioli, G.F. Relationship between the tumor necrosis factor receptor II (TNF-RII) gene polymorphism and sTNF-RII plasma levels in healthy controls and in rheumatoid arthritis. *Hum. Immunol.* **2004**, *65*, 1420–1426. [CrossRef] [PubMed]
23. González-Lara, L.; Batalla, A.; Coto, E.; Gómez, J.; Eiris, N.; Santos-Juanes, J.; Queiro, R.; Coto-Segura, P. The TNFRSF1B rs1061622 polymorphism (p.M196R) is associated with biological drug outcome in Psoriasis patients. *Arch. Dermatol. Res.* **2015**, *307*, 405–412.
24. Karban, A.S.; Okazaki, T.; Panhuysen, C.I.; Gallegos, T.; Potter, J.J.; Bailey-Wilson, J.E.; Silverberg, M.S.; Duerr, R.H.; Cho, J.H.; Gregersen, P.K.; et al. Functional annotation of a novel NFKB1 promoter polymorphism that increases risk for ulcerative colitis. *Hum. Mol. Genet.* **2004**, *13*, 35–45. [CrossRef] [PubMed]
25. Sheng, Y.; Jin, X.; Xu, J.; Gao, J.; Du, X.; Duan, D.; Li, B.; Zhao, J.; Zhan, W.; Tang, H.; et al. Sequencing-based approach identified three new susceptibility loci for Psoriasis. *Nat. Commun.* **2014**, *5*, 4331. [CrossRef]

26. Wang, W.; Zhu, Z.; Zhu, C.; Zheng, X.; Zuo, X.; Chen, G.; Zhou, F.; Liang, B.; Tang, H.; Wang, Z.; et al. A genetic variant rs1020760at NFKB1 is associated with clinical features of Psoriasis vulgaris in a Han Chinese population. *Ann. Hum. Genet.* **2016**, *80*, 197–202. [CrossRef]
27. González-Lara, L. Genetic Variants within the TNF-Alpha/TNFRSF1A, B/CARD14/NFKB Signaling Network and Psoriasis Risk. Ph.D. Thesis, Oviedo University, Oviedo, Spain, 2018.
28. Seidi, A.; Mirzaahmadi, S.; Mahmoodi, K.; Soleiman-Soltanpour, M. The association between NFKB1 -94ATTG ins/del and NFKB1A 826C/T genetic variations and coronary artery disease risk. *Mol. Biol. Res. Commun.* **2018**, *7*, 17–24.
29. Hong, M.; Ye, B.D.; Yang, S.-K.; Jung, S.; Lee, H.-S.; Kim, B.M.; Bin Lee, S.; Hong, J.; Baek, J.; Park, S.H.; et al. Immunochip meta-analysis of inflammatory Bowel disease identifies three novel loci and four novel associations in previously reported loci. *J. Crohns Colitis* **2018**, *12*, 730–741. [PubMed]
30. Stuart, P.E.; Nair, R.P.; Ellinghaus, E.; Ding, J.; Tejasvi, T.; Gudjonsson, S.A.; Li, Y.; Weidinger, S.; Eberlein, B.; Gieger, C.; et al. Genome-wide association analysis identifies three Psoriasis susceptibility loci. *Nat. Genet.* **2010**, *42*, 1000–1004. [CrossRef]
31. Capon, F.; Spencer, C.C.; Knight, J.; Weale, M.E.; Allen, M.H.; Barton, A.; Band, G.; Bellenguez, C.; Bergboer, J.G.; Palmer, C.N.; et al. A genome-wide association study identifies new Psoriasis susceptibility loci and an interaction between HLA-C and ERAP1. *Nat. Genet.* **2010**, *42*, 985–990.
32. Zuo, X.; Sun, L.; Yin, X.; Gao, J.; Sheng, Y.; Xu, J.; Zhang, J.; He, C.; Qiu, Y.; Wen, G.; et al. Whole-exome SNP array identifies 15 new susceptibility loci for Psoriasis. *Nat. Commun.* **2015**, *6*, 6793. [CrossRef]
33. Coto-Segura, P.; Coto, E.; González-Lara, L.; Alonso, B.; Gómez, J.; Cuesta-Llavona, E.; Queiro, R. Gene variant in the NF-κB pathway inhibitor NFKBIA distinguishes patients with Psoriatic arthritis within the spectrum of Psoriatic disease. *Biomed. Res. Int.* **2019**, *2019*, 1030256. [CrossRef]
34. Stuart, P.E.; Nair, R.P.; Tsoi, L.C.; Tejasvi, T.; Das, S.; Kang, H.M.; Ellinghaus, E.; Chandran, V.; Callis-Duffin, K.; Ike, R.; et al. Genome-wide association analysis of Psoriatic arthritis and Cutaneous Psoriasis reveals differences in their genetic architecture. *Am. J. Hum. Genet.* **2015**, *97*, 816–836. [CrossRef]
35. Johansen, C. IκBζ: A key protein in the pathogenesis of Psoriasis. *Cytokine* **2016**, *78*, 20–21. [CrossRef] [PubMed]
36. Tsoi, L.C.; Spain, S.L.; Ellinghaus, E.; Stuart, P.E.; Capon, F.; Knight, J.; Tejasvi, T.; Kang, H.M.; Allen, M.H.; Lambert, S.; et al. Enhanced meta-analysis and replication studies identify five new Psoriasis susceptibility loci. *Nat. Commun.* **2015**, *6*, 7001. [CrossRef] [PubMed]
37. Coto-Segura, P.; Gonzalez-Lara, L.; Gómez, J.; Eiris, N.; Batalla, A.; Gómez, C.; Requena, S.; Queiro, R.; Alonso, B.; Iglesias, S.; et al. NFKBIZ in Psoriasis: Assessing the association with gene polymorphisms and report of a new transcript variant. *Hum. Immunol.* **2017**, *78*, 435–440. [CrossRef]
38. Herrington, F.D.; Carmody, R.J.; Goodyear, C.S. Modulation of NF-κB signaling as a therapeutic target in autoimmunity. *J. Biomol. Screen.* **2016**, *21*, 223–242. [CrossRef] [PubMed]
39. Bertelsen, T.; Ljungberg, C.; Litman, T.; Huppertz, C.; Hennze, R.; Rønholt, K.; Iversen, L.; Johansen, C. IκBζ is a key player in the antipsoriatic effects of secukinumab. *J. Allergy Clin. Immunol.* **2020**, *145*, 379–390. [CrossRef]
40. Coto-Segura, P.; González-Fernández, D.; Batalla, A.; Gómez, J.; González-Lara, L.; Queiro, R.; Alonso, B.; Iglesias, S.; Coto, E. Common and rare CARD14 gene variants affect the antitumour necrosis factor response among patients with Psoriasis. *Br. J. Dermatol.* **2016**, *175*, 134–141. [CrossRef]
41. Bek, S.; Nielsen, J.V.; Bojesen, A.B.; Franke, A.; Bank, S.; Vogel, U.; Andersen, V. Systematic review: Genetic biomarkers associated with anti-TNF treatment response in inflammatory bowel diseases. *Aliment. Pharmacol. Ther.* **2016**, *44*, 554–567. [CrossRef] [PubMed]
42. Coto-Segura, P.; González-Lara, L.; Batalla, A.; Eiris, N.; Queiro, R.; Coto, E. NFKBIZ and CW6 in Adalimumab response among Psoriasis patients: Genetic association and alternative transcript analysis. *Mol. Diagn. Ther.* **2019**, *23*, 627–633. [CrossRef] [PubMed]
43. O'Rielly, D.D.; Rahman, P. Genetic, epigenetic and pharmacogenetic aspects of Psoriasis and Psoriatic arthritis. *Rheum. Dis. Clin. North. Am.* **2015**, *41*, 623–642. [CrossRef] [PubMed]
44. Batalla, A.; Coto, E.; González-Fernández, D.; González-Lara, L.; Gómez, J.; Santos-Juanes, J.; Queiro, R.; Coto-Segura, P. The Cw6 and late-cornified envelope genotype plays a significant role in anti-tumor necrosis factor response among psoriatic patients. *Pharmacogenet. Genom.* **2015**, *25*, 313–316.
45. Novelli, L.; Lubrano, E.; Venerito, V.; Perrotta, F.M.; Marando, F.; Curradi, G.; Iannone, F. Extra-articular manifestations and comorbidities in Psoriatic Disease: A journey into the immunologic crosstalk. *Front. Med. (Lausanne)* **2021**, *8*, 737079. [CrossRef] [PubMed]
46. Sundarrajan, S.; Arumugam, M. Comorbidities of Psoriasis—Exploring the links by network approach. *PLoS ONE* **2016**, *11*, e0149175. [CrossRef] [PubMed]
47. Koch, M.; Baurecht, H.; Ried, J.S.; Rodriguez, E.; Schlesinger, S.; Volks, N.; Gieger, C.; Rückert, I.-M.; Heinrich, L.; Willenborg, C.; et al. Psoriasis and cardiometabolic traits: Modest association but distinct genetic architectures. *J. Invest. Dermatol.* **2015**, *135*, 1283–1293. [CrossRef] [PubMed]
48. Coto, E.; Díaz-Corte, C.; Tranche, S.; Gómez, J.; Alonso, B.; Iglesias, S.; Reguero, J.R.; López-Larrea, C.; Coto-Segura, P. Gene variants in the NF-KB pathway (NFKB1, NFKBIA, NFKBIZ) and their association with type 2 diabetes and impaired renal function. *Hum. Immunol.* **2018**, *79*, 494–498. [CrossRef]

49. Fiordelisi, A.; Iaccarino, G.; Morisco, C.; Coscioni, E.; Sorriento, D. NFkappaB is a key player in the crosstalk between inflammation and cardiovascular diseases. *Int. J. Mol. Sci.* **2019**, *20*, 1599.
50. Coto, E.; Reguero, J.R.; Avanzas, P.; Pascual, I.; Martín, M.; Hevia, S.; Morís, C.; Díaz-Molina, B.; Lambert, J.L.; Alonso, B.; et al. Gene variants in the NF-KB pathway (NFKB1, NFKBIA, NFKBIZ) and risk for early-onset coronary artery disease. *Immunol. Lett.* **2019**, *208*, 39–43. [CrossRef]
51. Wang, Y.; Wu, B.; Zhang, M.; Miao, H.; Sun, J. Significant association between rs28362491 polymorphism in NF-kappaB1 gene and coronary artery disease: A meta-analysis. *BMC Cardiovasc. Disord.* **2020**, *20*, 278. [CrossRef]
52. Kracht, M.; Müller-Ladner, U.; Schmitz, M.L. Mutual regulation of metabolic processes and proinflammatory NF-κB signaling. *J. Allergy Clin. Immunol.* **2020**, *146*, 694–705.
53. Alonso, S.; Villa, I.; Fernández, S.; Martín, J.L.; Charca, L.; Pino, M.; Riancho, L.; Morante, I.; Santos, M.; Brandy, A.; et al. Multicenter study of secukinumab survival and safety in Spondyloarthritis and Psoriatic arthritis: SEcukinumab in cantabria and ASTURias study. *Front. Med. (Lausanne)* **2021**, *8*, 679009. [CrossRef] [PubMed]

International Journal of
Molecular Sciences

MDPI

Review

Towards Understanding the Genetic Nature of Vasovagal Syncope

Natalia Matveeva [1,2], Boris Titov [1,2], Elizabeth Bazyleva [1], Alexander Pevzner [1] and Olga Favorova [1,2,*]

1 National Medical Research Center for Cardiology, 121552 Moscow, Russia; natalijamat70@gmail.com (N.M.); titovborisvas@gmail.com (B.T.); dr.fedorova.ea@gmail.com (E.B.); avpevzner@mail.ru (A.P.)
2 Department of Molecular Biology and Medical Biotechnology, Pirogov Russian National Research Medical University, 117997 Moscow, Russia
* Correspondence: olga.favorova@gmail.com

Abstract: Syncope, defined as a transient loss of consciousness caused by transient global cerebral hypoperfusion, affects 30–40% of humans during their lifetime. Vasovagal syncope (VVS) is the most common cause of syncope, the etiology of which is still unclear. This review summarizes data on the genetics of VVS, describing the inheritance pattern of the disorder, candidate gene association studies and genome-wide studies. According to this evidence, VVS is a complex disorder, which can be caused by the interplay between genetic factors, whose contribution varies from monogenic Mendelian inheritance to polygenic inherited predisposition, and external factors affecting the monogenic (resulting in incomplete penetrance) and polygenic syncope types.

Keywords: syncope; vasovagal syncope; genetics; complex disorders; susceptibility; twin studies; family studies; candidate gene association studies; genome-wide studies

Citation: Matveeva, N.; Titov, B.; Bazyleva, E.; Pevzner, A.; Favorova, O. Towards Understanding the Genetic Nature of Vasovagal Syncope. *Int. J. Mol. Sci.* **2021**, *22*, 10316. https://doi.org/10.3390/ijms221910316

Academic Editor: Elixabet Lopez-Lopez

Received: 6 September 2021
Accepted: 23 September 2021
Published: 24 September 2021

Publisher's Note: MDPI stays neutral with regard to jurisdictional claims in published maps and institutional affiliations.

1. Introduction

Syncope, or fainting, is characterized by global cerebral hypoperfusion, transient loss of consciousness with disturbed postural tone, disturbance of the cardiovascular and respiratory systems, and spontaneous recovery back to the normal state [1].

A reduction in systemic blood pressure (BP) causing a decrease in cerebral blood flow plays a major role in the pathogenesis of syncope [2]. In turn, systemic BP depends on cardiac output and systemic peripheral vascular resistance [2,3]. The cardiac output and peripheral vascular resistance are affected by many factors such, as autonomic dysfunction, various cardiovascular (CV) diseases, decreased venous return, etc. The classification of syncope is based on these and other factors. According to the European Society of Cardiology Guidelines for the diagnosis and management of syncope [1], the following subtypes of syncope are distinguished: reflex syncope, cardiac syncope, and syncope due to orthostatic hypotension.

Approximately 40% of humans during their lifetime have transient loss of consciousness [4]; two-thirds of them are reflex syncope (also known as neurally mediated syncope) [5]. Vasovagal syncope (VVS) is the most common type of fainting in this group. A person develops VVS due to abnormal autonomic control of blood circulation, when sympathetic tone is decreased and the parasympathetic nervous system temporarily becomes overactive, thus resulting in arterial hypotension and cerebral hypoperfusion. VVS is often accompanied by bradycardia, and in some cases, by prolonged asystole [6]. VVS can be induced by various triggers such as orthostatic stress, exposure to emotional stress, medical manipulations, etc., and has such autonomic symptoms as hot flushes and nausea followed by dizziness and transient loss of consciousness. Although the outcome of VVS is favorable, this condition significantly worsens quality of life and can cause physical and mental injury. VVS more likely occurs at a young age, although cases when it first occurs in middle-aged and elderly patients have also been reported [7]. There are sex-specific differences in the prevalence and clinical manifestations of VVS: females are 50% more likely to have premonitory signs and symptoms of VVS than males [8].

A surrogate marker, the head-up tilt test (or prolonged passive head up tilt testing) has been used to diagnose VVS since the late 1980s. In this test, the patient is moved from the horizontal to vertical position using a special tilting table in order to simulate the neurally mediated reflex and induce syncope. The probability that VVS is induced during the head-up tilt test is 40–60%. The positive head-up tilt test response is associated with frequent VVS episodes and may be indicative of significant severity of orthostatic disorders [9]. Meanwhile, a negative head-up tilt test does not always mean that an individual cannot be diagnosed with neurally mediated syncope. This diagnosis should be suspected in individuals with typical clinical manifestations once all other reasons for transient loss of consciousness have been eliminated. The causes and the mechanism of VVS development have not been fully elucidated yet. This review describes and analyzes the publications focusing on the contribution of genetic components to the development of VVS. Having identified the genetic risk factors, one obtains efficient tools for performing a personalized prognosis of susceptibility to VVS. Furthermore, the molecular foundations of VVS pathogenesis can be revealed to develop new strategies for its prevention and management.

2. The Inheritance Pattern of VVS

The first step in studying hereditary diseases is often to perform a familial aggregation analysis. If familial aggregation is revealed, the patient's relatives have a higher risk of developing the disease compared to the average risk within a population, which is inversely proportional to the genetic distance from the proband. The earliest studies focused on the inheritance of VVS have identified familial aggregation of syncope. Kleinknecht and Lenz [10] reported that 66% of students that faint as a result of seeing blood or injuries on themselves or on others had at least one parent with VVS, while 41% of non-fainters had parents with VVS ($p < 0.01$). Among individuals with VVS unrelated to medical triggers, 94% of subjects had a family history of syncope [11].

In an early study, the family history of 30 children with VVS was examined [12]. For 24 of these children, their family histories were compared to those of their best friends (the control group). In these two groups, 90% of children with VVS and only 33% of children in the control group ($p < 0.01$) had at least one first degree relative also suffering from VVS. Furthermore, 37% of children with VVS had both a sibling and a parent with syncope vs. 4% of controls ($p < 0.05$). It was inferred that VVS is a complex disorder with both inherited genetic and external factors involved in its development.

In a study [13] involving 441 patients with VVS confirmed by the head-up tilt test, 19% of patients were found to have a familial history of this condition. It turned out that 37.2% of relatives of these patients also had VVS episodes. The researchers drew the conclusion that VVS undoubtedly has a genetic component and suggested that VVS either is an autosomal recessive disorder or its inheritance pattern is complex (does not strictly obey the Mendelian inheritance pattern).

Studying family history records identified families in which several generations suffered from VVS [14,15]. Thus, the study [15] presented a case report of a family consisting of three generations where all nine family members suffered from syncope. In this family, the inheritance pattern of VVS coincided with that expected in cases of autosomal dominant inheritance with incomplete penetrance.

Therefore, family aggregation studies of VVS inevitably demonstrated that probands with VVS were more likely to have a positive family history compared to the control group. A number of researchers have put forward a skeptical hypothesis that family aggregation might be observed randomly due to the high syncope rate among the population [16,17]; other scholars, however, do not support this point of view [18].

Twin studies allow one to examine the contribution of genetics to the development of a disease/phenotype by comparing the concordance in monozygotic and dizygotic twins. The first twin study focusing on syncope was conducted in 659 twin pairs from the Australian Twin Registry with respect to syncope related to blood/injury/injection fear [19]. A significant family aggregation of this type of syncope was observed; however, statistical methods could not discern contributions of the genetic component and overall external factors as the reason for fainting. In another study conducted on 51 monozygotic twin pairs from the Australian Twin Registry where at least one twin experienced syncope, a higher concordance among monozygotic twins compared to dizygotic ones was observed [20]. Significant effects were revealed both for syncope unrelated to external factors ($p = 0.018$) and for syncope related to typical vasovagal triggers (sight of blood, injuries, medical manipulations, standing too long in one place, or pain) ($p < 0.001$). The results of this study are also consistent with the assumption that VVS is a complex disorder, with both genetic and environmental factors contributing to its development. According to [20], the number of close relatives suffering from VVS complies with an autosomal dominant inheritance pattern in 7 out of 19 pairs of concordant monozygotic twins.

A total of 2,694,442 subjects from several Swedish nationwide registries were enrolled in a recent large-scale study [21], including 1,570,128 siblings, out of whom 24,020 subjects were twins; 264,244 subjects were half-siblings; and 1,044,546 subjects were cousins. The risk of syncope among relatives with VVS was maximal in twins and decreased systematically depending on the genetic distance from the proband.

The reported data are indicative of the role played by the genetic component in the development of VVS. However, as mentioned in the review by Sheldon and Sandhu [17], a significant number of cited studies have drawbacks such as the lack of clearly defined diagnostic criteria for the formation of initial study groups and/or control groups and the use of approaches based on self-assessment of patients whose recollections can be biased (especially with respect to the family history of syncope).

The effect of sex and age on the risk of VVS can be another reason for erroneous results. Indeed, a study enrolling 62 medical students and their families reported that females were more likely to experience syncope than males [22]. By the age of 30 years, the risk of syncope for females and males was 34% and 10%, respectively, if both their parents had no VVS episodes. This parameter increased to 48% and 28% for females and males, respectively, if one of the subjects' parents had VVS, and was as high as 78% (females) or 55% (males) if both parents had VVS. A positive maternal history of VVS increased the risk of syncope threefold in both male and female descendants, while a positive paternal history of VVS increased the risk only for male descendants. Similar data were obtained using the proportional hazards model [14]. It was shown that both male and female descendants with maternal history of VVS were more likely to experience syncope than those whose mother was a non-fainter, while the paternal history of syncopal episodes significantly increased the risk of VVS in sons but not in daughters. Interesting observations were made regarding the family history of three concordant monozygotic twin pairs with recurrent VVS [14,23]. The mother of a pair of twin girls also suffered from syncope [14], while both parents of the two pairs of twin boys were non-fainters [14,23]. The risk of VVS increases with age: the first syncopal episode occurs in most patients by age 30 years, followed by syncope recurrence over the next decades [7].

Hence, the individual susceptibility to VVS largely depends on one's sex and age, as well as on the sex of the parent who suffers from VVS. These data are consistent with the theory of a potential contribution of epigenetic factors to the development of VVS. Epigenetic regulation is not related to nucleotide sequence changes in the genome, but affects the transcriptional level of the genes important for phenotype development via covalent modification of DNA or histone proteins. Since sex hormones can modulate gene expression, the differences in prevalence of VVS in males and females can be attributed to the sex-hormone-dependent epigenetic mechanisms.

Overall, the observed complex inheritance pattern demonstrates that VVS has complex (multifactorial) origins: its development can be regulated by interplay between the genetic and epigenetic, as well as environmental factors. Importantly, environmental factors can modulate epigenetic processes [24], and these two mechanisms are not alternative in real practice. Meanwhile, taking into account the clinical heterogeneity of VVS and data variation in selected publications on family history, pedigrees, and results of twin studies, it is quite likely that the contribution of environmental factors and the inherited genetic component varies within a broad range. In turn, the genetic component can vary from monogenic Mendelian inheritance (with autosomal recessive or autosomal dominant patterns) to polygenic inherited predisposition.

What is the ratio between the different inheritance patterns of VVS? The findings reported in [25] provide a rough idea of this. Among the 44 multiplex families with VVS examined in [25], an autosomal dominant inheritance pattern was revealed in 6 families. The largest of these families included 30 subjects suffering from VVS in 3 generations; in the remaining 5 families, the number of subjects with VVS ranged from 4 to 14.

3. Candidate Gene Association Studies

The conventional approach to searching for genes involved in disease/phenotype development still remains relevant and is based on analyzing the association of individual candidate genes with the phenotype. An assumption that a gene is possibly associated with the phenotype is made based on the function of the gene product (the "phenotype-to-gene" approach). According to views on pathogenesis of VVS, one can expect that the genes whose products regulate the functioning of the autonomic nervous system and the cardiovascular system are mainly involved in syncope development. It is worth mentioning that genes are ascribed to either of these categories rather tentatively, since the autonomic nervous system regulates the function of all internal organs, including the heart and blood vessels, while variations in functioning of the cardiovascular system trigger a response from the nervous system. In most cases, researchers analyze polymorphic variants of genes, represented by single nucleotide polymorphisms (SNPs).

Data obtained from case-control studies on the association between carriership of polymorphic variants of a certain candidate gene and susceptibility to VVS are summarized in Table 1. The genotype frequencies in patients with VVS and non-fainting controls were compared in almost 50% of the studies. In the remaining studies, a surrogate marker was used instead of controls: patients susceptible and not susceptible to induction of VVS by the head-up tilt test were employed. Undoubtedly, this comparison can be reasonably performed, but it is not equivalent to direct comparison of the presence or absence of VVS. Indeed, as one can see in Table 1, the identified associations between a gene and the head-up tilt test response often mismatch the data obtained by comparing patients and controls.

Before we discuss the results obtained by assessing the involvement of individual genes in the development of VVS, we would like to point out the methodological flaws that some of the cited studies have. In a conventional association study, the groups being compared need to be characterized according to the ESC Guidelines for the diagnosis and management of syncope [4]. However, in actual practice, the study groups (and especially the control one) were often formed using questionnaire data only. The study group could include individuals with suspected VVS rather than those with a definitive diagnosis [26], or include patients with a history of both typical and atypical VVS [27]. In individual studies, patients and controls in the groups being compared were members of the same families, which does not meet the sample independence criterion [28]. Finally, the sample size was often insufficient: the number of probands with VVS ranged from 50 [29] to 347 individuals [30], while the number of controls ranged from 32 [31] to 150 individuals [32]. The publication that stands apart is study [33], where population-wide data from an earlier study [34] was used as the control group. Although a study containing large control groups is certainly appealing, this method generates doubts when taking into account the high population-wide frequency of VVS.

Table 1. Data on the association of polymorphic loci of candidate genes with vasovagal syncope (VVS).

Gene Symbol (Protein)	Rs Number (Amino Acid Substitution)	Ref.	Probands with VVS, N			Controls (without VVS, Ctrl), N	p Value (Compared Groups)	Note
			Total	Tilt +	Tilt −			
Genes of the adrenergic receptors								
ADRA1A (Alpha 1—adrenergic receptor)	rs1048101 (Arg347Cys)	[35]	89	89	0	40	**<0.001 (VVS vs. Ctrl)**	All controls were with a negative tilt test
		[28]	82	*	*	78	NS (VVS vs. Ctrl)	
		[36]	129	73	56	0	NS (Tilt+ vs. Tilt−)	
		[37]	134	88	46	0	NS (Tilt+ vs. Tilt−)	
	rs1383914 rs574584 rs573542	[37]	134	88	46	0	NS (Tilt+ vs. Tilt−)	
ADRB1 (Beta 1 adrenergic receptor)	rs1801253 (Arg389Gly)	[29]	50	33	17	0	**0.012 (Tilt+ vs. Tilt−)**	
		[38]	70	48	22	0	**0.001 (Tilt+ vs. Tilt−)**	
		[36]	129	73	56	0	NS (Tilt+ vs. Tilt−)	Also no association with the number of new syncope
		[39]	205	95	110	143	NS (VVS **vs.** C)	
		[28]	82	*	*	78	NS (VVS **vs.** C)	
		[37]	134	88	46	0	NS (Tilt+ vs. Tilt−)	
		[40]	123	123	0	0	**0.012 (Arg389Arg vs. Arg389Gly)**	Association with the number of syncope
	rs1801252 (Ser49Gly)	[36]	129	73	56	0	NS (Tilt+ vs. Tilt−)	Also no association with the number of new syncope episodes
		[28]	82	*	*	78	NS (VVS **vs.** Ctrl)	
		[37]	134	88	46	0	**0.02 (Tilt+ vs. Tilt−)**	
ADRB2 (Beta 2 adrenergic receptor)	rs1042713 (Gly16Arg)	[36]	129	73	56	0	NS (Tilt+ vs. Tilt−)	
		[37]	134	88	46	0	**0.04 (Tilt+ vs. Tilt−)**	
	rs1042714 (Gln27Glu)	[36]	129	73	56	0	NS (Tilt+ vs. Tilt−)	
		[37]	134	88	46	0	NS (Tilt+ vs. Tilt−)	

Table 1. *Cont.*

Gene Symbol (Protein)	Rs Number (Amino Acid Substitution)	Ref.	Probands with VVS, N			Controls (without VVS, Ctrl), N	*p* Value (Compared Groups)	Note
			Total	Tilt +	Tilt −			
ADRB3 (Beta 3 adrenergic receptor)	rs4994	[37]	134	88	46	0	NS (Tilt+ vs. Tilt−)	
Genes of the serotonin signaling								
SLC6A4 (Serotonin transporter)	rs25531	[41]	191	117	74	0	NS (Tilt+ vs. Tilt−)	
	rs4795541	[36]	129	73	56	0	NS (Tilt+ vs. Tilt−)	
	no rs; 43-bp insertion/deletion in the promoter region	[28]	82	*	*	78	NS (VVS **vs.** Ctrl)	There was a trend towards gender-specific differences in the effect of alleles ($p > 0.05$)
HTR1A (Serotonin 5-HT1A receptor)	rs6295	[28]	82	*	*	78	**0.005 (VVS vs. Ctrl, only males)**	
COMT (Catechol O-methyltransferase)	rs4680	[28]	82	*	*	78	**0.017 (VVS vs. Ctrl, with gender-specific allele effect)**	
Genes of the adenosine receptors								
ADORA2A (Adenosine A2A receptor)	rs5751876 (Tyr361Tyr)	[42]	105	52	53	121	**< 0.0001 (Tilt+ vs. Tilt−)** NS (VVS **vs.** Ctrl)	There was also an association with the frequency of syncope episodes
		[30]	347	207	140	83	NS (VVS **vs.** Ctrl) NS (Tilt+ vs. Tilt−)	An association was observed with heart rate in the early phase of tilt and during syncope
		[28]	82	*	*	78	NS (VVS **vs.** Ctrl)	
G protein signaling genes								
GNAS1 (G protein alpha subunit)	rs7121 C393T (silent mutation Ile131)	[43]	137	96	41	**0**	**<0.001 (Tilt+ vs. Tilt−)**	
		[44]	307	207	100	74	NS (VVS **vs.** Ctrl) NS (mild vs. malignant syncope)	
		[28]	82	*	*	78	NS (VVS **vs.** Ctrl)	

Table 1. *Cont.*

Gene Symbol (Protein)	Rs Number (Amino Acid Substitution)	Ref.	Probands with VVS, N			Controls (without VVS, Ctrl), N	*p* Value (Compared Groups)	Note
			Total	Tilt +	Tilt −			
GNB1 (G protein beta 1 subunit)	rs17363334 rs77354509 rs79516120	[45]	74	74	0	208	NS (VVS vs. Ctrl)	
GNB3 (G protein beta 3 subunit)	rs5443 C825T leads to alternative splicing with a loss of 41 amino acids	[27]	68	68	0	0	**<0.001 (typical vs. non-typical vasovagal history)**	56 patients with typical VVS history and 12—with non-typical VVS history
		[31]	213	*	*	32	NS (VVS **vs.** Ctrl)	All controls were with a negative tilt test
		[44]	307	207	100	74	NS (VVS **vs.** Ctrl) NS (mild vs. malignant syncope)	
		[46]	217	152	65	0	NS (Tilt+ vs. Tilt−)	
		[36]	129	73	56	0	NS (Tilt+ vs. Tilt−)	
		[26]	157	91	66	109	NS (VVS **vs.** Ctrl) NS (Tilt+ vs. Tilt−)	Patients with suspected VVS
GNG2 (G protein gamma 2 subunit)	no rs; c.87 + 34G > A	[45]	74	74	0	208	NS (VVS vs. Ctrl)	
RGS2 (G protein signaling regulator)	rs4606 C1114G	[46]	217	152	65	0	NS (Tilt+ vs. Tilt−)	
		[44]	307	207	100	74	NS (VVS **vs.** Ctrl) NS (syncope severity)	
		[47]	214	145	69	40	**0.04 (different number of syncope episodes)**	
		[32]	300	150	150	150	NS (VVS **vs.** Ctrl) NS (Tilt+ vs. Tilt−)	

Table 1. *Cont.*

Gene Symbol (Protein)	Rs Number (Amino Acid Substitution)	Ref.	Probands with VVS, N			Controls (without VVS, Ctrl), N	*p* Value (Compared Groups)	Note
			Total	Tilt +	Tilt −			
Genes of the potassium channels								
KCNJ5 (Inwardly rectifying potassium channel, subfamily J, member 5)	rs45516097	[45]	74	74	0	208	**0.001 (VVS vs. Ctrl)**	Minor allele T is less common in patients with VVS
	rs6590357 rs7118824 rs7118833 rs7102584 rs4937391	[45]	74	74	0	208	NS (VVS vs. Ctrl)	
KCNJ3 (Inwardly rectifying potassium channel, subfamily J, member 3)	rs16838016 rs3111033 rs17642086 rs80085601	[45]	74	74	0	208	NS (VVS vs. Ctrl)	
KCNH2 (Voltage-gated potassium channel subfamily H member 2)	rs1805123	[28]	82	*	*	78	NS (VVS vs. Ctrl)	
KCNE1 (Voltage-gated potassium channel subfamily E member 1)	rs1805127	[28]	82	*	*	78	NS (VVS vs. Ctrl)	
Genes encoding vasoactive proteins								
ACE (Angiotensin-converting enzyme)	rs4646994 insertion/deletion of the Alu repeat	[33]	165	165	0	>6000	NS (VVS vs. Ctrl)	Control data according to [34]
		[41]	191	117	74	0	NS (Tilt+ vs. Tilt−)	
AGT (Angiotensinogen)	rs699	[41]	191	117	74	0	NS (Tilt+ vs. Tilt−)	
AGTR1 (Angiotensin II receptor Type 1)	rs5186	[41]	191	117	74	0	NS (Tilt+ vs. Tilt−)	
eNOS (Endothelial NO synthase 3)	rs2070744 rs1799983	[28]	82	*	*	78	NS (VVS vs. Ctrl)	

Table 1. *Cont.*

Gene Symbol (Protein)	Rs Number (Amino Acid Substitution)	Ref.	Probands with VVS, N			Controls (without VVS, Ctrl), N	*p* Value (Compared Groups)	Note
			Total	Tilt +	Tilt −			
EDNRA (Endothelin type A receptor)	rs5333	[48]	107	58	49	208	NS (VVS vs. Ctrl) NS (Tilt+ vs. Tilt−)	
EDN1 (Endothelin 1)	rs1800997 insertion/ deletion 3A/4A	[48]	107	58	49	208	NS (VVS vs. Ctrl) **0.048 (Tilt+ vs. Tilt−)**	Allele 4A is associated with a positive tilt test
Other genes								
DBH (Dopamine beta hydroxylase)	rs1611115	[36]	129	73	56	0	NS (Tilt+ vs. Tilt−)	
CHRM2 (Muscarinic M2 receptor)	rs138806839 c.1114C > G	[45]	74	74	0	208	NS (VVS vs. Ctrl)	

Tilt+: VVS patients with a positive tilt test. Tilt−: VVS patients with a negative tilt test. Associations are considered significant at $p < 0.05$ and highlighted in bold. NS—non-significant. * Data not provided.

Since such neurotransmitters as norepinephrine, epinephrine, and serotonin are believed to play a major role in the development of syncope, genetic factors of susceptibility to VVS are primarily searched for among genes encoding receptors, carrier proteins, and enzymes partaking in the synthesis of these mediators.

The genes encoding adrenergic receptors are the most interesting candidate genes for VVS; the contribution of polymorphic variants of these genes was studied both to the development of VVS and head-up tilt test response. In the *ADRA1A* gene encoding the alpha-1A adrenergic receptor, the SNP rs1048101 (1039T > C) is responsible for a Cys347Arg substitution at the C-terminal end of alpha-1 adrenergic receptor; this substitution can affect receptor–protein interactions and, therefore, signal transduction from the receptor to the cell. Hernández-Pacheco et al. compared groups consisting of 89 tilt-positive patients and 40 healthy tilt-negative subjects without a history of VVS, heart or lung disease, and revealed a positive association between VVS and carriership of the C allele and the CC genotype (i.e., the presence of Arg347 in the protein) ($p < 0.001$) [35]. The authors suggest that Arg347 accelerates receptor internalization and therefore reduces the intracellular concentration of calcium ions, causing vasodilation and reducing venous return, thus increasing the risk of BP reduction and the development of VVS. In another study, a comparison of 82 patients with VVS with 79 healthy controls without structural or ECG cardiac abnormalities originating from one of nine families did not identify this association (this sample was used to analyze another 11 polymorphisms) [28]. The researchers put forward a hypothesis that rs1048101 (Arg347Cys) is associated with a positive head-up tilt test response rather than with VVS, although their samples were not characterized using the head-up tilt test. However, this assumption is not consistent with data obtained in studies where the polymorphic variants of the *ADRA1A* gene were compared in tilt-positive and tilt-negative patients with VVS. Thus, Sorrentino et al. [36] revealed no association between *ADRA1A* rs1048101 and head-up tilt response in 129 patients suffering from VVS without a history of cardiovascular disease or carotid sinus syndrome who were not taking medications affecting the cardiovascular system. A recent study observed differences in allele/genotype frequencies for neither rs1048101 nor other SNPs of the *ADRA1A* gene (rs1383914, rs574584, and rs573542) when comparing 88 tilt-positive and 46 tilt-negative patients with VVS [37].

Most of the studies listed in Table 1 focus on the contribution of variants of the beta 1 adrenergic receptor gene *ADRB1* to the development of VVS. This gene mediates positive chronotropic and inotropic effects in the cardiac muscle tissue and acts as a target for beta blockers. Researchers focused on two SNPs in the coding region of the *ADRB1* gene, rs1801253 (Arg389Gly) and rs1801252 (Ser49Gly), which affect the receptor function and its response to adrenoblockers. The Arg389Gly polymorphism resides in the C-terminal region of the beta 1 adrenergic receptor, affects its binding to G protein and, thereby, activation of adenylate cyclase. The Arg389 variant was shown to stimulate adenylate cyclase more efficiently and enhances signal transduction from adrenergic receptors compared to Gly389 [49]. The Ser49Gly polymorphism resides in the extracellular domain of the protein; the Gly49 variant enhances receptor desensitization after exposure to agonist and reduces receptor activity [50]. Comparison of patients suffering from VVS and the controls without previous history of fainting revealed no association between the SNP rs1801253 or SNP rs1801252 and VVS [28,39]. However, associations have been detected when comparing tilt-positive and tilt-negative patients. Thus, an association between the rs1801253(G) allele and positive head-up tilt test response ($p = 0.012$) [29] and, showing a good agreement with these data, an association between the CC genotype and negative head-up tilt test response ($p < 0.001$) were revealed [38]. However, a more recent study employing a larger sample replicated the data on association only between the SNP rs1801253 (but not SNP rs1801252) and the positive head-up tilt test response ($p = 0.02$) [37]. Sorrentino et al. [36] found no association between both of these polymorphic regions and head-up tilt test response. Special mention should be made of the study where the association of genotypic and phenotypic traits in 123 tilt-positive patients who had at least three syncopal episodes over one year was assessed [40]. Patients with the CC genotype (Arg389Arg) had a much higher number of syncopal episodes ($p = 0.012$); these patients also showed a better response to beta blocker therapy compared to those with the CG genotype (Arg389Gly) ($p < 0.001$).

For variants of the beta 2 adrenergic receptor gene, *ADRB2*, a comparison was performed only between tilt-positive and tilt-negative patients with VVS. No association of *ADRB2* rs1042713 or *ADRB2* rs1042714 with head-up tilt test response was found [36]. Márquez et al. [37] also found no association between rs1042714 and head-up tilt test response but observed this association for rs1042713 ($p = 0.04$). They also found an association between rs4994 in the *ADRB3* gene encoding beta 3 adrenergic receptor and the head-up tilt test response ($p = 0.03$).

Therefore, data on contribution of polymorphic variants of genes encoding alpha and beta adrenergic receptors to genetic susceptibility to VVS are rather controversial. When patients with VVS were compared to healthy controls, only an association between *ADRA1A* rs1048101 and VVS was observed [35], whereas no associations were detected when making other similar comparisons (including one more study for rs1048101 [28]). When analyzing the association between polymorphic variants of the adrenergic receptor genes and head-up tilt test response, an association was observed for 4 out of 13 comparisons.

The previously mentioned study [36] focused not only on SNPs of the adrenergic receptor genes, but also on the rs1611115 variant of the *DBH* gene encoding dopamine beta-hydroxylase, which catalyzes conversion of dopamine to norepinephrine. Comparison of tilt-positive and tilt-negative patients with VVS found no association between this SNP and head-up tilt test response (the data are provided in the "Other genes" section of Table 1).

There currently is no agreement regarding the role of serotonin in VVS development. An assumption was made that since serotonin is related to BP regulation and is found in the brain regions involved in the development of VVS, this neurotransmitter may contribute due to the antisympathetic effects mediated by the central nervous system [51]. Table 1 summarizes data on the association of head-up tilt test response and the *SLC6A4*, *HTR1A*, and *COMT* genes, whose products participate in the serotoninergic system. No differences in the carriership of polymorphic variants of rs25531 [41] and rs4795541 [36] in the serotonin transporter gene *SLC6A4* between tilt-positive and tilt-negative patients with VVS were found. Negative results were also received for the insertion/deletion polymorphism L/S

(43 bp Ins/Del) within the promoter of this gene for patients with VVS compared to healthy controls [28].

A study [28] conducted using familial data reported interesting findings on sex-related differences in the contribution of the genes of the serotoninergic system to the development of VVS. Association of the G allele of the SNP rs6295 (1019G > C) of the serotonin 1a receptor gene (*HTR1A*) with VVS was revealed only in males ($p = 0.005$). This polymorphic variant resides in the promoter region of the gene directly within the binding site of the transcription factor NUDR, which can affect the receptor expression level [52]. As reported in the study [28], the A allele of the SNP rs4680 (472G > A, Val158Met) of the *COMT* gene encoding catechol-O-methyl transferase related to a reduction in the enzyme level was associated with a lower risk of VVS in males and higher risk in females ($p = 0.017$). The *COMT* gene is known to have different effects on cerebral function and dysfunction in males and females, and is involved in sex-specific dimorphism of susceptibility to mental disorders [53]. It should be mentioned here that catechol-O-methyl transferase catalyzes the degradation of dopamine, as well as epinephrine, norepinephrine, and catechol estrogens; it is not directly related to the serotoninergic system. However, the reduced level of this enzyme increases the concentration of dopamine, which competes with serotonin for transport proteins capable of carrying monoamines, and disrupts serotoninergic regulation. Sheldon et al. [28] observed a similar trend of sex specificity for the L allele of the insertion/deletion polymorphism L/S in the *SLC6A4* gene; however, the differences did not reach statistical significance ($p = 0.059$). A conclusion has been drawn that males carrying any of the aforementioned three allelic variants of the genes involved in serotoninergic regulation are protected against VVS compared to females and to other males not carrying these allelic variants. Undoubtedly, these results need to be reproduced using independent samples.

A large group of candidate genes refers to the genes whose products are involved in functioning of the cardiovascular system (regulation of cardiac rhythm and vascular tone, as well as BP maintenance). Special focus is placed on the genes whose products participate in purinergic signal transduction (adenosine receptors), nitric oxide metabolism, functioning of potassium channels, and of the renin–angiotensin–aldosterone system.

The association of the SNP rs5751876 in the *ADORA2A* gene encoding adenosine A2A receptor with susceptibility to VVS, as well as to induction of VSS during the head-up tilt test, was investigated in three publications. The study [42], which included 105 patients with a history of at least two syncopal or presyncopal episodes over the preceding year and 121 healthy controls, identified an association of the CC genotype of rs5751876 with a positive head-up tilt test response in VVS patients ($p < 0.0001$), as well as with a high frequency of syncopal episodes ($p = 0.004$), however, no association with susceptibility to VVS has been found. The SNP rs5751876 is responsible for the synonymous substitution Tyr361Tyr. The researchers suggested that this polymorphism may affect gene expression level and protein folding. In a more recent study, analyzing a much greater number of patients (n = 347) with a history of at least one syncopal episode of unknown etiology and 85 controls without a history of syncopal episodes, no association of this SNP with head-up tilt test response or VVS was detected [30]. Sheldon et al. [28] also observed no association of the SNP rs5751876 with susceptibility to VVS.

Many receptors involved in signal transduction in patients suffering from VVS belong to the group of G protein-coupled receptors. The potential role of genes encoding G proteins in the formation of genetic predisposition to VVS has mainly been investigated by Lelonek et al. In the series of studies, they analyzed the SNP rs7121 of the *GNAS1* gene encoding protein G subunit alpha, the SNP rs5443 of the *GNB3* gene encoding protein G subunit beta 3, and the SNP rs4606 of the *RGS2* gene encoding the G protein signaling regulator 2; these SNPs have also been analyzed by other researchers. A comparison of VVS patients with healthy controls revealed that none of these genes are associated with VVS [31,46,47]. Other researchers obtained similar results for the *GNAS1* [28], *GNB3* [26], and *RGS2* genes [32]; the sample size of the group analyzed in [32] was rather large (300 children with VVS and 150 healthy children). Furthermore, no association of the SNPs rs17363334, rs77354509, or rs79516120 of the *GNB1* gene encoding G protein subunit beta 1 and c.87 + 34G > A of the *GNG2* gene encoding G protein subunit gamma 2 with VVS was detected [45]. No association between polymorphisms of the *GNAS1* and *GNB3* genes and syncope severity was revealed in patients with a history of more than three syncopal episodes over the preceding two years [44]. Meanwhile, polymorphisms in the *GNB3* gene were found to be associated with a history of VVS (typical VVS vs. atypical VVS) ($p < 0.001$) [27]; polymorphisms in the *RGS2* gene were found to be associated with the number of syncopal episodes ($p = 0.04$) [47]. Out of five studies focusing on the association between the G protein genes (*GNAS1*, *GNB3* or *RGS2*) and head-up tilt test response, such an association was found only in study [43] for the SNP rs7121 of the *GNAS1* gene ($p < 0.001$).

Hence, the best reproducibility of findings has been achieved by analyzing the association between G protein genes and VVS. These data allow one to infer that the analyzed polymorphic variants of the G protein genes have no significant contribution to VVS susceptibility.

Potassium channels are involved in cardiac rhythm and vascular tone regulation. An association between the SNP rs45516097 of *KCNJ5* (a gene encoding one of the potassium channel proteins) and VVS was identified when compared 74 tilt-positive patients without cardiac, endocrine, or neurological disorders and 208 healthy blood donors ($p = 0.001$) [45]. No association with VVS was observed for the SNPs rs6590357, rs7118824, rs7118833, rs7102584, and rs4937391 of the *KCNJ3* gene. Negative results were also obtained in this study for the SNP rs138806839 of the *CHRM2* gene encoding acetylcholine receptors M2 (muscarinic receptors) capable of modulating muscarinic potassium channels (the data are provided in the "Other genes" section of the Table 1). Furthermore, no association of *KCNH2* rs1805123 and *KCNE1* rs1805127 with VVS was found [28].

Endothelial nitric oxide synthase (eNOS) plays a crucial role in the regulation of vascular tone, blood flow, and BP. However, there were no association between the polymorphic regions of the *eNOS* gene and VVS [28].

Association of VVS with polymorphic variants of other candidate genes whose products are involved in regulation of vascular tone and BP (the SNPs rs4646994 of the *ACE* gene encoding angiotensin-converting enzyme [33,41], rs699 of the angiotensin gene *AGT*, rs5186 of the *AGTR1* gene encoding angiotensin II [41], rs10478694 of the endothelin 1 gene *EDN1* [48], and rs5333 of the endothelin receptor type A gene *EDNRA* [48]) have also been studied, but no association with VVS has been revealed. However, the 4A polymorphism of the *EDN1* gene (rs10478694) was shown to be associated with a positive head-up tilt test response ($p = 0.048$). Carriers of this allele have increased endothelin-1 production. Since endothelin-1 has a vasoconstrictive effect and is expected to prevent VVS, researchers suggest that its involvement in the development of VVS is not confined to affecting vascular tone and is possibly related to a different mechanism [48].

Therefore, studies focusing on the role of individual candidate genes in VVS patients are rather controversial, which can be explained by small sample size and ethnic differences in the groups being compared, as well as the fact that it is rather labor-intensive to diagnose this condition. Furthermore, patients with VVS subdivided into tilt-positive or tilt-negative were used as controls in a number of studies, or patients with typical and atypical syncope were compared. A plausible reason for non-reproducibility of the results is that the control groups were formed from conditionally healthy individuals without any prior diagnostics, which is especially important because of the high frequency of VVS (up to 25%) in the general population. Meanwhile, the association data for a number of genes (such as serotoninergic system genes, potassium channel genes, most vascular-tone-regulating genes, etc.) have been obtained in a single study and need to be reproduced. Taking into account the complex pathogenesis and heterogeneous clinical course of VVS, it is fair to assume that this disorder is polygenic. However, none of the studies analyzed the overall contribution of candidate genes to its development.

4. Genome-Wide Studies

Unlike the "candidate gene" approach when a hypothesis regarding potential involvement of a gene in phenotype is put forward according to its nature and function of the gene product (the "phenotype-to-gene" approach), the genome-wide searching for genes involved in disease/phenotype development employs panels of genetic markers with known chromosomal localization. These panels can be used to identify the genomic regions where these genes localize, and then search for the genes directly associated with the phenotype within these genomic loci is performed. This approach can be defined as the "genome-to-gene" approach. Polymorphic variants distributed over the entire genome more or less uniformly serve as genetic markers. Most typically, those are SNPs. Insertions/deletions (indels) and mini- or microsatellite repeats are also analyzed rather commonly. Microarrays and other modern technologies allow one to simultaneously identify as many as several dozen thousand to several million polymorphisms within a single sample.

Genome-wide studies were originally applied to analyze the linkage between a disease and chromosomal loci in families where several members had this disease. Later, genome-wide association studies (GWAS) were used as a more powerful tool for studying human genetic architecture.

The first genome-wide study of VVS was conducted by linkage analysis in a large family presumably with autosomal dominant inheritance (30 affected individuals with VVS over three generations) using microarray-based SNP genotyping data [25]. In this family, significant linkage to the locus on chromosome 15q26 was revealed for VVS; the logarithm of odds score was 3.28. Sequencing of the *SLCO3A1*, *ST8SIA2* and *NR2F2* candidate genes residing within the linkage interval detected no mutations. Linkage to the chromosome 15q26 region was excluded in two additional large families, suggesting that different genes may be relevant in the development of VVS in these families.

Demir et al. [54] conducted the first GWAS of VVS based on comparing the copy number variations (CNV) of repetitive genomic regions sized from one thousand to several million base pairs. Due to the greater genome coverage ensured by CNV compared to SNP markers, CNV is an important source of genetic variability and is regarded as an alternative type of DNA marker. The small study involved 16 subjects from four families with familial VVS; 13 of these subjects had a history of recurrent syncope and positive head-up tilt test response, while 3 subjects had no history of syncope. Twenty-six CNV variants whose presence in the genome differed significantly in patients with VVS compared to healthy subjects ($p < 0.05$) was revealed. In patients with VVS, the CNV segments were longer than those in healthy controls. However, the copy number presumably is not directly related to the pathophysiology of syncope, since no identical CNVs specific for individuals diagnosed with VVS have been identified.

The accuracy of assessments made in association studies directly depends on the number of DNA samples being studied. It is noteworthy that both patients and controls need to belong to the same ethnic group. A large-scale GWAS for syncope and collapse was conducted in 2020 [55]. This study used UK Biobank metadata [56], containing detailed information on the health status and genotyping results of more than 500,000 subjects. A total of 805,426 SNPs were examined as potential genetic markers. A British population was enrolled in the study: a group consisting of 9163 patients with syncope or collapse according to the International Classification of Disease (ICD-9, code 780.2, and ICD-10, code R55) and 399,798 healthy controls.

A new locus on chromosome 2q32.1 associated with VVS was identified at a significance level complying with modern requirements of GWAS results (the Bonferroni-adjusted p value for 1,000,000 comparisons needs to be $<5 \times 10^{-8}$). The p value for the lead SNP rs12465214 was 5.8×10^{-15}; for the other four SNPs in this locus (rs7593266, rs17582219, rs12621296, and rs2219224), the p value ranged from 1.0×10^{-9} to 3.3×10^{-8}. Carriership of the rs12465214*C risk allele was characterized by a hazard ratio of 1.13 (95% confidence interval 1.10–1.17) [55].

Results of this study were validated using the Danish Neonatal Screenings Biobank database [57]. Samples from 2352 subjects suffering from syncope and 51,929 controls were selected for GWAS analysis, which confirmed the association of the rs12465214*C allele with syncope ($p = 8.82 \times 10^{-6}$).

Functional annotation identified 26 genes in the locus associated with a risk of syncope [55]; the researchers believe that the *ZNF804A* gene encoding zinc finger protein 804A is the most plausible candidate gene associated with the risk of developing VVS. The SNP rs12465214 resides at a distance of approximately 250 kb from this gene and affects its expression as shown by the transcriptome analysis also conducted in [55]. Using quantitative polymerase chain reaction (q-PCR), the authors demonstrated that the *ZNF804A* gene was preferentially expressed in the brain, cerebral arteries, and endocrine tissue rather than in the cardiac muscle tissue [55].

The *ZNF804A* gene is known to be associated with schizophrenia and bipolar disorder [58,59]. Its protein product regulates processing of mRNA precursors and expression of genes, associated with synaptic transmission and development of the nervous system [60]. Based on these data, the search for associations of the VVS-associated SNPs rs12465214, rs7593266, rs17582219, rs12621296 and rs2219224 with schizophrenia was performed [55] using the GWAS data for schizophrenia [61]. However, none of these SNPs were associated with schizophrenia at a nominal significance level [55].

Enhanced expression in patients with VVS was shown for other genes encoding zinc finger proteins (ZNF28, ZNF845 and ZNF146) in a study involving a Chinese cohort of children with VVS- and age-matched controls [62].

Therefore, convincing findings on the association of the rs12465214*C risk allele residing on chromosome 2 in close proximity to the *ZNF804A* gene encoding zinc finger protein 804A have been obtained and validated using GWASs on two large ethnically homogeneous cohorts [55]. This protein contains an N-terminal C2H2-type zinc finger domain (Cys2-His2) [63]. Such domains are typical of classical transcription factors [64].

However, this study has serious limitations related to the fact that the definition of the disease was given using the ICD-10 code "syncope and collapse," without differentiating between syncope subtypes. Since VVS is the most common type of syncope [65], there is hope that the identified associations describe the genetic nature of this specific subtype. However, this assumption needs to be verified experimentally.

5. Conclusions

The question regarding the genetic nature of VVS was raised in the late 1980s, and is still far from being fully answered.

Data on the inheritance of VVS accumulated thus far are indicative of its familial aggregation; the genetic component can vary from a monogenic Mendelian inheritance pattern to polygenic inherited predisposition. The observed genetic heterogeneity can be correlated with both the diversity of VVS triggers and its clinical heterogeneity.

Both genetic and external factors contribute to the development of VVS; they can play a role in monogenic (resulting in incomplete penetrance) and polygenic syncope. The effect of external and epigenetic factors makes polygenic VVS a conventional complex (multifactorial) disorder. Interplay between these factors is responsible for the complex inheritance pattern. Since the effect of external factors can modulate epigenetic processes, these mechanisms cannot be fully delineated.

While the familial analysis data are rather convincing, the conventional approach to searching for the genes involved in syncope development by analyzing the association with individual candidate genes (selected according to VVS pathophysiology and function of the gene product) has not yielded unambiguous results yet. Although the range of candidate genes is rather broad, genes whose association with VVS that can be confirmed using independent samples or at least obtained by comparing large cohorts of patients and controls sufficient to perform robust statistical analysis are yet to be discovered.

The "genome-to-gene" approach has been used to analyze genetic architecture in a few studies and has yielded fundamentally important results. The genome-wide linkage analysis of VVS in several multiplex families with autosomal dominant inheritance patterns revealed the mutant locus 15q26 in only one family [25]. These data argue convincingly in favor of the genetic heterogeneity of VVS. In 2020, a GWAS involving large heterogeneous study groups (patients with syncope and collapse vs. healthy controls) revealed a single polymorphic variant (rs12465214) associated with the disease at a genome-wide significance level in the locus 2q32.1, which was later validated using an independent sample [55]. At first glance, this finding is not consistent with the genetic heterogeneity of VVS, to say nothing of the group of syncope and collapse patients that is more heterogeneous in terms of clinical characteristics. Further research is needed to resolve this controversy. Today, it is only fair to state with a high degree of probability that the phenomenon of observing a single syncope-associated locus (while its genetic heterogeneity has been established) is attributed to the key functions of the gene closely linked to the rs12465214.

Hadji-Turdeghal et al. [55] believe that the association of the SNP rs12465214 with syncope is most likely due to the fact that it resides in close proximity to the *ZNF804A* gene in locus 2q32.1. The C2H2 transcription factor containing zinc finger protein 804A (ZNF804A) encoded by the *ZNF804A* gene can play a pivotal role in the protein–protein interaction network by participating in various regulatory and signaling pathways involved in syncope development.

We believe, however, that it is not the only possible interpretation of the results. When discussing localization of the lead SNP associated with syncope, rs12465214, on chromosome 2, Hadji-Turdeghal et al. [55] classified it as an intergenic variant. However, the SNP rs12465214 actually resides in the shared intron of two non-coding RNA genes, *LOC105373776* and *LOC102724340*, in the positive and negative DNA strands, respectively. Although the functions of these genes are unknown, we presume that these very genes might be responsible for the role of the 2q32.1 region as a locus associated with the risk of syncope development. As reported in the NCBI database [66], these genes produce several transcript variants which are long non-coding RNAs (lncRNAs). Residing in the intron, the SNP rs12465214 can affect splicing of the transcripts of these genes, thus altering the composition of lncRNAs. According to modern views, lncRNAs play a crucial role in regulating of the expression of numerous genes in various biological and pathophysiological contexts (in particular, in neuronal dysfunction and immune response) [67]. It is quite reasonable to assume that lncRNAs are involved in the formation of the genetic architecture of syncope.

Identifying genetic and epigenetic factors involved in VVS has proven to be a promising field of research, not only to improve knowledge of risk factors for VVS, which could be of help for prevention, but also to improve the understanding of the pathophysiology of syncope subtypes, and to optimize and personalize the treatment of patients with syncope in the future.

Author Contributions: Conceptualization, N.M. and O.F.; writing—original preparation, B.T.; visualization, B.T. and E.B.; writing—review and editing, O.F. and A.P. All authors have read and agreed to the published version of the manuscript.

Funding: This study was conducted as a part of the Research & Development project No. 121031300196-1 of the National Medical Research Center for Cardiology of Russian Ministry of Health.

Institutional Review Board Statement: Not applicable.

Informed Consent Statement: Not applicable.

Data Availability Statement: Not applicable.

Acknowledgments: The authors would like to thank N.B. Baulina and A.V. Favorov for their valuable advice.

Conflicts of Interest: The authors declare no conflict of interest. The funders had no role in the design of the study; in the collection, analyses, or interpretation of data; in the writing of the manuscript, or in the decision to publish the results.

References

1. Brignole, M.; Moya, A.; de Lange, F.J.; Deharo, J.C.; Elliott, P.M.; Fanciulli, A.; Fedorowski, A.; Furlan, R.; Kenny, R.A.; Martín, A.; et al. ESC Scientific Document Group. 2018 ESC Guidelines for the diagnosis and management of syncope. *Eur. Heart. J.* **2018**, *39*, 1883–1948. [CrossRef] [PubMed]
2. Van Dijk, J.G.; van Rossum, I.A.; Thijs, R.D. Timing of Circulatory and Neurological Events in Syncope. *Front. Cardiovasc. Med.* **2020**, *7*, 36. [CrossRef] [PubMed]
3. Fu, Q.; Levine, B.D. Pathophysiology of neurally mediated syncope: Role of cardiac output and total peripheral resistance. *Auton. Neurosci.* **2014**, *184*, 24–26. [CrossRef] [PubMed]
4. Ganzeboom, K.S.; Mairuhu, G.; Reitsma, J.B.; Linzer, M.; Wieling, W.; van Dijk, N. Lifetime cumulative incidence of syncope in the general population: A study of 549 Dutch subjects aged 35–60 years. *J. Cardiovasc. Electrophysiol.* **2006**, *17*, 1172–1176. [CrossRef]
5. Brignole, M.; Ungar, A.; Casagranda, I.; Gulizia, M.; Lunati, M.; Ammirati, F.; Del Rosso, A.; Sasdelli, M.; Santini, M.; Maggi, R.; et al. Syncope Unit Project (SUP) investigators. Prospective multicentre systematic guideline-based management of patients referred to the Syncope Units of general hospitals. *Europace* **2010**, *12*, 109–118. [CrossRef]
6. Varosy, P.D.; Chen, L.Y.; Miller, A.L.; Noseworthy, P.A.; Slotwiner, D.J.; Thiruganasambandamoorthy, V. Pacing as a Treatment for Reflex-Mediated (Vasovagal, Situational, or Carotid Sinus Hypersensitivity) Syncope: A Systematic Review for the 2017 ACC/AHA/HRS Guideline for the Evaluation and Management of Patients with Syncope: A Report of the American College of Cardiology/American Heart Association Task Force on Clinical Practice Guidelines and the Heart Rhythm Society. *Circulation* **2017**, *136*, e123–e135. [CrossRef]
7. Sheldon, R.S.; Grubb, B.P.; Olshansky, B.; Shen, W.K.; Calkins, H.; Brignole, M.; Raj, S.R.; Krahn, A.D.; Morillo, C.A.; Stewart, J.M.; et al. 2015 heart rhythm society expert consensus statement on the diagnosis and treatment of postural tachycardia syndrome, inappropriate sinus tachycardia, and vasovagal syncope. *Heart Rhythm.* **2015**, *12*, e41–e63. [CrossRef]
8. Romme, J.J.; Van Dijk, N.; Boer, K.R.; Dekker, L.R.; Stam, J.; Reitsma, J.B.; Wieling, W. Influence of age and gender on the occurrence and presentation of reflex syncope. *Clin. Auton. Res.* **2008**, *18*, 127–133. [CrossRef]
9. Brignole, M.; Menozzi, C.; Del Rosso, A.; Costa, S.; Gaggioli, G.; Bottoni, N.; Bartoli, P.; Sutton, R. New classification of haemodynamics of vasovagal syncope: Beyond the VASIS classification. Analysis of the pre-syncopal phase of the tilt test without and with nitroglycerin challenge. Vasovagal Syncope International Study. *Europace* **2000**, *2*, 66–76. [CrossRef]
10. Kleinknecht, R.A.; Lenz, J. Blood/injury fear, fainting and avoidance of medically-related situations: A family correspondence study. *Behav. Res. Ther.* **1989**, *27*, 537–547. [CrossRef]
11. Kleinknecht, R.A.; Lenz, J.; Ford, G.; DeBerard, S. Types and correlates of blood/injury-related vasovagal syncope. *Behav. Res. Ther.* **1990**, *28*, 289–295. [CrossRef]
12. Camfield, P.R.; Camfield, C.S. Syncope in childhood: A case control clinical study of the familial tendency to faint. *Can. J. Neurol. Sci.* **1990**, *17*, 306–308. [CrossRef]
13. Newton, J.L.; Kenny, R.; Lawson, J.; Frearson, R.; Donaldson, P. Newcastle Cohort. Prevalence of family history in vasovagal syncope and haemodynamic response to head up tilt in first degree relatives: Preliminary data for the Newcastle cohort. *Clin. Auton. Res.* **2003**, *13*, 22–26. [CrossRef]
14. Márquez, M.F.; Urias, K.I.; Hermosillo, A.G.; Jardón, J.L.; Iturralde, P.; Colín, L.; Nava, S.; Cárdenas, M. Familial vasovagal syncope. *Europace* **2005**, *7*, 472–474. [CrossRef] [PubMed]
15. Newton, J.L.; Kerr, S.; Pairman, J.; McLaren, A.; Norton, M.; Kenny, R.A.; Morris, C.M. Familial neurocardiogenic (vasovagal) syncope. *Am. J. Med. Genet. A* **2005**, *133*, 176–179. [CrossRef] [PubMed]
16. Olde Nordkamp, L.R.; Wieling, W.; Zwinderman, A.H.; Wilde, A.A.; van Dijk, N. Genetic aspects of vasovagal syncope: A systematic review of current evidence. *Europace* **2009**, *11*, 414–420. [CrossRef] [PubMed]
17. Sheldon, R.S.; Sandhu, R.K. The Search for the Genes of Vasovagal Syncope. *Front. Cardiovasc. Med.* **2019**, *6*, 175. [CrossRef] [PubMed]
18. Klein, K.M.; Berkovic, S.F. Genetics of vasovagal syncope. *Auton. Neurosci.* **2014**, *184*, 60–65. [CrossRef]
19. Page, A.C.; Martin, N.G. Testing a genetic structure of blood-injury-injection fears. *Am. J. Med. Genet.* **1998**, *81*, 377–384. [CrossRef]
20. Klein, K.M.; Xu, S.S.; Lawrence, K.; Fischer, A.; Berkovic, S.F. Evidence for genetic factors in vasovagal syncope: A twin-family study. *Neurology* **2012**, *79*, 561–565. [CrossRef]
21. Fedorowski, A.; Pirouzifard, M.; Sundquist, J.; Sundquist, K.; Sutton, R.; Zöller, B. Risk Factors for Syncope Associated with Multigenerational Relatives with a History of Syncope. *JAMA Netw. Open.* **2021**, *4*, e212521. [CrossRef]
22. Serletis, A.; Rose, S.; Sheldon, A.G.; Sheldon, R.S. Vasovagal syncope in medical students and their first-degree relatives. *Eur. Heart J.* **2006**, *27*, 1965–1970. [CrossRef]
23. Arikan, E.; Yeşil, M.; Apali, Z.; Postaci, N.; Bayata, S. Vasovagal syncope in monozygotic twins. *Anadolu. Kardiyol. Derg.* **2009**, *9*, 61–62. [PubMed]
24. Szyf, M.; McGowan, P.; Meaney, M.J. The social environment and the epigenome. *Environ. Mol. Mutagen.* **2008**, *49*, 46–60. [CrossRef] [PubMed]

25. Klein, K.M.; Bromhead, C.J.; Smith, K.R.; O'Callaghan, C.J.; Corcoran, S.J.; Heron, S.E.; Iona, X.; Hodgson, B.L.; McMahon, J.M.; Lawrence, K.M.; et al. Autosomal dominant vasovagal syncope: Clinical features and linkage to chromosome 15q26. *Neurology* **2013**, *80*, 1485–1493. [CrossRef] [PubMed]
26. Evin, L.; Mitro, P.; Habalova, V.; Simurda, M.; Muller, E.; Murin, P. Beta 3 subunit of G-protein and its influence on autonomic nervous system in patients with vasovagal syncope. *Bratisl. Lek. Listy.* **2016**, *117*, 142–147. [CrossRef]
27. Lelonek, M.; Pietrucha, T.; Stanczyk, A.; Goch, J.H. Vasovagal syncope patients and the C825T *GNB3* polymorphism. *Anadolu. Kardiyol. Derg.* **2007**, *7*, 206–208. [PubMed]
28. Sheldon, R.; Rose, M.S.; Ritchie, D.; Martens, K.; Maxey, C.; Jagers, J.; Parboosingh, J.; Gerull, B. Genetic Association Study in Multigenerational Kindreds with Vasovagal Syncope: Evidence for Involvement of Sex-Specific Serotonin Signaling. *Circ. Arrhythm. Electrophysiol.* **2019**, *12*, e006884. [CrossRef]
29. Márquez, M.F.; Hernández-Pacheco, G.; Hermosillo, A.G.; Gómez, J.R.; Cárdenas, M.; Vargas-Alarcón, G. The Arg389Gly beta1-adrenergic receptor gene polymorphism and susceptibility to faint during head-up tilt test. *Europace* **2007**, *9*, 585–588. [CrossRef] [PubMed]
30. Mitro, P.; Habalova, V.; Evin, L.; Muller, E.; Simurda, M.; Slaba, E.; Murin, P.; Valocik, G. Gene Polymorphism of the Adenosine A2a Receptor in Patients with Vasovagal Syncope. *Pacing Clin. Electrophysiol.* **2016**, *39*, 330–337. [CrossRef]
31. Lelonek, M.; Pietrucha, T.; Matyjaszczyk, M.; Goch, J.H. C825T G-protein beta3 subunit gene polymorphism, tilt test results and point score in syncopal patients. *Clin. Auton. Res.* **2008**, *18*, 158–161. [CrossRef] [PubMed]
32. Chen, T.T.; Huang, Y.J.; Wang, J.Y.; Zhang, G.Q.; Xu, M.; Huang, M. Association between C1114G polymorphism in the regulator of G protein signaling 2 and vasovagal syncope in children. *Zhonghua Er Ke Za Zhi Chin. J. Pediatrics* **2018**, *56*, 856–860. [CrossRef]
33. Newton, J.L.; Donaldson, P.; Parry, S.; Kenny, R.A.; Smith, J.; Gibson, A.M.; Morris, C. Angiotensin converting enzyme insertion/deletion polymorphisms in vasovagal syncope. *Europace* **2005**, *7*, 396–399. [CrossRef] [PubMed]
34. Keavney, B.; McKenzie, C.; Parish, S.; Palmer, A.; Clark, S.; Youngman, L.; Delépine, M.; Lathrop, M.; Peto, R.; Collins, R. Large-scale test of hypothesised associations between the angiotensin-converting-enzyme insertion/deletion polymorphism and myocardial infarction in about 5000 cases and 6000 controls. International Studies of Infarct Survival (ISIS) Collaborators. *Lancet* **2000**, *355*, 434–442. [CrossRef]
35. Hernández-Pacheco, G.; González-Hermosillo, A.; Murata, C.; Yescas, P.; Espínola-Zavaleta, N.; Martínez, M.; Serrano, H. Arg347Cys polymorphism of α1a-adrenergic receptor in vasovagal syncope. Case-control study in a Mexican population. *Auton. Neurosci.* **2014**, *183*, 66–71. [CrossRef] [PubMed]
36. Sorrentino, S.; Forleo, C.; Iacoviello, M.; Guida, P.; D'Andria, V.; Favale, S. Lack of association between genetic polymorphisms affecting sympathetic activity and tilt-induced vasovagal syncope. *Auton. Neurosci.* **2010**, *155*, 98–103. [CrossRef]
37. Márquez, M.F.; Fragoso, J.M.; Pérez-Pérez, D.; Cázares-Campos, I.; Totomoch-Serra, A.; Gómez-Flores, J.R.; Vargas-Alarcón, G. Polymorphisms in β-adrenergic receptors are associated with increased risk to have a positive Head-Up Tilt Table test in patients with vasovagal syncope. *Rev. Investig. Clin.* **2019**, *71*, 124–132. [CrossRef]
38. Hernández-Pacheco, G.; Serrano, H.; Márquez, M.F.; Hermosillo, A.G.; Pérez-Vielma, N.; Sotomayor, A.; Ferreira-Vidal, A.D.; Salas-Silva, E.; Cárdenas, M. Estudio genético del síncope vasovagal asociado al polimorfismo Arg389Gly del receptor adrenérgico beta1 [Genetic study of the vasovagal syncope associated to the Arg389Gly polymorphism of the beta1 adrenergic receptor]. *Arch. Cardiol. Mex.* **2008**, *78*, 134–138. [PubMed]
39. Zelazowska, M.; Lelonek, M.; Fendler, W.; Pietrucha, T. Arg389Gly β1-adrenergic receptor polymorphism and susceptibility to syncope during tilt test. *Arch. Med. Sci.* **2014**, *10*, 240–245. [CrossRef]
40. Atici, A.; Rasih-Sonsoz, M.; Ali-Barman, H.; Durmaz, E.; Demirkiran, A.; Gulsen, K.; Elitok, A.; Onur, I.; Sahin, I.; Kaya Bilge, A. The role of Beta-1 receptor gene polymorphism in Beta-Blocker therapy for vasovagal syncope. *Rev. Investig. Clin.* **2020**, *72*, 300–307. [CrossRef]
41. Mudrakova, K.; Mitro, P.; Salagovic, J.; Habalova, V.; Kirsch, P.; Tkac, I. Gene polymorphisms of renin angiotensin system and serotonin transporter gene in patients with vasovagal syncope. *Bratisl. Lek. Listy.* **2009**, *110*, 73–76.
42. Saadjian, A.Y.; Gerolami, V.; Giorgi, R.; Mercier, L.; Berge-Lefranc, J.L.; Paganelli, F.; Ibrahim, Z.; By, Y.; Guéant, J.L.; Lévy, S.; et al. Head-up tilt induced syncope and adenosine A2A receptor gene polymorphism. *Eur. Heart J.* **2009**, *30*, 1510–1515. [CrossRef]
43. Lelonek, M.; Pietrucha, T.; Matyjaszczyk, M.; Goch, J.H. Mutation T/C,Ile 131 of the gene encoding the alfa subunit of the human Gs protein and predisposition to vasovagal syncope. *Circ. J.* **2008**, *72*, 558–562. [CrossRef]
44. Lelonek, M.; Pietrucha, T.; Matyjaszczyk, M.; Goch, J.H. Genetic insight into syncopal tilted population with severe clinical presentation. *Auton. Neurosci.* **2009**, *147*, 97–100. [CrossRef]
45. Holmegard, H.N.; Benn, M.; Mehlsen, J.; Haunsø, S. Genetic variation in the parasympathetic signaling pathway in patients with reflex syncope. *Genet. Mol. Res.* **2013**, *12*, 2601–2610. [CrossRef] [PubMed]
46. Lelonek, M.; Pietrucha, T.; Matyjaszczyk, M.; Goch, J.H. A novel approach to syncopal patients: Association analysis of polymorphisms in G-protein genes and tilt outcome. *Europace* **2009**, *11*, 89–93. [CrossRef]
47. Lelonek, M.; Pietrucha, T.; Matyjaszczyk, M.; Goch, J.H. Polymorphism C1114G of gene encoding the cardiac regulator of G-protein signaling 2 may be associated with number of episodes of neurally mediated syncope. *Arch. Med. Res.* **2009**, *40*, 191–195. [CrossRef]
48. Sorrentino, S.; Forleo, C.; Iacoviello, M.; Guida, P.; D'Andria, V.; Favale, S. Endothelin system polymorphisms in tilt test-induced vasovagal syncope. *Clin. Auton. Res.* **2009**, *19*, 123–129. [CrossRef]

49. Mason, D.A.; Moore, J.D.; Green, S.A.; Liggett, S.B. A gain-of-function polymorphism in a G-protein coupling domain of the human beta1-adrenergic receptor. *J. Biol. Chem.* **1999**, *274*, 12670–12674. [CrossRef] [PubMed]
50. Rathz, D.A.; Brown, K.M.; Kramer, L.A.; Liggett, S.B. Amino acid 49 polymorphisms of the human beta1-adrenergic receptor affect agonist-promoted trafficking. *J. Cardiovasc. Pharmacol.* **2002**, *39*, 155–160. [CrossRef] [PubMed]
51. Matzen, S.; Secher, N.H.; Knigge, U.; Pawelczyk, J.; Perko, G.; Iversen, H.; Bach, F.W.; Warberg, J. Effect of serotonin receptor blockade on endocrine and cardiovascular responses to head-up tilt in humans. *Acta Physiol. Scand.* **1993**, *149*, 163–176. [CrossRef] [PubMed]
52. Theodorakis, G.N.; Markianos, M.; Zarvalis, E.; Livanis, E.G.; Flevari, P.; Kremastinos, D.T. Provocation of neurocardiogenic syncope by clomipramine administration during the head-up tilt test in vasovagal syndrome. *J. Am. Coll. Cardiol.* **2000**, *36*, 174–178. [CrossRef]
53. Harrison, P.J.; Tunbridge, E.M. Catechol-O-methyltransferase (COMT): A gene contributing to sex differences in brain function, and to sexual dimorphism in the predisposition to psychiatric disorders. *Neuropsychopharmacology* **2008**, *33*, 3037–3045. [CrossRef] [PubMed]
54. Demir, E.; Hasdemir, C.; Ak, H.; Atay, S.; Aydin, H.H. Genome-Wide Association Study of Copy Number Variations in Patients with Familial Neurocardiogenic Syncope. *Biochem. Genet.* **2016**, *54*, 487–494. [CrossRef]
55. Hadji-Turdeghal, K.; Andreasen, L.; Hagen, C.M.; Ahlberg, G.; Ghouse, J.; Bækvad-Hansen, M.; Bybjerg-Grauholm, J.; Hougaard, D.M.; Hedley, P.; Haunsø, S.; et al. Genome-wide association study identifies locus at chromosome 2q32.1 associated with syncope and collapse. *Cardiovasc. Res.* **2020**, *116*, 138–148. [CrossRef] [PubMed]
56. Bycroft, C.; Freeman, C.; Petkova, D.; Band, G.; Elliott, L.T.; Sharp, K.; Motyer, A.; Vukcevic, D.; Delaneau, O.; O'Connell, J.; et al. The UK Biobank resource with deep phenotyping and genomic data. *Nature* **2018**, *562*, 203–209. [CrossRef]
57. Pedersen, C.B.; Bybjerg-Grauholm, J.; Pedersen, M.G.; Grove, J.; Agerbo, E.; Bækvad-Hansen, M.; Poulsen, J.B.; Hansen, C.S.; McGrath, J.J.; Als, T.D.; et al. The iPSYCH2012 case-cohort sample: New directions for unravelling genetic and environmental architectures of severe mental disorders. *Mol. Psychiatry* **2018**, *23*, 6–14. [CrossRef]
58. Chang, H.; Xiao, X.; Li, M. The schizophrenia risk gene ZNF804A: Clinical associations, biological mechanisms and neuronal functions. *Mol. Psychiatry* **2017**, *22*, 944–953. [CrossRef]
59. Williams, H.J.; Norton, N.; Dwyer, S.; Moskvina, V.; Nikolov, I.; Carroll, L.; Georgieva, L.; Williams, N.M.; Morris, D.W.; Quinn, E.M.; et al. Fine mapping of ZNF804A and genome-wide significant evidence for its involvement in schizophrenia and bipolar disorder. *Mol. Psychiatry* **2011**, *16*, 429–441. [CrossRef] [PubMed]
60. Chapman, R.M.; Tinsley, C.L.; Hill, M.J.; Forrest, M.P.; Tansey, K.E.; Pardiñas, A.F.; Rees, E.; Doyle, A.M.; Wilkinson, L.S.; Owen, M.J.; et al. Convergent Evidence That ZNF804A Is a Regulator of Pre-messenger RNA Processing and Gene Expression. *Schizophr. Bull.* **2019**, *45*, 1267–1278. [CrossRef] [PubMed]
61. Ripke, S.; Neale, B.M.; Corvin, A.; Walters, J.T.; Farh, K.-H.; Holmans, P.A.; Lee, P.; Bulik-Sullivan, B.; Collier, D.A.; Huang, H.; et al. Biological insights from 108 schizophrenia-associated genetic loci. *Nature* **2014**, *511*, 421–427. [CrossRef]
62. Huang, Y.J.; Zhou, Z.W.; Xu, M.; Ma, Q.W.; Yan, J.B.; Wang, J.Y.; Zhang, Q.Q.; Huang, M.; Bao, L. Alteration of gene expression profiling including *GPR174* and *GNG2* is associated with vasovagal syncope. *Pediatr. Cardiol.* **2015**, *36*, 475–480. [CrossRef] [PubMed]
63. O'Donovan, M.C.; Craddock, N.; Norton, N.; Williams, H.; Peirce, T.; Moskvina, V.; Nikolov, I.; Hamshere, M.; Carroll, L.; Georgieva, L.; et al. Molecular Genetics of Schizophrenia Collaboration. Identification of loci associated with schizophrenia by genome-wide association and follow-up. *Nat. Genet.* **2008**, *40*, 1053–1055. [CrossRef] [PubMed]
64. Matthews, J.M.; Sunde, M. Zinc fingers—Folds for many occasions. *IUBMB Life* **2002**, *54*, 351–355. [CrossRef]
65. Soteriades, E.S.; Evans, J.C.; Larson, M.G.; Chen, M.H.; Chen, L.; Benjamin, E.J.; Levy, D. Incidence and prognosis of syncope. *N Engl J. Med.* **2002**, *347*, 878–885. [CrossRef] [PubMed]
66. Available online: https://www.ncbi.nlm.nih.gov/snp/rs12465214?vertical_tab=true#variant_details (accessed on 23 September 2021).
67. Statello, L.; Guo, C.J.; Chen, L.L.; Huarte, M. Gene regulation by long non-coding RNAs and its biological functions. *Nat. Rev. Mol. Cell Biol.* **2021**, *22*, 96–118. [CrossRef]

International Journal of
Molecular Sciences

MDPI

Review

Role of Virus-Induced Host Cell Epigenetic Changes in Cancer

Valeria Pietropaolo [1,*], Carla Prezioso [1,2] and Ugo Moens [3,*]

1 Department of Public Health and Infectious Diseases, "Sapienza" University, 00185 Rome, Italy; carla.prezioso@uniroma1.it
2 IRCSS San Raffaele Roma, Microbiology of Chronic Neuro-Degenerative Pathologies, 00161 Rome, Italy
3 Molecular Inflammation Research Group, Department of Medical Biology, Faculty of Health Sciences, University of Tromsø—The Arctic University of Norway, 9037 Tromsø, Norway
* Correspondence: valeria.pietropaolo@uniroma1.it (V.P.); ugo.moens@uit.no (U.M.)

Abstract: The tumor viruses human T-lymphotropic virus 1 (HTLV-1), hepatitis C virus (HCV), Merkel cell polyomavirus (MCPyV), high-risk human papillomaviruses (HR-HPVs), Epstein-Barr virus (EBV), Kaposi's sarcoma-associated herpes virus (KSHV) and hepatitis B virus (HBV) account for approximately 15% of all human cancers. Although the oncoproteins of these tumor viruses display no sequence similarity to one another, they use the same mechanisms to convey cancer hallmarks on the infected cell. Perturbed gene expression is one of the underlying mechanisms to induce cancer hallmarks. Epigenetic processes, including DNA methylation, histone modification and chromatin remodeling, microRNA, long noncoding RNA, and circular RNA affect gene expression without introducing changes in the DNA sequence. Increasing evidence demonstrates that oncoviruses cause epigenetic modifications, which play a pivotal role in carcinogenesis. In this review, recent advances in the role of host cell epigenetic changes in virus-induced cancers are summarized.

Keywords: chromatin remodeling; circular RNA; DNA methylation; histone modification; non-coding RNA; oncogenes; tumor suppressor genes; tumor virus

Citation: Pietropaolo, V.; Prezioso, C.; Moens, U. Role of Virus-Induced Host Cell Epigenetic Changes in Cancer. *Int. J. Mol. Sci.* **2021**, *22*, 8346. https://doi.org/10.3390/ijms22158346

Academic Editor: Elixabet Lopez-Lopez

Received: 13 July 2021
Accepted: 2 August 2021
Published: 3 August 2021

Publisher's Note: MDPI stays neutral with regard to jurisdictional claims in published maps and institutional affiliations.

1. Introduction

Viruses are infectious agents that can cause malignant and non-malignant diseases. Approximately 15% of all human cancers have a viral etiology and six human viruses are firmly associated with cancer [1]. They include the RNA viruses human T-lymphotropic virus 1 (HTLV-1) and hepatitis C virus (HCV), and the DNA viruses Merkel cell polyomavirus (MCPyV), high-risk human papillomaviruses (HR-HPVs), Epstein-Barr virus or human herpes virus-4 (EBV/HHV-4), Kaposi's sarcoma-associated herpes virus or human herpesvirus-8 (KSHV/HHV-8) and hepatitis B virus (HBV) [2–4]. Despite their differences in structure and genome, all human tumor viruses apply the same mechanisms to induce oncogenesis. They convey the hallmarks of cancer on the host cell. Human viral oncoproteins will cause cells to evade growth suppression and apoptosis, to sustain proliferation and immortalization, to induce mutations and genome instability, to promote chronic inflammation, invasion/metastasis and angiogenesis, to escape immune destruction, and to deregulate cellular energetics [5,6]. Many of these processes are brought about by virus-mediated changes in gene expression because viral oncoproteins can directly modulate gene expression by activating transcription factors, inhibiting transcriptional repressors, and acting as transcription factors [5,6]. Oncoviruses can also affect cellular gene expression by epigenetic mechanisms, including modifying host DNA methylation, inducing chromatin remodeling, expressing viral-encoded non-coding RNAs such as microRNAs, long non-coding RNAs (lncRNAs) and circular RNAs (circRNAs), and changing cellular non-coding RNAomics [7].

It is very difficult to study the epigenetic changes in virus-induced cancer cells for several reasons. Tumors are usually not detected in an early stage and tumor cells represent end products rather than initiation products. Moreover, oncoviruses have often a very long incubation time and virus-induced tumors often occur several decades after the original infection [8–10]. It is challenging to differentiate between an epigenetic change that is directly due to viral infection, due to the host antiviral response or due to a subsequent downstream effect of the transformation process [11]. In vitro infection studies with human oncoviruses may give an idea of the initial epigenetic changes triggered by viral infection, but for oncoviruses such as HPV, MCPyV and HBV good cell systems are lacking.

Viruses also employ epigenetic changes to regulate their life cycle. This review focuses predominantly on the role of virus-induced epigenetic modifications of the host cell in carcinogenesis. The reader is referred to excellent reviews that expound how epigenetic changes modulate the viral life cycle replication [12–15].

2. Oncoviruses and Host Cell DNA Methylation

2.1. The Cellular DNA Methylation Machinery

DNA methylation occurs at cytosine residues in CpG dinucleotides and is a fundamental mechanism in silencing gene transcription and is catalyzed by a family of DNA methyltransferases (DNMTs). DNMT3A and DNMT3B are responsible for establishing DNA methylation. DNMT3L is catalytically inactive but stimulates the enzymatic activity of DNMT3A/3B. DNMT1 is responsible for maintaining the DNA methylation pattern. Erasing DNA methylation is executed by the demethylating enzymes ten-eleven translocation (TET), activation-induced cytidine deaminase (AICDA) and thymine DNA glycosylase (TDG). Methylation of DNA reduces gene expression, whereas demethylation has the opposite effect. Methylation of DNA can prevent transcription regulatory proteins to bind or allow proteins with high affinity for methylated CpG to bind. There are three families of such proteins: methyl-CpG-binding domain (MBD), ubiquitin-like, containing PHD and RING finger domain (UHRF), and Zinc-finger domain. The MBD family comprises MeCP2, MBD1, MBD2, MBD3, and MBD4. The UHRF family contains UHRF1 and UHRF2, and the last family includes Kaiso, Zinc finger and BTB domain containing 4 (ZBTB4) and ZBTB38 [16,17]. MeCP2 and MBD2 act as transcription repressors by recruiting histone deacetylases (HDACs), the nucleosome remodeling complex (NuRD), and the transcriptional repressor switch independent 3A (SIN3A) [18,19]. However, both MeCP2 and MBD2 were shown to function as transcriptional activators [20,21]. The other CpG binding proteins have been less studied.

Aberrant methylation is associated with diseases, including cancer [22,23]. Induction of de novo (de)methylation is one of the common mechanisms used by all human tumor viruses to alter host cell gene expression. Remarkably, virus-induced (de)methylation is non-random and occurs at CpG islands of specific genes, whose role in cancer has been well-established. This will be discussed for each human tumor virus in Sections 2.2–2.8 and the effects of viral oncoproteins on enzymes involved in CpG methylation are summarized in Table 1.

Table 1. Effects of viral oncoproteins on DNA methylating/demethylating enzymes. See text for details.

Viral Oncoprotein	DNA Methylation/Demethylation Enzymes	References
HTLV-1		
Tax	MDB2 recruitment	[24,25]
Unknown	Increased DNMT1 and DNMT3B levels	[26]
HCV		
Core protein	Increased DNMT1 and DNMT3B levels	[27–30]
MCPyV	Unknown	
HR-HPV		
E6	Increased DNMT1 level	[31,32]
E7	Increased DNMT1 level and activity	[31,32]

Table 1. *Cont.*

Viral Oncoprotein	DNA Methylation/Demethylation Enzymes	References
EBV		
LMP1	Increased DNMT1, DNMT3A, and DNMT3B levels and activity	[33–36]
LMP1	Increased recruitment of DNMT1 to promoters	[37]
LMP1	Decreased DNMT1 level	[38]
LMP2A	Increased DNMT1 and DNMT3A levels	[39,40]
LMP2A	Decreased TET1 and TET2 levels	[41]
EBNA3C	Increased DNMT3A level	[42]
Unknown	Increased DNMT3A level	[38]
Unknown	Decreased DNMT3B level	[38]
KSHV		
LANA	Increased recruitment of DNMT3A to promoters	[43–45]
LANA	Increased recruitment of MeCP2 to promoters	[46]
vIRF1	Increased DNMT1 level	[47,48]
vIL6	Increased DNMT1 level	[49]
HBV		
HBx	Releasing DNMT3A from promoters	[50]
HBx	Increased DNMT1 level and recruitment	[51]
HBx	Increased DNMT3A level and recruitment	[52–56]
HBx	Decreased DNMT3B level	[52,54–56]
HBx	Recruitment of MeCP2	[52]
HBx	Decreased DNMT3A level	[52]
HBx	Decreased DNMT3L level	[50]
HBx	Increased recruitment of MeCP2	[50]
HBx	Increased recruitment of MBD1	[54]

2.2. HTLV-1 and Host Cell DNA Methylation

The retrovirus HTLV-1 infects 10–20 million people worldwide, but only 3–5% of infected individuals will develop adult T-cell leukemia-lymphoma (ATL) 30–50 years after initial infection [57,58]. HTLV-1 is also linked to a neurodegenerative disease called tropical spasticparaparesis/HTLV-I–associated myelopathy [59]. The viral proteins Tax and basic zipper (HBZ) are crucial for tumorigenesis [60–62]. However, not all ATL tumor cells express Tax and during the late stage of leukemogenesis, Tax expression is frequently inactivated through several mechanisms such as loss of or DNA hypermethylation of the 5′ long terminal repeat (LTR) or nonsense, insertion or deletion mutations in the *Tax* gene, suggesting that the Tax protein is not essential for the maintenance of ATL [63]. HBZ is transcribed as an antisense transcript of the HTLV-1 provirus and is constitutively expressed in all ATL cases [64].

The integrated HTLV-1 genome is often hypermethylated. Tax was able to increase the transcriptional activity of HLTLV-1 LTR even when heavily methylated [25]. Stimulation of hypermethylated LTR by Tax required association with MDB2. Tax and MBD2 possibly target other methylated sequences and activate transcription from methylated promoters. Indeed, Tax:MBD2 could activate methylated cAMP-response element (CRE) containing promoters [25], suggesting that Tax may induce expression of cellular CRE containing promoters, even if they are hypermethylated. Genome-wide analysis has identified approximately 4000 CRE-containing promoters in the human genome [65], whose expression may be affected by Tax independently of their methylation state.

Methylation analysis of ATL genomes showed prominent CpG hypermethylation and hypomethylation in comparison with controls [66–69]. This altered methylation pattern was associated with transcriptional silencing and upregulation of cellular gene expression. Kruppel-like factor 4 (*KLF4*) and early growth response 3 (*EGR3*) were among the genes that were hypermethylated. Ectopic expression of KLF4 and EGR3 in ATL cell induced apoptosis, indicating that hypermethylated-mediated silencing of these genes enables ATL cell to escape from cell death [70]. Transcription factor-encoding genes and Major histocompatibility complex class I (*MHC-I*) genes were also hypermethylated. This may result in altered gene expression and may help ATL cells to evade the immune system [68,69]. Hypomethylated genes in ATL cells included PR/SET domain 16 (*PRDM16*), resulting in elevated expression of the protein encoded by the *PRSM16* gene, transcription factor *MEL1*. Overexpression of this protein is associated with leukemogenesis [67]. The *FOX3P* locus was found to be hypomethylated in cells from ATL patients and higher FOX3P protein levels were observed [71]. Tax was previously shown to reduce, whereas HBZ increased FOX3P expression [72,73]. However, Tax and HBZ levels did not relate to hypomethylation status of the FOX3P locus, suggesting that hypomethylation was not induced by HTLV-1 [71].

The mechanisms by which HTLV-1 enforces DNA methylation are incompletely understood. Although DNMT1 and DNMT3B were upregulated in HTLV-1 transformed T cells, not all cells expressed Tax, suggesting a Tax-independent mechanism [26]. The promoter of the tumor suppressor gene Src homology-2-containing protein tyrosine phosphatase (*SHP-1*) gene is hypomethylated in ATL cells and SHP-1 expression is lost. The authors showed that Tax repressed SHP-1 expression by recruiting HDAC1, but whether demethylation of the promoter depended on Tax was not investigated [24]. The tumor suppressor gene N-myc downregulated gene 2 (*NDRG2*) is frequently downregulated in ATL. Tax indirectly contributed to repression of this promoter by increasing the expression of enhancer of zeste homolog 2 (EZH2), a histone methyltransferase. Overexpression of EZH2 suppressed transcription of *NDRG2* via DNA methylation and trimethylation of histone 3 at lysine 27 (H3K27me3) [74]. Both examples suggest that Tax indirectly can modulate DNA methylation. Tax may induce irreversible changes in DNA methylation during the initial phase of HTLV-1 infection and this may explain why constitutive Tax expression is not required in ATL. Tax was shown to interact with coactivator associated arginine methyltransferase 1 (CARM1 or PRMT4), and this stimulated histone H3 methylation [75]. A possible role of HBZ in DNA methylation has not been divulged. Importantly, aberrant DNA methylation in ATL cells may not only be caused by HTLV-1 because aging and cancer are closely related to aberrant DNA methylation. The long incubation time of ATL and the prolonged life span of these cells might be predisposing factors for perturbed DNA methylation [76,77].

2.3. HCV and DNA Methylation

HCV is a (+) RNA virus belonging to the family Flaviviridae and is one of the leading causes of hepatocellular carcinoma (HCC). The viral genome is translated into a polypeptide of approximately 3000 amino acids that is cleaved by viral-encoded and cellular proteases to generate structural and non-structural proteins [78]. In vitro studies and transgenic animal models have shown that the viral proteins NS3, NS5A, and the core protein have oncogenic properties [6,78–80].

The methylation landscape of HCV-positive HCC tissues differs from non-tumor controls and a correlation between HCV infection and aberrant methylation of genes such as *CDKN2A* (cyclin-dependent kinase inhibitor 2A), *CDH1* (cadherin 1), *SOCS1* (suppressor of cytokine signaling 1), *RASSF1A* (Ras associated domain family member 1), APC (adenomatous polyposis coli protein), *GSTP1* (glutathione S-transferase Pi 1), *STAT1* (Signal transducer and activator of transcription 1), and *PRDM2* (PR/SET domain 2) in HCV-positive HCC has been established. Hampered expression of these genes contributes to cancer by promoting cell proliferation, mobility and invasion, and immune evasion [27,29,81–84]. The core protein seems to be implicated in HCV-induced DNA methylation because DNMT1 and DNMT3B levels were enhanced in HCV core protein expressing HepG2 cells and in Huh-7 cells compared to control cells [27–30]. The exact mechanisms by which the core protein induces expression of DNMT1 and DNMT3B is unknown but required activation of the STAT pathways by this viral protein [30]. Another possible mechanism, which is applied by the HBX protein of HBV (see Section 2.8), is through the retinoblastoma (pRb)/E2F pathway [53]. The *DNMT1* gene is an E2F1 target gene and the core protein has been shown to phosphorylate pRb, resulting in activation of E2F1-dependent transcription.

2.4. MCPyV and Host Cell DNA Methylation

MCPyV is the most recently identified virus to be linked to a human cancer. It is associated with about 80% of Merkel cell carcinoma (MCC), a rare, but aggressive cutaneous malignancy. The MCPyV genome is always integrated in all virus-positive MCCs examined [9,85]. MCPyV is a non-enveloped virus belonging to the *Polyomaviridae* family [86]. The viral oncoproteins are large tumor antigen (LT) and small tumor antigen (sT). In vitro and animal studies and the detection of sT in the absence of LT in some MCC indicate that sT may be more involved in the oncogenic process, whereas LT is required to sustain the tumor cell growth [85,87].

The DNA methylomes of MCPyV-negative and MCPyV-positve MCCs display significant differences in several genes that are associated with cancer. Frequent occurrence of *RASSF1A* promoter hypermethylation was observed in MCPyV-positive MCC [88]. DNA methylation examination of MCPyV-positive and MCPyV-negative MCC specimens showed that 54% had hypermethylation of the *RASSF1A* promoter and 22% of the *CDKN2A* promoter, whereas the promoters of the tumor suppressor genes fragile histidine triad diadenosine triphosphate (*FHIT*), tumor promoter p73 (*TP73*), and protein tyrosine phosphatase receptor type G (PTPRG) had no or infrequent hypermethylation. However, no significant correlation between viral infection and hypermethylation was observed, indicating that MCPyV infection may not induce DNA hypermethylation of these promoters [88]. Hypermethylation of the promoters of the *RASSF2*, *RASSF5A*, *RASSF5C* and *RASSF10* and the *TERT* gene (encoding telomerase reverse transcriptase) was frequently detected in MCCs compared to normal skin samples, but again no correlation with MCPyV infection was found [89,90]. The promoter of the *RB1* gene (encoding retinoblastoma protein pRb) was hypermethylated in MCCs compared to normal skin samples, but the pattern of hypermethylation of the RB1 promoter was similar in all MCCs independent of the MCPyV status [91]. MCPyV LT can inactivate pRb through interacting with the protein, suggesting the hypermethylation of the RB1 gene to inactivate expression is superfluous. However, the polyomavirus SV40 LT can both bind pRb and induced hypermethylation of the RB1 promoter in diffuse large B-cell type lymphomas [92]. This illustrates that LT of different polyomaviruses can possess distinct functions. The *INK4A-ARF* (*CDNK2A*) locus and *DUSP2* (dual specificity phosphatase 2) gene were found to be frequently hypermethylated in MCC tumors, but the viral status in these tumors was not specified, so that a possible role for MCPyV in hypermethylation cannot be determined [93,94]. In another study, no difference in *INK4A-ARF* methylation was found between virus-positive and virus-negative MCC tumors [95]. Hypomethylation of the *PTCH1* gene (encoding the Patched 1) and the gene for Atonal BHLH transcription factor 1 (*ATOH1*) was detected in both virus-negative and virus-positive MCC cell lines [96,97]. MCC is considered a neuroendocrine tumor and repressor element 1 silencing transcription factor (REST) is a key regulator in neuronal programs. Moreover, REST can act as an oncogene in neural cells and a tumor suppressor in non-neural cells. Therefore, Chteinberg et al. investigated the expression of REST in MCC. REST protein was not detected in any of the examined MCPyV-negative and MCPyV-positive tumors and MCPyV-negative and MCPyV-positive cell lines, but no hypermethylation of the *REST* promoter was observed in all tissues and cell lines, indicating that silencing of *REST* is not caused by hypermethylation and occurred independently of the virus status. The authors speculated that miR-9, which is upregulated in MCCs and targets the 3′ untranslated region of *REST* mRNA, may prevent REST synthesis [98]. The loss of O6-methylguanine-DNA methyltransferase expression has been associated with a wide variety of cancers. The O6-methylguanine-DNA methyltransferase promoter was hypermethylated in six MCPyV-positive MCC cell lines, but hypomethylated in 18 MCC tissues with unknown viral status [99]. This finding emphasizes that caution is warranted when comparing results from tumor cell lines and tumor tissue.

In conclusion, aberrant DNA methylation of cancer-related genes is common in both MCPyV-negative and MCPyV-positive MCCs and does not seem to be provoked by MCPyV infection. Viral-independent modification of host DNA methylation was further confirmed in a study that showed that DNA methylation in MCC tissues was significantly lower as compared to the patients' chronological age. The accelerated DNA methylation in patients was irrespective of the viral presence [100]. Although SV40 LT can upregulate the expression of DNMT3B, thereby contributing to the oncogenic phenotype in a lung cancer model [101], it is not recognized whether MCPyV LT can affect the expression levels or activity of specific DNMTs. A recent study demonstrated a correlation between MCPyV and the methylation pattern in MCC. The authors found that the programmed cell death 1 (*PDCD1*) promoter was hypomethylated in 42 out of 69 MCCs tissues and hypomethylation was significantly more frequent in virus-positive tumors. Virus-positive MCC patients with hypomethylated *PDCD1* promoter had a better prognosis than those with high *PDCD1* methylation [102]. Further studies are required to establish whether MCPyV infection has an effect on host DNA methylation.

2.5. *High-Risk (HR) HPV and Host Cell DNA Methylation*

Human papillomaviruses (HPV) are non-enveloped viruses with a circular dsDNA genome of approximately 8000 base-pairs [103]. More than 200 different types of HPV have been isolated and several of them, so called high risk HPV (HR-HPV) are associated with anogenital and oropharyngeal cancers [104]. HR-HPV are responsible for >99% of cervical cancer cases, with HPV16 (55% of all cases) and HPV18 (15% of all tumors) the two most common types [105]. In the USA about 40–80% of oropharyngeal cancers are positive for HR-HPV, whereas in Europe the incidence varies between 15% and 90%, with >90% of the cases containing HPV16 [106]. The main oncoproteins are E5, E6 and E7 (for a recent review see [107]).

Methylome analyses of HPV-positive cancers revealed differences in DNA methylation compared to matching normal tissue or HPV-negative tumors and transfection studies have confirmed that the E6 and E7 oncoproteins provoked hypermethylation tumor suppressor genes and hypomethylation of proto-oncogenes [31,32,108–114]. Both these viral proteins have been shown to upregulate the expression of DNMT1. E7 does so by derepressing E2F through sequesting pRb, whereas E6 inactivates p53, which abrogates the interaction of p53 with transcription factor Sp1 on the DNMT1 promoter. As the p53:Sp1 complex represses the *DNMT* promoter, E6 releases the repression by appropriating p53 [31,32]. Furthermore, E7 associates with DNMT1 and stimulates its activity [32]. Increased expression of DNMT3B was reported in non-smoking female lung cancer patients with HPV16 or HPV18 positive tumors, but the role of E6 and E7 was not investigated [115]. The mechanism(s) by which HR-HPV provoke hypomethylation of the host genome remain enigmatic. In conclusion, HPV-mediated changes in DNA methylation affects the expression of several cellular genes and has been proven to stimulate cell proliferation, cell survival, adhesion and migration [32,114].

2.6. *EBV and Host Cell DNA Methylation*

EBV or HHV4 is an enveloped virus with a dsDNA genome of around 170 kilobase-pairs. More than 90% of the world population have lifelong infection with this virus. EBV is associated with Burkitt's lymphoma, Hodgkin's disease, primary effusion lymphoma (PEL), nasopharyngeal carcinoma lymphoma, gastric carcinoma, but also with non-malignant diseases, including infectious mononucleosis [3,5,116]. EBV-induced cancer has an incidence of about 1 in 200,000 per year. The major EBV oncoprotein is LMP1, but other viral proteins including LMP2A, EBNA1, EBNA2, EBNA3 and EBNA-LP, and viral RNA transcripts (see further) are implicated in EBV-induced tumorigenesis [3,6,117].

EBV-associated cancers such as gastric cancer, nasopharyngeal carcinoma and Burkitt's lymphoma are characterized by extensive hypermethylation of the host DNA compared with non-infected tumors and cell culture studies have illustrated that EBV infection induces de novo methylation [45,111,118–122]. Many of the genes whose expression is affected by EBV-induced methylation code for proteins involved in cell cycle control, signaling pathways, apoptosis, invasion and migration [45,111,122,123]. Some of these genes will be discussed, as well as the viral proteins involved in their methylation.

LMP1 induces hypermethylation of the *CDH1* promoter and downregulation of cadherin 1 by augmenting the expression and activity of DNMT1, 3A and 3B [33]. Loss of function of the *CDH1* gene contributes to cancer progression by increasing proliferation, invasion, and metastasis [124]. The gene for tumor suppressor *RASSF10*, which encodes a protein that inhibits cell proliferation, invasion, and migration and induces apoptosis was hypermethylated in EBV-positive gastric cancer compared to EBV-negative gastric cancers. The authors demonstrated that LMP1 promoted DNMT1 expression, which was responsible for hypermethylation of the *RASSF10* gene. Overexpression of LMP1 in human gastric adenocarcinoma AGS cells stimulated migration, invasion and cell colony formation and this was counteracted when RASSF10 was co-expressed. Xenograft studies with LMP1 and LMP1 plus RASSF10 cells confirmed that RASSF10 thwarted the LMP1-malignant phenotype. These results suggest that LMP1-mediated methylation and silencing of the *RASSF10* gene plays a role in EBV-induced oncogenesis [125]. Other studies confirmed that LMP1 upregulates DNMT1, DNMT3A and DNTM3B. LMP1-induces DNMT1 expression dependent on activation of the c-Jun N-terminal kinase (JNK)/AP1 pathway, whereas DNMT3A and DNMT3B were induced via the NFκB pathway [34,35]. LMP2A increased expression of DNMT1 via STAT3 and DNMT3A via the mitogen-activated protein kinase (MAPK) pathway and downregulated the expression of the demethylating enzymes TET1 and TET2 [39–41]. However, in germinal center B-cells, presumptive progenitors of Hodgkin's lymphoma, EBV infection resulted in downregulation of DNMT1 and DNMT3B and upregulation of DNMT3A and the authors found that LMP1 is responsible for downregulation of DNMT1, while the mechanism for DNMT3A and DNMT3B remains unknown as ectopic expression of LMP1 or of LMP2A had no effect on DNMT3A and DNMT3B levels [38]. LMP2A caused hypermethylation of the phosphatase and tensin homolog (*PTEN*) gene through stimulation of DNMT1 in a STAT3-dependent manner [39]. EBNA3C, another EBV protein, could induce hypermethylation of the *RASSF1A* promoter by enhancing DNMT3A expression. This epigenetic modification results in decreased RASSF1A expression, leading to increased cell proliferation [42]. Finally, EBV-mediated methylation also affects genes whose products are involved in histone modification and chromatin remodeling. LMP1 could recruit DNMT1 to the promoter of the lysine-specific demethylase 2b (*KDM2B*) and trigger hypermethylation. KDM2B demethylates histone 3 at lysine 4 (H3K4me3). H3K4me3 is commonly associated with active transcription and demethylation will result in transcriptional silencing [37]. Thus, EBV-provoked changes in the host DNA methylation can contribute to virus-induced tumorigenesis.

2.7. KSHV and Host Cell DNA Methylation

KSHV or HHV8 is the causative agent of Kaposi sarcoma and associated with the lymphoproliferative disorders, multicentric Castleman's disease and PEL [126,127]. No individual KSHV gene product appears to transform primary human cells by itself, but several viral proteins and non-coding RNAs have been shown to play a pivotal role in the pathogenesis of KSHV-associated tumors [6,128]. The viral proteins latency-associated nuclear antigen (LANA), vCyclin, and viral FLICE inhibitory protein (vFLIP) drive cell proliferation and prevent apoptosis, while viral interleukin 6 (vIL6), vGPCR, and ORFK1 contribute to angiogenesis and inflammation [127].

CpG methylation analysis of the human DNA in KSHV-infected cells and KSHV-associated PELs revealed both hyper- and hypomethylated promoters compared with KSHV-negative lymphoma BJAB cells. Genes encoding proteins involved in cell cycle control, signaling pathways and metastasis were differently methylated in the KSHV-positive cells and tumors compared to control cells [111,129]. Some of the genes that were hypermethylated in KSHV-infected PEL cell lines included *CDNK2A*, *CDH1* and *CDH13* (cadherin 1 and 13), *LDHB* (lactate dehydrogenase B), *HLTF* (helicase like transcription factor, a member of the chromatin remodeling SWI/SNF family), *CCND2* (cyclin D2). The authors showed that KHSV LANA recruited DNMT3A to chromatin, and induced hypermethylation and transcriptional inactivation of these genes [43–45]. LANA may not only repress transcription of cellular genes by inducing hypermethylation, but it may potentiate transcriptional inhibition through recruiting the transcriptional repressor methyl CpG binding protein 2 (MeCP2), which interacts with LANA [46]. Moreover, LANA could inhibit the promoter of the TGF-β type II receptor (*TGFBR2*) through inducing hypermethylation of Sp1 binding sites, thereby preventing Sp1 binding. Epigenetic silencing of this promoter contributed to the pathogenesis of KSHV-associated tumors [130]. Two other KSHV proteins interfere with DNA methylation. vIRF1 could upregulate DNMT1 expression in a STAT3-dependent manner and by inhibiting p53 [47,49]. vIL6-induced modifications in DNA methylation promoted proliferation and migration of endothelial cells [47]. Another group showed that the vIL6/STAT3/DNMT1 axis was involved in silencing expression of caveolin 1, which promoted cell proliferation, invasion and angiogenesis of endothelial cells [49]. The mechanism by which KSHV achieves hypomethylation of the host DNA is not known. Taken together, these results indicate that KSHV-triggered DNA methylation play a role in KSHV-associated cancers.

2.8. HBV and Host Cell DNA Methylation

It is estimated that more than 250 million people globally are chronically infected with HBV, and each year around 800,000 patients died from HBV- and HCV-related HCC. Of these, approximately 50% of are caused by HBV [131]. HBV-induced hepatocarcinogenesis occurs due to viral genome integration causing mutations and through the actions of the viral proteins, predominantly HBx (also referred to as pX), but the surface proteins preS and S also contribute to tumor development as shown by in vitro and animal studies. The mechanisms by which HBV induces HCC have been comprehensively reviewed by others [3,5,6,78,132].

Comparing the DNA methylation profile of HBV-associated HCC and HBV-negative tumors or healthy adjacent liver tissue, HBV-infected and non-infected cells, and HBx transgenic mouse model and control mice disclosed differentially methylation. Several cellular promoters were hypermethylated in the presence of HBV or HBx, including the promoters of the genes encoding cyclin-dependent protein kinases inhibitors $p21^{CIP1/WAF1}$ (*CDKN1A*), $p14^{ARF}$ (*CDKN2A*) and $p14^{INK4B}$ (*CDKN2B*), cadherin 1, RASSF1A, the spleen associated tyrosine kinase SYK (*SYK*), GSTP1, the protein phosphatase 1 regulatory subunit 13B (*PP1R13B*), the tumor promotor p53 binding protein 2 (*TP53BP2*), and insulin like growth factor binding protein 3 (*IGFBP3*) [52,53,82,84,133–138]. These proteins are involved in cell cycle control, apoptosis, migration and invasion, indicating that HBV-induced silencing of these genes play a role in HCC. Some CpG islands of genes associated with HBV-induced tumorigenesis were significantly hypomethylated in transgenic mice with liver-specific HBx-expression compared to wild-type animals, illustrating that HBV infection can also upregulate gene expression by demethylating their DNA [50].

HBV seems to affect DNA methylation by several mechanisms. One study showed that HBx could cause hypomethylation through releasing DNMT3A from promoters [51]. HBx also upregulated expression of DNMT1 and DNMT3A, but repressed DNMT3B expression in liver cell lines [52]. HBx upregulated DNMT1 expression by repressing $p16^{INK14A}$, resulting in activation of the cyclin-dependent kinase 4/6-pRb-E2F1 pathway, and ultimately in stimulation of DNMT1 expression [53]. Moreover, HBx was shown to downregulate miR-152 and miR-101, which target DNMT1 mRNA and DNMT3A mRNA, respectively, thereby increasing the levels of DNMT1 and DNMT3A [55,56]. Another study demonstrated that HBx could recruit MeCP2, which repressed transcription [52]. HBx was found to modestly suppress DNMT3A expression in mouse liver, and to cause a strong decrease in DNMT3L levels. The latter has no methyltransferase activity but stimulates the enzymatic activity of DNMT3A. The authors also showed that HBx stimulated recruitment of HDAC1 [50]. The reason for the antagonistic effect of HBx on DNMT3B expression in liver cells and in liver is not known. Other studies demonstrated that HBx did not directly influence the expression of DNMT1 and DNMT3A and of MeCP2 and MBD1, but increased their recruitment to promoters, as was shown for the *PP1R13B* and *TP53BP2* promoters [54]. Similar to the other human tumor viruses, HBV infection alters the methylation profile of the host cell DNA, resulting in up- and downregulation of cancer-related genes, which can contribute to HBV-induced hepatocarcinogenesis.

3. Oncoviruses and Chromatin Remodeling

3.1. Histone Modification and Chromatin Remodeling Machinery

Host cell DNA is packed and present in a highly organized structure called chromatin, which is a complex of DNA, histones and other proteins. Chromatin is a dynamic structure that regulates the accessibility of DNA for transcription, replication, DNA repair and recombination. Nucleosomes are the basic units of chromatin and consist of two copies of the canonical histones H2A, H2B, H3 and H4 around which DNA is twisted. The linker histone H1 is interspersed between nucleosomes. Posttranslation modifications (PTMs) of histones will affect the chromatin structure and hence the accessibility of the DNA. The most studied and best understood histone PTMs are acetylation of lysine (K) and methylation of lysine and arginine (R) residues, and phosphorylation of serine (S), threonine (T) and tyrosine (Y) [139–141]. Acetylation is a reversible process and is catalyzed by an histone acetylase (HAT), while an histone deacetylase (HDAC) will reverse acetylation. Acetylation of histones will neutralize the positive charges of K residues, thereby disrupting the interaction with e.g., the negative phosphate groups of the DNA. Acetylation of histones is associated with transcriptional activity, and HDAC acts as a transcriptional repressor. Multiple methylation events can occur at the same K or R residue in histones. H3K4me3 is associated with transcriptional activity, whereas high methylation levels of histone 3 at K9 and K27 and of histone 4 at K20 (H4K20me) are typical for transcriptionally repressed chromatin. Lysine methyltransferases (KMTs) and lysine demethyltransferases (KDMs) add or remove methyl groups. Phosphorylation of histones adds negative charges that undoubtfully influence chromatin structure, but the precise role of this PTM in transcription is less understood. Histone PTMs will affect nucleosome–DNA interactions, as well as histone–histone interactions and interactions with other proteins such as histone chaperones [141,142]. Histone modifying enzymes often exist in multisubunit complexes. For example, the polycomb repressive complex 2 (PRC) includes either enhancer of zeste homolog 1 (EZH1) or EZH2, and the proteins embryonic ectoderm development (EED), suppressor of zeste 12 homolog (SUZ12) and retinoblastoma-binding protein RbAp46 or RbAp48. PRC2 catalyzes H3K27me3 by the enzymatic activity of EZH1 or EZH2 [143].

Another mechanism to change the chromatin structure is by chromatin remodelers [143,144]. ATP-dependent remodelers use ATP to remodel the chromatin. Four major families of ATP-dependent remodeling complexes exist: switching defective/sucrose nonfermenting (SWI/SNF), imitation switch (ISWI), chromodomain helicase DNA-binding protein (CHD), and inositol requiring 80 (INO80). All these complexes consist of multiple proteins [145].

Perturbed histone and modifications and remodeling of chromatin are pivotal events in oncogenesis [146]. In the next section we will discuss how tumor viruses can induce histone modifications and chromatin remodeling and how this may contribute to tumorigenesis. The effects of viral oncoproteins on histone modifying enzymes and proteins of chromatin remodeling complexes are summarized in Table 2.

Table 2. Effects of viral oncoproteins on histone modifying enzymes and protein chromatin remodeling complexes. See text for details.

Viral Oncoprotein	Histone Modifying Enzyme and Chromatin Remodeling Protein	References
HTLV-1		
Tax	Recruitment of p300/CBP	[147,148]
Tax	Recruitment of HDAC1	[149]
Tax	Recruits SIRT1	[150]
Tax	Increased EZH2 level and interaction with EZH2	[151]
Tax	Recruitment of SWI/SNF components BRG1, BAF53, BAF57, BAF155	[152]
Tax	Recruitment of CARM1/PRMT4	[75]
HBZ	Sequestering of p300/CBP	[147]
HBZ	Inhibition of KAT7 activity	[153]
HBZ	Impaired recruitment of KMT1A	[154]
HBZ	Displacement of BRF1	[152]
Unknown	Increased levels of the PRC2 complex proteins	[151]
Unknown	Recruitment of CTCF	[155]
HCV		
Core protein	Increased HDAC1 level	[156]
Core protein	Degradation of the PRC1 component RNF2	[157]
Unknown	Inactivation of PRMT1	[158]
MCPyV		
sT	Recruitment EP400 HAT and chromatin remodeling complex	[159]
sT	Increased expression *KDMA1* gene	[159]
HR-HPV		
E6	Inhibition of p300/CBP activity	[160]
E6	Stimulation proteasomal degradation of TIP60	[161]
E6	Increased EZH2 level	[162]
E6	Stimulation proteasomal degradation of KDM5C	[163]
E6	Inhibition of CARM1/PRMT4 activity	[164]
E6	Inhibition of PRMT1 activity	[164]
E6	Inhibition of KMT5A	[164]
E6/E7	Decreased level of the PRC1 protein BMI1	[165]
E7	Stimulation of p300/CBP activity	[166]
E7	Inhibition of PCAF/KAT2B activity	[167]
E7	Sequestering the NuR complex components Mi2β, HDAC1, and HDAC2	[168]
E7	Increased activity of BRG1	[169]
E7	Increased EZH2 level	[162]
E7	Increased KDM6A and KDM6B levels	[170]
Unknown	Increased HDAC1 and HDAC2 levels	[171]
EBV		
LMP1	Increased KDM6 levels	[172]
EBNA2	Stimulation of p300/CBP and PCAF/KAT2B activities	[173]
EBNA2	Recruitment of the chromatin remodeling complex INO80	[174]
EBNA3C	Inhibition of p300/CBP activity	[175]
EBNA3C	Recruitment HDAC1 and HDAC2	[176]
EBNA-LP	Recruitment of the chromatin remodeling complex INO80	[174]
BZLF1	Recruitment of p300/CBP	[177]
BZLF1	Recruitment of chromatin remodeling proteins SNF2h and INO80	[178]
BRLF1	Recruitment of p300/CBP	[179]
Unknown	Increased level of the SNF2 member LSH	[180]
KSHV		
LANA	SAP30	[181]
LANA	KMT1A/SUV39H1	[182]
LANA	Increased EZH2 level	[183]
LANA	Recruitment of KMT2F	[184]
LANA	Recruitment of BRD/BET	[185]

Table 2. *Cont.*

Viral Oncoprotein	Histone Modifying Enzyme and Chromatin Remodeling Protein	References
vIRF	Inhibits p300/CBP activity	[186]
vIRF3	Prevents nuclear export HDAC5	[187]
Rta	Recruitment of p300/CBP	[188]
Rta	Recruitment of BRG1	[188]
vIL6	Increased EZH2 level	[183]
vFLIP	Increased EZH2 level	[183]
HBV		
HBx	Recruitment of p300/CBP	[189,190]
	Recruitment of HDAC1	[50]
	Increased HDAC1, HDAC2, and HDAC3 levels and activities	[191,192]
	Increased SETDB1 level	[193]
	Increased EZH2 level	[56,194,195]
	Increased SMYD3 level	[196]
	Increased PRC2 activity	[192]
	Increased KDM1A activity	[192]
	Complex formation with RelA, EZH2, TET2, and DNMT3L	[197]
	Stabilization of WDR5 and recruitment to chromatin	[198]

3.2. HTLV-1 and Histone Modification and Chromatin Remodeling

HTLV-1 infection can affect histone acetylation as demonstrated for the $p21^{CIP1}/^{WAF1}$ encoding gene. Expression of this cyclin-dependent kinase inhibitor was upregulated in HTLV-1 infected cells and it was shown that histone H4, but not histone H3 was acetylated [199]. Both Tax and HBZ have been shown to be involved in the regulation of histone acetylation. Tax could bind CREB-binding protein (CBP) and its paralog p300, as well as HDAC1, whereas HBZ sequestered p300/CBP [147–149]. Competition between HBZ and Tax for p300/CBP disrupted the interaction of Tax with p300/CBP and abrogated Tax-induced stimulation the HTLV-1 promoter [147]. As not all ATLs express Tax, but do express HBZ, HBZ may usurp p300/CBP, thereby reducing expression of cellular genes [200–202]. HBZ bound to and repressed activity of another HAT, lysine acetyltransferase 7 (KAT7 alias HBO1), which acetylates histones H3 and H4 [153]. Protein levels of the HDAC sirtuin 1 (SIRT1) were higher in ATL cells compared to healthy peripheral blood mononuclear cells (PBMC). Interestingly, SIRT1 inhibitors induced apoptosis of ATL cells, suggesting an anti-apoptotic action of SIRT1 [203]. The mechanism for upregulation of SIRT1 in ATL cells is not known, but SIRT1 has been shown to interact with Tax and to suppress HTLV-1 gene expression [150]. These findings suggest that interfering with HDAC and HAT may be important in the development of HTLV-1 associated ATL.

Altered histone methylation may also contribute to HTLV-1-induced cancer. The H3K27me3 pattern in ATL cells was different from normal CD4+ T cells, indicating that HTLV-1 reprograms the H3K27me3 profile. H3K27me-silenced genes included genes whose products are involved in control of cell proliferation, cell migration, transcriptional regulation, immune response and cellular metabolism [151,204]. Fujikawa and colleagues reported that the expression of all proteins that constitute the PRC2 complex were upregulated in ATL cells compared to normal CD4+ T cells, whereas downregulated genes included tumor suppressor genes, genes encoding transcription factors, histone demethylases, and other epigenetic modifiers [151]. Tax-dependent immortalized cells showed H3K27me3 reprogramming that was significantly similar to that of ATL cells, suggesting that changes in the H3K27me3 landscape are at least partially dependent on Tax. Indeed, Tax, but not HBZ, stimulated EZH2 promoter activity in a MAPK- and NFκB-dependent manner, increased EZH2 protein levels and interacted with EZH2. Moreover, the authors showed that inhibition of EZH2 prevented Tax-dependent growth and immortalization of Tax-transfected PBMC [151]. Taken together, Tax/EZH2-dependent epigenetic modifications contribute to altered gene expression and to the survival of HTLV-1-infected cells. Tax protein induced transcription of the Ellis Van Creveld 1 (*EVC1*) and *EVC2* genes though stimulating histone H3 acetylation and H3K4me3 [205]. The EVC1 and EVC2 proteins are positive modulators of the Hedgehog signaling pathway and aberrant activation of the Hedgehog signaling is an oncogenic pathway in many types of cancer [206]. Mukai and Ohshima demonstrated that HBZ interacted with centromere protein B (CENP-B), a protein that enhances H3K9me3 by recruiting the histone methyltransferase KMT1A/SUV39H1. The interaction between HBZ and CENP-B impaired recruitment of KMT1A and significantly reduced the amount of H3K3me3 [154]. Transcription of the BCL2 like 11 (*BCL2L11*) gene, which encodes the proapoptotic protein BCL2 interacting mediator of cell death (BIM), was decreased in ATL cells compared to HTLV-negative T cell lines and normal PBMC. Ectopic expression of HBZ in T cells inhibited transcription of the BCL2L11 gene. The authors showed that HBZ-mediated repression of *BCL2L11* transcription involved inactivation of the transcription factor Forkhead box O3A (FOXO3A), hypermethylation, upregulation of H3K9me2 and H3K27me3, and reduced acetylation of histone H3. HBZ-mediated silencing of BIM expression led to decreased apoptosis and may thus contribute to HTLV-1 induced oncogenesis [207].

Two studies demonstrated that HTLV-1 could induce chromatin remodeling. The integrated HTLV-1 genome bound CCCTC-binding factor (CTCF), a chromatin remodeling protein and regulator of transcription. Recruitment of CTCF by HTLV-1 provirus may spread abnormalities in the chromatin structure of host cells, thereby affecting gene expression [155]. Mass spectrophotometry and immunoprecipitation studies showed that Tax could interact with the SWI/SNF components BRM/SWI2-related gene (BRG1) and the BRG-associated factors BAF53, BAF57, and BAF155. Tax recruited BRG1, the ATPase subunit of the SWI/SNF chromatin remodeling complex, to the HTLV-1 promoter and cellular promoters and induced acetylation of histone H4, thereby stimulating the HTLV-1 promoter activity [152]. Interestingly, HBZ displaced BRG1 from the HTLV-1 promoter. Similar to p300/CBP, Tax and HBZ compete for BRG1, thereby activating or repressing promoters. The opposite roles of Tax and HBZ in viral expression may be important for maintaining viral latency and persistence, which may ultimately lead to the development of ATL [208].

3.3. HCV and Histone Modification and Chromatin Remodeling

HCV can modulate histone acetylation as shown for secreted frizzled related protein 1 (*SFRP1*) promoter. The core protein was shown to downregulate SFRP1 expression by an epigenetic mechanism. The core protein increased the levels of DNMT1 and HDAC1 and stimulated their binding to the *SFRP1* promoter. This resulted in hypermethylation and reduction in histone H3 acetylation. Silencing of SRFP1 led to deregulated activation of the Wnt signaling pathway and may thus contribute to HCC-induced HCC [156].

HCV infection is associated with changes in histone methylation. Ectopic expression of the entire HCV polypeptide resulted in a significant loss of H4K16ac, H4R3me2, and H4K20me3, and was correlated with the altered expression of genes important in hepatocarcinogenesis such as avian myelocytomatosis viral oncogene homolog (*c-MYC*), *PTEN*, *CDH1*, epidermal growth factor (*EGF*), *CDKN2A*, and *IGFBP3* [158]. Increased protein phosphatase A catalytic subunit alpha (PPP2CA) levels and reduced H4R3me2 were observed in HCV-positive HCC tumor samples compared to matching non-tumor liver tissue. The authors showed that altered H4R3me2 was caused by PPP2CA-mediated inactivation of protein arginine methyltransferase 1 (PRMT1) [158]. HCV infection of the Huh7.5 cell line resulted in significant enrichment of the transcriptional active chromatin labels H3K9ac and H3K4me, and of the transcriptional silent chromatin marker H3K9me3, but not of H3K27me3. Infection of primary human hepatocytes or the Huh7.5 cell line was associated with reprogrammed gene expression, which can be linked to HCV pathogenesis [209]. The authors also demonstrated that once epigenetic changes had occurred, this specific gene expression pattern is maintained in cells cured for HCV infection by direct acting antivirals treatment. Thus, the presence of the virus seems no longer required for its oncogenic effects on the host cells, supporting a hit-and-run mechanism. HCV can also alter the ubiquitination pattern of histones and this may affect transcription as exemplified for several homeobox (*HOX*) genes. Kasai et al. reported that the expression of several *HOX* genes was induced in HCV infected or core protein expressing cells. HCV and core protein stimulated *HOX* gene expression by impairing histone H2A monoubiquitination via degradation of PRC1 component E3 ligase RNF2 (ring finger protein 2) [157]. As HOX proteins are associated with tumorigenesis, HCV-regulated expression of these genes may contribute to HCV-induced hepatocarcinogenesis.

3.4. MCPyV and Histone Modification and Chromatin Remodeling

The LTs of the murine and SV40 polyomaviruses were found to bind to, and to upregulate the expression and the activity of p300/CBP [210–213]. Whether MCPyV LT possesses similar properties has not been investigated. Busam and colleagues evidenced a strong reduction of H3K27me3 staining in virus-positive MCCs compared with virus-negative tumors. This observation suggests that epigenetic deregulation may play a role in the pathogenesis of Merkel cell polyomavirus associated MCC, but the mechanism for MCPyV-induced reduction in H3K27me and the biological significance remain to be solved [214]. Cheng and coworkers showed that sT interacted with MYCL and together they recruited the EP400 HAT and chromatin remodeling complex and bound to specific cellular promoters to stimulate their activity. One of the upregulated genes was *KDMA1*, indicating that sT may affect histone methylation. sT:MYCL:EP400 complex formation was required to transform IMR90 human diploid fibroblasts, suggesting that complex formation is important in the development of MCPyV-positive MCC [159].

3.5. HR-HPV and Histone Modification and Chromatin Remodeling

Several studies have shown that HATs and HDACs can play a role in HR-HPV associated cancers. Expression levels of HDAC1 and HDAC1 were increased in invasive HPV-positive cervical cancers compared normal epithelium and inversely correlated with $p21^{CIP1/WAFf1}$ levels. RNA interference-mediated silencing of HDAC2 in HPV18 positive HeLa cells increased expression of the $p21^{CIP1/WAFf1}$ tumor suppressor and stimulated apoptosis [171]. It is not known whether HPV oncoproteins promote HDAC1/2 expression, but it could be a strategy of the virus to prevent apoptosis. E6 of HR-HPV16, but not of LR HPV6, binds and inhibits HAT activity of p300 and CBP, whereas binding of E7 to p300/CBP stimulated their activity [160,166]. E7 also interacted with lysine acetyltransferase 2B (KAT2B; also known as p300/CBP-associated factor PCAF) and reduced its ability to acetylate histones in vitro [167]. The interaction of E6 and E7 with these HATs has been demonstrated to downregulate expression of interleukin 8 (IL-8), which is a chemotactic factor for immune cells. Hence, E6/E7-mediated downregulation of IL-8 may help HPV-infected cells to evade the immune system. The HAT TIP60, which acetylates histone H4, was targeted for proteasomal degradation by E6 and reduced acetylation of histone H4 was observed in HPV-positive cell lines compared to control cells [161]. TIP60 also helps to recruit the transcriptional repressor bromodomain containing 4 (BRD4) and is involved in DNA damage response and apoptosis. Hence, E6-induced TIP60 destabilization may relieve gene expression, abrogate DNA repair, and prevent apoptotic pathways, thereby contributing to HPV-induced carcinogenesis [215].

HR-HPV E7 was shown to interact with Mi2β, HDAC1 and HDCA2, which are constituents of the NuRD complex, a CHD chromatin remodeling complex. HPV E7 could through this interaction downregulate expression of proteins involved in immune responses and promote cell growth [168,216]. Furthermore, E7 binds BRG1, a component of the chromatin remodeling SWI/SNF complex. This interaction overcomes repression of the FBJ murine osteosarcoma viral oncogene homolog (*c-FOS*) gene transcription. Hence, E7-mediated upregulation of c-FOS protein levels may contribute to deregulation of cell cycle control [169].

HR-HPV can affect histone methylation by several mechanisms. The PRC2 complex mediates H3K27me3, which is associated with transcriptional repression. Subsequently, PRC1 binds to H3K27me-marked chromatin and further silences gene expression by monoubiquitinating lysine 119 of histone H2A. PRC2 contains the histone methyltransferase EZH2, which catalyzes mono-, di-, and trimethylation of H3 [217,218]. Perturbed H3K27me is a common histone modification in many different cancers, including HPV-positive cancers [146,219]. HPV16 E6/E7 transformed primary human skin fibroblasts had increased expression of EZH2 and reduced global H3K27me3 levels compared to normal keratinocytes. Increased EZH2 levels and the loss of H3K27me3 was also observed in HP16-positive high-grade cervical intraepithelial lesions compared to matched normal tissue. E6 and E7 were shown to stimulate expression of EZH2. E6 enhanced the levels of transcription factor FOXM1, whereas E7 activated E2F1 by binding pRb. FOXM1 and E2F1 bind the EZH2 promoter and enhance transcription [162]. Furthermore, it has been shown that p53 represses expression of EZH2, suggesting that increased expression of EZH2 may be mediated through E6-mediated loss of p53 [165]. It is somewhat paradoxical that the HPV oncoproteins upregulate expression of EZH2, while a decrease in H3K27me is observed. One explanation is that KDM6A and KDM6B, which demethylate H3K27me3, were also upregulated in E6/E7 transformed primary human skin fibroblasts cells and these may counteract the effect of EZH2. Reduced H3K27me3 and increased EZH2, KDM6A, KDM6B levels were also observed in primary human foreskin keratinocytes expressing HPV16 E7 compared to control cells [170]. The PRC1 protein B lymphoma murine leukemia virus insertion region 1 (BMI1), which recognizes H3K27me3 and stabilizes this repressive methylation mark, was downregulated in E6/E7 transformed cells [165]. This may also explain the diminished H3K27me3 levels, despite increased EZH2 levels. Moreover, phosphorylation of EZH2 by AKT negatively regulates EZH2′s enzymatic activity and E6/E7 induces EZH2 phosphorylation by AKT [165], so that the levels of EZH2 may be high, but the protein is inactive. E6/E7 modulation of EZH2, BMI1, and KDM6A levels resulted in significantly reduced H3K27me3 levels of the promoters of *HOX* genes. In accordance with cervical cancer, expression of these genes was upregulated in the E6/E7 transformed fibroblasts and in E7-expressing keratinocytes cells compared to control cells [165,170]. E6 stimulates *hTERT* promoter activity by increasing H3K4me3 and H3K9ac, which are transcription activation modifications, and decreasing methylation of the transcription repressive modification H3K9me2 [163]. HPV16-positive CaSki cervical cancer cells had lower levels of KDMC5 than HPV-negative C33A cervical cancer cells. E6 was shown to interact with histone H3K4 demethylase KDM5C and promote proteasomal degradation. The authors demonstrated that CaSki cells, which overexpressed KDMC5, grew slower and invasion and migration were reduced compared to control cells. A mouse xenograft model showed that tumors derived from CaSki-KDMC5 cells grew more slowly than CaSki-derived tumors [220]. E6 could inhibit the enzymatic activity of CARM1 (as known as PRMT4), PRMT1, and the lysine methyl-transferase KMT5A. Inhibition of the methyltransferase activity of these enzymes hampered histone methylation at p53-responsive promoters and prevented the binding of p53, hence suppressing p53-mediated transcription [164].

In conclusion, changes in histone acetylation and methylation resulted in dysregulation of cellular gene expression and may contribute to HPV-induced oncogenesis.

3.6. EBV and Histone Modification and Chromatin Remodeling

Increased histone acetylation and increased cellular gene expressed were observed in EBV-transformed lymphoblastoid cell lines compared to control cells [175]. EBNA2 was shown to interact with and stimulate the activity of the HATs p300, CBP, and KAT2B/PCAF, suggesting a role for EBNA2 in regulating histone acetylation [173]. EBNA3C bound p300 but interacted with also HDAC1 and HDAC2 and downregulated EBNA2-induced HAT activity [175,176]. This suggests that EBNA3C may counteract the EBNA2-induced histone acetylation by sequestering p300 and recruiting HDAC. However, EBNA2 and EBNA3C are not typically expressed in EBV-positive Burkitt's lymphoma, gastric cancer and most nasopharyngeal carcinomas, suggesting that their role in epigenetic changes in the cancer cell may be limited. Two viral proteins that can interfere with histone acetylation are BRLF1 and BZLF1, which were found to recruit CBP [177,179]. The human genome contains almost 200,000 putative BZLF1 binding sites, suggesting that appropriation of CBP by BZLF1 may repress transcription. Indeed, induced expression of BZLF1 in EBV-negative cells caused only minor, whereas overexpression of BZLF1 in latently infected B cells provoked profound reduction in gene expression and decreased open chromatin structure ([221] and references therein).

EBV infection was also associated with changes in histone methylation. EBV infection of nasopharyngeal epithelial cells reduced the transcriptional activation mark H3K4me3 and enhanced the suppressive mark H3K27me3 at the promoter regions of several genes, including 16 DNA damage repair genes. The reduced DNA repair ability in EBV-infected nasopharyngeal epithelial cells may play an important role in nasopharyngeal carcinoma [222]. Infection of B cells with EBV resulted in a loss of H3K9me3, H3K27me3, and H4K20me3, histone markers that are associated with histone condensation. Reduction of these markers was linked to increased chromatin accessibility and gene expression, including genes involved in hallmarks of cancer such as cell cycle regulation and apoptosis, and was associated with transformation. Similar decrease in H3K9me3, H3K27me3, and H4K20me3 patterns was also obtained with LMP1 and EBNA2 deficient mutant viruses, suggesting that these proteins are not required [223]. Histone modification and chromatin remodeling seems also involved in EBV-induced pathogenesis. Schaeffner and her coworkers reported that the EBV transcription factor BZLF1 interacted with the chromatin remodeling proteins SNF2h and INO80 and this led to increased chromatin accessibility on the EBV genome [178]. EBNA-LP and EBNA2 could also associate with the INO80 complex [174]. Whether the interaction of these viral proteins with chromatin remodeling complexes affects the chromatin structure of host cells was not investigated. Another study showed that EBNA2:SNF complex was recruited to the cellular Fc fragment of IgE receptor II (*FCER2* or *CD23*) promoter [224]. It was previously demonstrated that EBNA2 stimulates CD23 expression [225], suggesting the EBNA2-mediated recruitment of SNF may be involved. The SNF2 member lymphoid-specific helicase (LSH) is overexpressed in EBV-positive nasopharyngeal tumor samples compared to EBV-negative samples, but the biological relevance was not investigated [180].

Taken, together, EBV-induced histone modifications and chromatin remodeling may be a potential cancer driver in EBV-related tumors.

3.7. KSHV and Histone Modification and Chromatin Remodeling

KSHV-infected cells displayed changes in the level of H3K27me3 at promoters of genes encoding proteins relevant in KSHV-induced carcinogenesis such as vascular endothelial growth factor (VEGF), p53, and toll-like receptors (TLRs) [226]. Several KSHV proteins have been shown to interfere with histone modifying enzymes and proteins of chromatin remodeling complexes. Viral interferon regulatory factor (vIRF) was shown to interact with the HATs p300 and CBP and inhibited their activity. These interactions resulted in altered chromatin structure and reduced gene expression [186]. HDAC5 lacks enzymatic activity but can be phosphorylated and transported to the cytoplasm. This will ultimately lead to anti-angiogenic gene expression [227]. It was demonstrated that vIRF3 interacted with HDAC5 and prevented nuclear export, thereby contributing to virus-induced lymphoangiogenesis [187]. Another viral protein, Rta, could also recruit CBP, as well as the SWI/SNF complex through interaction with the BRG1 subunit, and the transcriptional regulatory complex TRAP/Mediator. However, the effect on cellular gene expression in KSHV-induced oncogenesis remains to be determined [188]. LANA could interact with SAP30 (Sin3-associated protein), a component of the HDAC complex and with histone methyltransferase KMT1A/SUV39H1 and heterochromatin protein 1 to induce H3K9 methylation [181,182]. LANA, vIL6, and vFLIP stimulated EZH2 expression via the NFκB pathway. KSHV induced expression of the H3K27-specific methyltransferase EZH2 of the PRC2 complex promoted production of the proangiogenic factor ephrin-B2, indicating that EZH2 is essential for KSHV-induced angiogenesis [183,186]. Moreover, LANA was found to associate with H3K4 methyltransferase KMT2F/SETD1A and to bind the members of the chromatin modulator family BRD/BET [184,185], indicating that LANA can modify chromatin structure. However, LANA chromatin-immunoprecipitation techniques showed that LANA predominantly bound to sites that were already in an open chromatin formation and most transcription of the genes located close to LANA binding sites did not change significantly. However, LANA may induce gene-specific chromatin changes as demonstrated for some interferon gamma (IFNγ)-responsive genes [128]. LANA was found to induce sumoylation of Sp100, a component of ND10 nuclear bodies, resulting in release from chromatin and this coincided with acquisition of H3K27me3 marks [228]. KDM6B is overexpressed in several EBV-positive tumors and KDM6B expression was induced in LMP1-transfected in germinal centre B cells [172]. In conclusion, several KSHV proteins may induce histone modifications and chromatin rearrangements, thereby contributing to oncogenesis.

3.8. HBV and Histone Modification and Chromatin Remodeling

HBx protein of HBV was shown to activate or repress cellular gene expression. This opposite effect depended on whether HBx attracted HATS or HDACs to the promoter. HBx stimulated CRE binding protein (CREB)-dependent transcription by recruiting p300/CBP. Induction of CREB target genes may play a role in the development of HCC associated with HBV infection [189]. HBx also increased histone acetylation on the DNMT1, DNMT3A and DNMT3B promoters, thereby increasing their expression (see Section 2.8). This suggests that HBx stimulated HAT binding to these promoters [52]. HBx was shown to bind p300/CBP and to stimulate transcription of the *IL-8* and proliferating cell nuclear antigen (*PCNA*) genes. IL-8 possesses mitogenic, motogenic and angiogenic properties, whereas PCNA is implicated in DNA synthesis. Increased expression of these proteins may represent key steps in neoplastic transformation by HBV [190]. On the other hand, HDAC1, HDAC2, and HDAC3 expression was increased in HBV-positive HCCs, in HBx-expressing cells, and in the liver of HBx transgenic mice compared to matching non-tumor tissue, control liver cells, and wild-type mice, respectively [191]. HBx was shown to interact with HDAC1 and HDAC2, and HBx-induced stabilization of hypoxia-inducible factor 1 alpha (HIF-1α), a key regulator in tumor growth, angiogenesis and metastasis of HCC, involved deacetylation by HDAC1 [191,229].

HBx-caused changes in histone methylation is mediated by different enzymes. HBx stimulated the expression of the histone lysine 9-specific methyltransferase SETDB1, leading to the release of transcriptionally silenced HBV genome [193]. The effect on cellular gene expression was not examined, but upregulated expression of SETDB1 was significantly associated with HCC disease progression, cancer aggressiveness, and poorer prognosis of HCC patients [230]. HBx upregulated EZH2 expression by reducing levels of miR-101, which targets EZH2 transcripts, and by inhibiting pRb, resulting in E2F1 mediated transcription of the EZH2 gene. Furthermore, HBx increased the half-life of EZH2 [56,194,195]. HBx augmented the expression of the H3K4-specific methyltransferase set and mynd domain containing (SMYD3) and this resulted in increased transcription of the *c-MYC* proto-oncogene [196]. HBx upregulated expression of the polo like kinase 1 (PLK1). This serine/threonine kinase blocks the repressive effect of PRC2 and the transcription repression complex composed of lysine demethylase 1A (KDM1A), the co-repressor CoRest, HDAC1, and HDAC1 [192]. The KDM1A/CoREST/HDAC1/2 complex enzymatically removed histone acetylations and H3K4 methylations [231]. PLK1-mediated inhibition of PRC2 and KDM1A/CoREST/HDAC1 has been shown to stimulate the Wnt signaling pathway by increasing β-catenin expression and to promote the progression of hepatocellular carcinoma [232]. HBx was found to form a complex with the p65 subunit of NFκB, EZH2, TET2, and DNMT3L and to cause activation of the epithelial cell adhesion molecule (*Ep-CAM*) promoter [197]. HBx was shown to promote H3K4me3 by preventing proteasomal degradation of WD repeat domain 5 protein (WDR5), which is a core subunit of the H3K3 methyltransferase complex, and by recruiting this protein to chromatin. Silencing WDR5 expression reduced tumor formation of HBx expressing cell implanted in nude mice. These results suggest that HBx mediates its oncogenic effect in a WDR5-dependent manner [198].

Taken together, these findings emphasize an important role of HBV-induced histone modifications in the development of HCC.

4. Oncoviruses and microRNA

4.1. microRNA Biogenesis and Functions

MicroRNAs are short, non-coding RNAs that are involved in the regulation of gene expression. Most miRNA genes are transcribed by RNA polymerase II and generate an immature precursor pri-miRNA, which is processed by the RNase III enzymes Drosha and Dicer to produce mature microRNA of 21–23 bases. The mature miRNA is incorporated into the RNA-inducing silencing complex (RISC), which binds to complementary or quasi complementary sequences in the 3′ untranslated region of target mRNAs and induces their degradation or prevents their translation [233]. MicroRNAs play a pivotal role in developmental and cellular processes, but also in cancer [234]. Transcription of miRNA encoding regions is regulated by additional transcription factors and repressors, but also by DNA methylation and chromatin remodeling of their promoters. The role of some microRNAs in virus-positive cancers is outlined below and summarized in Table 3.

Table 3. Some of the microRNAs affected by human tumor viruses. See text for details.

Virus	miR	Viral Protein	Expression	Target	Effect	References
HTLV-1	miR-150	Tax, HBZ	Down	STAT1	↑ proliferation; evade immune clearance	[235,236]
	miR-223	Tax, HBZ	Down	STAT1	↑ proliferation; evade immune clearance	[235,236]
	miR-17	HBZ	Up	HSSB2	↑ proliferation; genome instability	[237]
	miR-21	HBZ	Up	HSSB2	↑ proliferation; genome instability	[237]
	miR-23b	HBZ	Up	HSSB2	↑ proliferation; genome instability	[237]
	miR-27b	HBZ	Up	HSSB2	↑ proliferation; genome instability	[237]
HCV	miR-30c	Core	Down	SNAI1	↑ EMT; ↓ apoptosis	[238]
	miR-122	Core	Down	HCV DNA	↑ viral replication	[238]
	miR-124	Core	Down	SMYD3, EZH2 TERT	↑ migration and invasion Histone modification; ↑ proliferation	[238,239]
	miR 138	Core	Down	WNT1	↑ immortality	[238]
	miR-152	Core	Down	SNAI2	↑ proliferation	[238]
	miR-203	Core	Down	PTEN	↑ EMT; ↓ apoptosis	[238]
	miR-21	Core	Up	IFNAR1	↑ proliferation; ↑ invasion	[238]
	miR-93	Core	Up	MCL1	evade antiviral effect	[238]
	miR-193b	Core	Up	FAS, TERT	↓ apoptosis	[238]
	miR-196b	Core	Up	TLR3, TLR7	↓ apoptosis; ↑ proliferation	[238]
	miR-758	Core	Up	ND *	Immune evasion	[238]
MCPyV	miR-M1	ND *	Up	LT, SP100	Immune evasion	[240,241]
	miR-30a-3p	ND	Up	ATG7, SQSTM1	Suppression of autophagy	[242]
	miR-30a-5p	ND	Up	ATG7, SQSTM1	Suppression of autophagy	[242]
	miR-34a	ND	Up	ND	ND	[243]
	miR-375	ND	Up	ATG7, SQSTM1, LDHB	Suppression of autophagy; ↑ proliferation	[242] [244]
HR-HPV	miR-21	E6, E7	Up	CCL20	↑ proliferation; ↑ migration; ↓ apoptosis	[108]
	miR34a	E6	Down	cyclinD, BCL2	↑ proliferation; ↓ apoptosis	[108,245]
	miR-107	ND	Down	MCL1	Evade antiviral effect	[114]
	miR-124	ND	Down	IGFBP7	↑ proliferation; ↑ migration	[246]
	miR-155	ND	Up	LKB1	↑ proliferation	[245]
EBV	BART2-5p	ND	Up	MCIB	Immune evasion	[247–249]
	BART5-5p	ND	Up	PUMA	↓ apoptosis	[247–249]
	BART9	ND	Up	BIM	↓ apoptosis	[247–249]
	BART11	ND	Up	BIM	↓ apoptosis	[247–249]
	BART11	ND	Up	BIM	↓ apoptosis	[247–249]
	BART15	ND	Up	NLRP3	↑ inflammation	[247–249]
	miR-146-5p	EBNA2	Up	KDM2	Histone modification	[222]
	miR-155	LMP		FOXO3a	↑ proliferation	[248]

Table 3. *Cont.*

Virus	miR	Viral Protein	Expression	Target	Effect	References
KSHV	miR-K12-1	ND	Up	$p21^{CIP1/WAF1}$	↑ proliferation	[250–252]
	miR-K10	ND	Up	BCLAF1	↓ apoptosis	[250–252]
	miR-K12-5	ND	Up	BCLAF1	↓ apoptosis	[250–252]
	miR-K12-9	ND	Up	BCLAF1	↓ apoptosis	[250–252]
	miR-K12-1	ND	Up	Caspase 3	↓ apoptosis	[250–252]
	miR-K12-3	ND	Up	Caspase 3	↓ apoptosis	[250–252]
	miR-K12-4-3	ND	Up	Caspase 3	↓ apoptosis	[250–252]
	miR-K12-4-5	ND	Up	Rbl2	↓ DNA methylation	[253]
	miR-K12-11	ND	Up	Jarid	Histone modification	[253]
	miR-21	K15	Up	ND	↑ migration; ↑ angiogenesis	[248,252]
	miR-31	K15	Up	ND	↑ migration; ↑ angiogenesis	[248,252]
	miR-221	LANA, kaposin	Down	ETS2	↑ migration	[248,252]
	miR-222	LANA, kaposin vFLIP ND	Down	ETS1	↑ migration	[248,252]
	miR-146a		Up	CXCR4	↑ release KSHV-infected endothelial cells	[248,252]
	miR-132		Up	p300	↓ antiviral immunity	[254]
HBV	HBV-miR-2	ND	Up	TRIM35, RAN	↑ migration and invasion	[255]
	HBV-miR-3	ND	Up	PP1A, PTEN	↑ proliferation; ↑ invasion	[256]
	miR-10	HBx	Up	EPHA4	↑ EMT	[257]
	miR-132	HBx	Down	AKT	↑ proliferation	[257]
	miR-143	HBx	Up	FNDC3B	↑ migration, evasion, and metastasis	[257]
	miR-193b	HBx	Up	MCL1	Evade antiviral effect	[257]

↑ = increased; ↓ = decreased; * ND: not determined.

4.2. HTLV-1 and microRNA

No HTLV-1-encoded microRNA has been described so far, but HTLV-1 can alter the expression levels of cellular microRNAs. HTLV-1-transformed cells and ATL-derived cell lines had reduced levels of miR-150 and miR-223. STAT1, whose mRNA is a direct target for these miRNAs, was upregulated in HTLV-1-transformed and ATL cells and was required for the proliferation of these cells. MHC-I levels were also increased in these cells and enhanced MHC-I expression helped the tumor cell to avoid immune clearance [235]. STAT1 has been found to play a role in chromatin decondensation of the MHC locus [236], which may explain concomitant increased expression of both proteins. The mechanisms by which HTLV-1 repressed miR-150 and miR-223 expression are incompletely understood, but Tax, as well as HBZ could increase the expression and activity of E2F1, which is a repressor of the miR-223 promoter [258–260]. The HTLV-1 HBZ protein was also shown to affect microRNA levels. HBZ upregulated miR-17, miR-21, miR23b, and miR-27b by a posttranscriptional maturation mechanism. These microRNAs target mRNA of the nucleic acid binding protein 1 (*NABP1*) gene encoding the ssDNA binding protein HSSB2. Silencing of this DNA repair factor stimulated cell proliferation and genomic instability, indicating that HTLV-1 infection may trigger proliferation and genomic instability by the HBZ/miR-17+miR-21/HSSB2 axis [237].

4.3. HCV and microRNA

HCV does not seem to encode viral microRNA probably because of its cytoplasmic location, which deprives the virus from nuclear proteins, such as RNA polymerase II and Drosha, required for microRNA biogenesis. However, comparative microRNAome profiling of HBV-associated HCCs and HBV-negative HCCs, and of HepG2 hepatocytes stably transfected and full-length HCV genome and control cells demonstrated that HCV elicited changes in cellular miRNA expression [238,261–263]. MicroRNAs including miR-30c, miR-122, miR-124, miR-138, miR-152, and miR-203 were downregulated, whereas miR-21, miR-93, 193b, miR-196a, and miR-758 were upregulated. These microRNAs were shown to regulate cell proliferation, invasion and migration, immune evasion, immortalization and cell survival. The core protein was demonstrated to be responsible for modulating the expression of these microRNA. One modus operandi of core protein-mediated microRNA repression was by inducing methylation of microRNA genes such as the miR-124 gene.

The transcript of the SMYD3 protein, a protein that stimulates migration and invasion, was shown to be a direct target of miR-124. Hence, the core protein can stimulate tumor migration and invasion by DNMT1/methylation-mediated inhibition of miR-124 expression, and consequently preventing miR-124-induced silencing of SMYD3 [262]. EZH2 was shown to be also a direct target of miR-124 and a significant inverse correlation between miR-124 and EZH2 mRNA levels was measured in HCC tissues [239]. This finding suggests that HCV core protein can affect H3K27me3 through a miR-124/EZH2 pathway. Another mechanism by which the core protein affected microRNA levels was by suppressing the activity of Dicer, thereby interfering with the biogenesis of microRNAs [264]. The non-structural proteins NS3, NS4A, NS4B, and NS5A also affected the expression of cellular microRNAs that stimulate proliferation, cell survival, migration and invasion, and immune evasion [238]. The mechanisms by which these HCV proteins modify microRNA expression remains to be determined.

4.4. MCPyV and microRNA

MCPyV encodes a microRNA, referred to as miR-M1, which negatively regulates the expression of LT, a viral protein involved in transcription and replication of the MCPyV genome [240,265]. This viral-encoded microRNA is predicted to regulate viral replication and promote immune evasion [240,241]. Ectopic expression of miR-M1 resulted in significant differentially expressed genes compared to control cells, especially genes whose proteins are involved in the immune response, but also in cell motility [241]. One of the confirmed miR-M1 targets was the transcript for SP100, a protein involved in antiviral immunity. MiR-M1-mediated silencing of SP100 resulted in reduced secretion of C-X-C- motif chemokine ligand 8 (CXCL8) and attenuated neutrophil migration in cell culture. These in vitro data suggest a role for miR-M1 in aiding MCPyV-positive MCCs to escape the immune system. However, deep sequencing analysis showed that very low miR-M1 levels are detectable in less than 50% of MCPyV-positive MCC tumors and undetectable in the majority of MCC tumors, jeopardizing miR-M1′s biological significance in tumorigenesis [265,266]. Minimizing the levels of miR-M1 allows the infected cell to produce more LT transcripts that can be translated into the LT oncoprotein.

Comparative microRNAome studies between virus-positive and virus-negative MCC cell lines and tumors have identified several cellular microRNAs whose expression is associated with the MCPyV status (for a recent review see [267]). These included miR-203, miR-30a-3p, miR-769-5p, miR-34a, miR-30a-5p and miR-375 [267,268]. MiR-375 has been most extensively studies and its serum level correlates with tumor burden, demonstrating that miR-375 serum levels can be considered a valid surrogate biomarker of tumor burden in MCC patients [243,269]. However, the function of miR-375 in MCC is controversial. Abraham and colleagues described the involvement of miR-375 in neuroendocrine differentiation and knockdown of miR-375 in virus-positive cell lines did not alter their growth properties [270]. Recently, Kumar and colleagues found that MCPyV T-antigens and the MCPyV-regulated miRNAs miR-375, miR-30a-3p and miR-30a-5p suppressed autophagy by targeting multiple autophagy genes, thereby protecting MCC cells from autophagy-associated cell death [242]. LDHB is a target of miR-375. This enzyme catalyzes the conversion of lactate to pyruvate and NAD+ to NADH and is known to play important roles in cancer cell growth and progression [271,272]. In another paper, Kumar and colleagues reported that LDHB expression was inversely correlated with miR-375 levels in MCC cells and LDHB was found to have distinct roles in MCPyV positive and MCPyV negative MCC cells. In virus-associated MCC cells, inhibition of miR-375 expression reduced cell growth and induced apoptosis, and silencing of LDHB restored cell growth caused by miR-375 inhibition. An opposite effect was observed in MCPyV negative cell lines were silencing of LDHB reduced cell growth [244]. MiR-375 expression seems to be activated by transcription factor ATOH1 [96]. However, ATOH1 is downregulated during MCC progression, whereas another study demonstrated that expression of ATOH1 was increased in advanced MCCs MCPyV associated carcinogenesis [273,274]. Interestingly, ATOH1 expression is induced by ectopic expression of truncated forms of LT (which are expressed in MCPyV-positive MCCs) in fibroblasts [96]. Another study questioned the role of miR-375 in MCPyV-associated MCC. Highly effective miR-375 knockdown in virus-positive MCC cell lines did not significantly modify the cell viability, morphology and oncogenic signaling pathways [275]. Enrichment of miR-375 in extracellular vescicles has been described, suggesting a role of this microRNA in intercellular communication of MCC. Becker and his group showed that extracellular vesicle-mediated transmission of miR-375 to fibroblasts caused phenotypic changes toward cancer-associated fibroblasts. This observation suggests that miR-375 may contribute to generating a tumor microenvironment [276].

A subset of miRNAs associated with tumor metastasis and MCC-specific survival has been identified. Functionally, overexpression of miR-203 was able to inhibit cell growth, to induce cell cycle arrest, and to regulate survivin expression in MCPyV negative-MCC cells, but not in MCPyV-positive MCC cells. These findings reveal a mechanism for survivin expression regulation in MCC cells and offer insights into the role of miRNAs in MCC tumorigenesis [268].

MCPyV has also been detected in other cancer types, including non-small cell lung cancer [277]. Lasithiotaki et al. demonstrated overexpression of miR-21 and miR-376c in MCPyV-positive compared MCPyV-negative non-small cell lung cancers, whereas miR-145 levels were higher in the virus-negative tumor samples [278].

In conclusion, the MCPyV-encoded microRNA miR-M1 does not seem to be involved in MCC, but MCPyV infection modifies cellular microRNA expression, which may play a role in tumorigenesis and the tumor microenvironment.

4.5. HR-HPV and microRNA

MicroRNA prediction algorithms have been used to forecast putative HPV16- and HPV18-encoded miRNAs [279,280]. By using Northern blotting, a weak hybridization signal corresponding to mature HPV18-miR-LCR3 was detected in the HPV16-positive CaSki cell line [279]. This putative HPV miRNA has high sequence identity to cellular miR-466. Possible targets are genes encoding proteins involved in proliferation, transcription, signaling pathways. Whether HPV18-miR-LCR3 is a truly HPV-encoded miRNA remains to be established. The group of Auvinen identified and validated the expression of 5 HPV16-encoded microRNAs (HPV16-miR-H1, H2, H3, H5 and H6) in HPV-positive cell lines and cervical cancers. In all cases, HPV16-miRs were expressed at low levels [280,281]. Among the putative targets were mRNAs encoding proteins involved in focal adhesion, cell migration, cell proliferation and tumor suppressors [280].

Several studies have shown that HR-HPV positive tumors and cell lines expressing the HR-HPV oncoproteins E5, E6 or E7 have altered cellular microRNAomics compared to control tissue and cells. Upregulation and downregulation of cellular microRNAs have been observed. The microRNAs dysregulated in HPV-positive cervical cancers are involved in cell proliferation, cell survival, angiogenesis, invasion, and migration underscoring their role in HR-HPV pathogenesis (Table 3; [108,109,114,245,282]).

One mechanism by which HPV affected microRNAs expression was by modifying the promoter methylation pattern of the genes encoding microRNAs [246]. For example, no methylation of miR-124 promoter was found in normal cervical cancer, whereas hypermethylation level of the miR-124 promoter increased with the cancer grade [246]. Methylation of the miR-124 promoter was increased and levels of this microRNA were decreased in human foreskin keratinocytes immortalized with either HPV16 or HPV18. Concordantly, levels of insulin like growth factor binding protein 7 (IGFBP7), whose mRNA is a target for miR-124, were increased. Furthermore, ectopic expression of miR-124 in HPV16-positive SiHa and CaSki cervical cancer cell lines reduced their proliferation rate and migration capacity. These results support a role for silencing miR-124 in HPV-mediated cervical carcinogenesis. HPV-induced hypermethylation of miRNA promoters is mediated by increased DNMT1 expression and activity by E6 and E7 as discussed in Section 2.4. HR-HPV infection was also associated with reduced methylation of microRNA genes, but the mechanism by which HR-HPV decreases microRNA promoter methylation is not known [108,109,114,245,282]. Another mechanism by which HR-HPV affected microRNA expression is through targeting cellular proteins involved in the transcription of microRNA genes. HR-HPV E6 induced degradation of p53 and E7 appropriated pRb, which altered the transcription levels of microRNA-encoding genes [114]. A third mode of disturbing microRNA levels is by interfering with the biogenesis of microRNAs. HR-HPV E6 and E7 could altered the expression of microRNA processing proteins, including Drosha and Dicer and different expression of these proteins was observed in HPV-induced cancers compared to normal tissue. Dysregulation of microRNA processing proteins perturbed miRNA biogenesis and affected translation of their target mRNAs [108,283,284].

4.6. EBV and microRNA

The EBV BHRF1 cluster and the BamHI-A rightward transcript (BART) clusters 1 and 2 encode >40 mature miRs, which can regulate host and viral gene expression. These viral miRs are crucial for EBV-associated tumorigenesis by e.g., inhibiting apoptosis, immune evasion, and cell growth [45,247–249]. For example, EBV miR-BART2-5p silences MHC class I polypeptide-related sequence B (MCIB) expression to inhibit natural killer cell recognition and activation, allowing immune evasion of the EBV-positive tumor cell. Other EBV miRNAs that have a predicted role in immune evasion, include miR-BHRF1-3 (target is CXCL11, a T cell attracting chemokine), miR-BART15 (target is the inducer of pro-inflammatory cytokines NLR family pyrin domain containing 3; NLRP3 or cryopin), and miR-BART5-5p (represses the expression of the pro-apoptotic protein p53 upregulated modulator of apoptosis; PUMA). EBV miR-BART9, miR-BART 11 and miR-BART 12 inhibit apoptosis by repressing expression of BIM [247–249].

EBV infection also altered the expression of host cell miRNAs. Comparison of the microRNAomes from EBV-positive nasopharyngeal tissue and non-tumor tissue disclosed several cellular miRNAs that were upregulated, but also many were downregulated. One of the cellular miRNAs induced by EBV is miR-155, an oncomir crucial for B cell transformation and proliferation [248]. The microRNA profile of EBV-positive gastric cancers and EBV-positive lymphomas also displayed differentially expressed host cell microRNAs compared with virus-negative tissue. Again, these microRNAs target transcripts of proteins involved in apoptosis, immune evasion, cell proliferation, invasion and metastasis, hinting to a crucial role in the carcinogenesis of these EBV associated tumors [248,249,285,286]. EBV induced chromatin changes can also be mediated by microRNA. The EBV protein EBNA2 was found to induce miR-146-5p, which targets KDM2 mRNA [222].

The mechanisms by which EBV modulate microRNA expression have been less studied but may include changes in DNA methylation and chromatin of the microRNA genes induced by viral proteins as discussed in Sections 2.6 and 3.6. EBV can also affect the biogenesis of microRNAs as shown for EBV miR-BART6-5p, which targets Dicer mRNA [248].

4.7. KSHV and microRNA

KSHV produces 25 mature microRNAs and more than 2000 host transcripts that encode proteins associated with KSHV pathogenesis can be directly targeted by these viral microRNAs [250,287,288]. The functions of KSHV microRNAs have been extensively studied and showed that they perturbed expression of host proteins, which are involved in angiogenesis, proliferation, cell survival, migration and invasion, and immune evasion [45,247,248,250,251,288]. A few examples are mentioned. KSHV miR-K12-1 helped evading cell cycle arrest by silencing $p21^{CIP1/WAF1}$ expression. KSHV miR-K12-5, miR-K12-9 and miR-K10a/b targeted the pro-apoptotic protein Bcl-2-associated factor 1 (BCLAF1), whereas miR-K12-1, miR-K12-3, and miR-K12-4-3p suppressed caspase 3 expression. These microRNAs allowed the virus to avoid apoptosis. KSHV evaded the innate immune system by miR-K12-5- and miR-K12-9-mediated reduction of myeloid differentiation primary response 88 (MYD88) and interleukin-1 receptor-associated kinase 1 (IRAK1), respectively.

Finally, KSHV microRNAs promoted angiogenesis by downregulating the levels of the anti-angiogenic factor thrombospondin, SH3 domain binding glutamate-rich protein (SH3BGR) and CD82 [247,248,252,288,289]. KSHV-encoded microRNAs were demonstrated to play a role in DNA methylation because infection with a mutant virus unable to express KSHV microRNAs resulted in almost complete loss of DNA methylation. Possible mechanisms could be through miR-K12-4-5p, a KSHV microRNA that prevented synthesis of the DNMT repressor Rbl2, and via miR-K12-11, which targets the PRC2 component Jarid2 [253]. Jarid 2 was also shown to function as a tumor suppressor and regulator of B-cell survival. Hence, KSHV miR-K12-11-mediated inhibition of Jarid2 may contribute to KSHV-induced malignant transformation [290].

The role of KSHV-provoked dysregulated expression of host cell microRNAs in cancer has been extensively reviewed [248,251,252]. We will briefly mention some examples. The viral protein K15 was shown to induce expression of cellular miR-21 and miR-31, which promoted cell migration, angiogenesis, and lymphangiogenesis. The viral proteins LANA and Kaposin B repressed expression of cellular miR-221 and miR-222, which resulted in increased cell migration. vFLIP upregulated miR-146a levels in an NFκB-dependent manner. This host cell microRNA silenced C-X-C motif chemokine receptor 4 (CXCR4), which promoted the premature release of KSHV-infected endothelial cell progenitors into the blood stream [248,252]. Similar to KSHV-encoded microRNAs, KSHV- induced host cell microRNAs could exert an effect on chromatin structure. KSHV was found to upregulate cellular miR-132, which targeted the HAT p300 mRNA [254]. These findings underscore a role for viral and cellular microRNA in KSHV-associated cancer.

4.8. HBV and microRNA

HBV encodes two viral miRNAs: HBV-miR-2 and HBV-miR-3. HBV-miR-2 may act as an oncomiR because it was found to promote cell growth, migration and invasion by downregulating the expression of the E3 ubiquitin-protein ligase tripartite motif containing 35 (TRIM35) and upregulating protein levels the GTPase RAN. TRIM35 is a proapototic protein and can inhibit the Warburg effect, whereas RAN is involved in nucleocytoplasmic transport, but also in metastasis. HBV-miR-3 enhances cell invasion and proliferation by e.g., silencing PP1A and PTEN [255,256].

Several studies showed a role for HBx in up- and downregulating the expression of cellular microRNAs, including miR-10, miR-132, miR-143, and miR-193b. This has been the topic of excellent reviews [190,255,257,263,291,292]. HBx modulates microRNA expression by inducing epigenetic changes in microRNA-encoding genes or modulating expression of genes whose products are involved in microRNA biogenesis. HBx can affect DNA methylation, histone acetylation and histone methylation as discussed in Sections 2.7 and 2.8, which will affect transcription of the microRNA-encoding region. HBx can also affect the affinity of transcription factors involved in transcription of microRNA genes. For example, HBx can interfere with p53 sequence-specific DNA binding of and inhibit p53′s transcriptional activity [132], stabilize c-MYC [293], and activate NFκB-mediated transcription [294]. These three transcription factors have been shown to affect transcription of microRNA genes [295–297]. Moreover, HBx can repress Drosha expression leading to dysregulation of microRNA biogenesis [298]. MicroRNAs modulated by HBx were demonstrated to target genes that encode proteins involved in cell cycle progression, cell survival, immune evasion, invasiveness and migration, and angiogenesis [292]. Thus, dysregulation of microRNA expression is a pivotal mechanism by which HBV promotes hepatocellular carcinogenesis.

5. Oncoviruses and Long Non-Coding RNAs

5.1. Long Non-Coding RNA Biogenesis and Functions

Long non-coding RNAs (lncRNAs) are a heterogeneous group of RNAs that are more than 200 nucleotides long and are not translated into functional proteins. Most lncRNAs are generated by RNA polymerase II and can contain a 5′ cap and 3′ polyA tail. So far, ~18,000 lncRNA genes have been identified in the human genome, but their number is still increasing. The lncRNA genes produce almost 50,000 transcripts, but many remain to be annotated [299]. LncRNAs can act as guides for proteins, including chromatin-modifying complexes and transcriptional activators or repressors. They can also sequester microRNAs and can by binding mRNA, regulate splicing and stability, editing and subcellular localization. LncRNAs can also associate with DNA and regulate histone modification and DNA methylation. Moreover, lncRNAs can induce structural changes in proteins. Therefore, lncRNAs play crucial roles in gene expression, but they are also important in maintaining chromosome integrity. LncRNAs are crucial for normal cellular processes, but there is clear evidence that they are involved in cancer [300,301]. Some examples of lcnRNA and their role in virus-positive cancers are discussed below and are summarized in Table 4.

Table 4. Human tumor virus and lncRNAs with their targets and known functions in virus-induced cancer. See text for details.

Virus	lncRNA	Expression	Viral Protein	Target	Effect	References
HTLV-1	HBZ antisense	Up	ND *	*CCR4*, *E2F1*, and *survivin* genes	↑ proliferation; ↓ apoptosis	[302]
	ANRIL	Up	Tax, HBZ	Recruits EZH2 and p65	↑ proliferation; ↓ apoptosis	[303]
HCV	UCA1	Up	ND	miR-203 sponge → increased SNAI2	↑ EMT	[304]
	PVT1	Up	ND	Recruits EZH2	↑ proliferation	[304,305]
	AK021443	Up	ND	ND	↑ proliferation	[304,305]
	LINC01419	Up	ND	ND	↑ proliferation	[304,305]
	HULC	Up	ND	ND	↑ proliferation	[304,305]
	AF070632	Down	ND	ND	metabolism	[304,305]
MCPyV	HNGA1	Up	ND	miR-375 sponge	↑ proliferation; ↑ glycolysis	[306]
HR-HPV	HOTAIR	Up	ND	miR-23b and miR-143-3p sponge; Recruits EZH2; increased expression VEGF and MMP9	↑ angiogenesis; ↑ invasion	[245]
	TMPOP2	Up	E6	miR-139 and miR-375 sponge → ↑ E6 and ↑ E7; inhibition of E-cadherin	↑ proliferation; ↑ invasion; ↑ angiogenesis	[307]
	CCDST	Down	E6, E7	Increased DHX9 level	↑ mobility; ↑ angiogenesis	[308]
	FANCI-2	Up	E6, E7	ND	ND	[309]
EBV	EBER1	Up	ND	ND	Immune evasion; ↓ apoptosis	[45,310,311]
	EBER2	Up	ND	Recruits PAX5	↓ apoptosis	[45,310,311]
	BART	Up	ND	↓ IFNB1 and ↓ CXCL8	Immune evasion	[312–315]
	BHLF1	Up	ND	Viral replication and latency	Persistent infection	[312–315]
	MALAT1	Up	ND	miR-124 and miR-195 sponge	↑ proliferation; ↑ invasion	[313]
	HOTAIR	Up	ND	miR-34a, miR-217 and miR-618 sponge	↑ angiogenesis; ↑ invasion and migration	[314]
	H19	Up	ND	miR-141 and miR-630 sponge	↓ apoptosis; ↑ invasion and migration	[314]
KSHV	PAN	Up	ND	Binds IRF4, histone demethylases, EZH2, and SUZ12	Immune evasion; chromatin modification	[316]
	H19	Up	ND	miR23b, miR-34a, miR124 sponge	Tumor progression and metastasis	[314]
	MALAT1	Up	ND	miR-124 and miR-195 sponge	↑ proliferation; ↑ invasion	[314]
	KIKAT	Up	ND	Interaction with KDM4A; ↑ ATOM	↑ angiogenesis; ↑ migration	[317]
HBV	HEIH	Up	HBx	Recruits EZH2	↑ proliferation; ↑ invasion; ↓ apoptosis	[318–320]
	UCA1	Up	ND	Recruits EZH2; miR-216b sponge	↑ proliferation; ↑ invasion; ↓ apoptosis	[318–324]
	HOTAIR	Up	ND	Recruits EZH2	↑ proliferation; ↑ invasion; ↓ apoptosis	[318–321]
	HULC	Up	HBx	Recruits EZH2; miR-186 and miR-372 sponge	↑ proliferation; ↑ invasion; ↓ apoptosis	[318–322]
	LINC00152	Up	HBx	Recruits EZH2	↑ proliferation; ↑ invasion; ↓ apoptosis	[319,325]
	ANRIL	Up	ND	Recruits PRC2; miR-122-5p sponge	↑ proliferation; ↑ invasion; ↓ apoptosis	[319]
	HBx-LINE1	Up	ND	Activates WNT pathway	↑ proliferation; ↑ invasion	[318,319,321,326]

↑ = increased; ↓ = decreased; * ND: not determined.

5.2. HTLV-1 and lncRNA

HTLV-1 produces the antisense mRNA HBZ that is inefficiently polyadenylated and as a result the minor fraction of properly polyadenylated HBZ mRNA is transported to the cytoplasm and translated into the HBZ protein, while the majority of aberrant polyadenylated antisense mRNA is retained in the nucleus and acts as lncRNA. Nuclear HBZ mRNA could bind to the promoters of the cellular genes, including the genes encoding

C-C chemokine receptor type 4 (CCR4) and E2F1, and enhanced transcription of these genes, resulting in stimulation of proliferation of HTLV-1-infected cells. HTLV-1 antisense mRNA also promoted expression of survivin [302]. The exact mechanism by which HBZ mRNA exerts its transcriptional regulatory functions are unknown but altered gene expression by this lcnRNA can contribute to HTLV-1-induced oncogenesis.

Comparing the levels of cellular lncRNA in ATL cells, HTLV-1-infected cell lines and control cells revealed upregulation of lncRNAs ANRIL (antisense noncoding RNA in the INK4 locus), H19, and SAF (Fas-Antisense) and slight downregulation of HOTAIR (HOX antisense intergenic RNA) and TUSC7 (tumor suppressor candidate 7) by HTLV-1 [303]. The authors showed that enhancement of ANRIL expression depended on transcription factor E2F1. The exact mechanism by which HTLV-1 regulates ANRIL expression is not known, but Tax has been shown to increase expression and activation of E2F1, whereas HBZ abrogated the interaction between pRb and HDAC3, thereby activating E2F1 [260,327]. Knockdown of ANRIL in ATL cells impaired proliferation and provoked apoptosis. Tumor growth of xenografted ANRIL knockout cells was reduced compared to wild-type cells in mice. ANRIL could form a complex with EZH2 and p65 and enhanced the binding of p65 to NFκB-responsive promoters, whereas ANRIL also formed a complex with EZH2 and repressed $p21^{CIP1/WAF1}$ expression by increasing H3K27me of the *CDKN1A* promoter [303]. In conclusion, HTLV-1-encoded lncRNA and HTLV-1-induced cellular lncRNAs are involved in processes controlling cell proliferation and cell survival and may contribute to HTLV-1 associated leukemogenesis.

5.3. HCV and lncRNA

So far, no HCV-encoded lncRNAs have been identified. However, it was shown that the 5′ untranslated region could be processed by the cellular endoribonuclease XRN1, generating subgenomic viral RNAs that are not translated and therefore may act as viral lncRNAs [328]. The functions of these subgenomic viral RNAs remain to be determined.

Results from several studies comparing HCV-positive HCC with healthy liver tissue showed that several lncRNAs have significantly different expression levels [291,304,305,329]. Several of these HCV-induced lncRNA affect the viral life cycle and are beyond the scope of this review [305,329]. However, other HCV-induced lncRNAs are related to HCC, and while the function of most of these lncRNAs remains elusive, the role of some lncRNAs in HCV-related HCC has been addressed. LncRNA urothelial carcinoma associated 1 (UCA1) is involved in epithelial-to-mesenchymal transmission through sponging miR-203, resulting in increased SNAI2 expression levels. PVT1 (plasmacytoma variant translocation 1) lncRNA recruits EZH2, which facilitates histone modification. PVT1 could promote HCC cell proliferation by stabilizing nucleolar protein 2 and by downregulating transcription of the proto-oncogene c-myc. AK021443, LINC01419 (or PRLH1 for p53-regulated lncRNA for homologous recombination repair 1), and HULC (highly upregulated in liver cancer) lncRNAs are upregulated and stimulate cell proliferation or metastasis. AF070632 and aHIF (antisense to hypoxia-inducible factor 1 alpha) lncRNAs are downregulated. The former is involved in cell metabolism, while the function of the latter is unknown [304,305].

5.4. MCPyV and lncRNA

The existence of putative MCPyV-encoding lncRNAs and the effect of MCPyV infection on host cell lncRNA expression have not been addressed. One study in head and neck squamous cell carcinoma cells demonstrated that miR-375 silenced the expression of the glucose transporter protein solute carrier family 2 member 1 (SCL2A1 or glucose transporter type 1; GLUT1) by targeting SCL2A1 mRNA. The head and neck squamous cell carcinoma glycolysis-associated 1 (HNGA1) lnc RNA was upregulated and HNGA1 functioned as a sponge for miR-375. Ectopic expression of lcnRNA HNGA1 in head and neck squamous cell carcinoma cells stimulated cell proliferation and glycolysis, and accelerated tumor growth in xenograft mouse [306]. The expression level of lcnRNA HNGA1 in MCC has to the best of our knowledge not been examined, but miR-375 is upregulated in MCC (see Section 4.4), suggesting that little or no HNGA1 lcnRNA is present in MCC tumors. MCPyV sT was shown to upregulate the expression of the glucose transporters SLC2A1 and SLC2A3 in normal human fibroblasts and SCL2A1 is significantly expressed in MCC [330–332]. These findings suggest that MCPyV uses an HNGA1-independent mechanism to upregulate expression of SLC2A1.

5.5. HPV and lncRNA

Several lncRNAs previously found to be involved in cancer are also differentially expressed in cervical cancers compared to control samples. RNA interference-mediated silencing of E7 in HPV18-positive HeLa cells or expression of E6 in primary human keratinocytes resulted in altered expression of several lncRNAs compared to control cells [114,245,309,333,334]. One of the upregulated lncRNA in cervical cancers is HOTAIR. This lncRNA could sponge miR-23b and miR-143-3p, could recruit EZH2 and promoted expression of VEGF and matrix metalloproteinase 9 (MMP9), thus stimulating carcinogenic processes [245]. Another lncRNA that was upregulated in cervical cancers is thymopoietin pseudogene 2 (TMPOP2) lncRNA. p53 bound the promoter region of the *TMPOP2* gene and inhibited its expression. E6 released p53-mediated inhibition of TMPOP2 expression. Interestingly, overexpression of TMPOP2 enhanced E6 and E7 protein levels because TMPOP2 sponged miR-375 and miR-139, which target E6/E7 mRNA [307]. Moreover, TMPOP2 inhibited E-cadherin expression by recruiting EZH2. Thus, enhanced expression of TMPOP2 seems to play an important role in HPV-induced tumorigenesis. E6 and E7 downregulated expression of cervical cancer DHX9 suppressive transcript (CCDST) lncRNA, which resulted in increased DExH-box helicase 9 (DHX9) protein levels, thereby accelerating cell mobility and angiogenesis [308]. E7 and to a lesser extent E6 elevated Fanconi anemia complementation group 1–2 (FANCI-2) lncRNA levels and this was dependent on YY1 binding sites in the promoter region of *FANC-2*. E6 and E7 were found to reduce miR-29a levels, which targets transcription factor YY1 and E7 altered the activity of YY1 [309]. The exact role of FANCI-2 in HPV-induced cancer remains elusive. Altered lncRNA expression was also observed in HPV-positive head and neck squamous carcinomas compared with normal tissue [333]. In conclusion, affecting cellular lncRNA expression may be a mechanism that contributes to HPV-induced carcinogenesis.

5.6. EBV and lncRNA

EBV encodes two small non-coding RNAs EBV-encoded RNA1 (EBER1) and EBER2 of 167 and 172 nucleotides long, respectively. Both are RNA polymerase III transcripts and they are present in high copy numbers (~10^6 for EBER1 and ~2.5×10^5 for EBER2) in infected cells. Although shorter than classical lncRNAs, EBERs are considered lncRNAs. EBER2 acts as a guide RNA to recruit transcription PAX5 to viral target sites, where the complex suppresses transcription. The role of EBER1 remains poorly understood, but both EBER1 and EBER2 play a role in suppressing the innate immune system, avoiding apoptosis and activating the oncogenic phosphatidylinositol 3-kinase (PI3K)-Akt signaling pathway [45,310,311]. A recent study reported that extracellular vesicles could transmit EBERs from EBV-positive nasopharyngeal carcinoma cells to endothelial cells and promote angiogenesis through upregulation of vascular cell adhesion molecule 1 (VCAM1) expression via TLR3/retinoic acid-inducible gene 1 protein (RIG-1)/MAPK pathway [335]. Taken together, EBER1 and EBER2 play a crucial role in EBV-induced tumorigenesis. EBV encodes other lncRNAs, including BART transcripts, and the EBV BamHI leftward reading frame 1 (BHLF1) region. The latter encodes also and circular RNA (see Section 6.6). BART lncRNAs are involved in the epigenetic regulation of host gene expression and were demonstrated to inhibit expression of interferon beta 1 (*IFNB1*) and *CXCL8* genes. BHLF1 lncRNAs promote EBV replication but may also contribute to viral latency [312–314].

Comparison of EBV-positive tumors with control cells identified several cellular lncRNAs that were differentially expressed. Some of these cellular lncRNAs, such as H19, HOTAIR, and metastasis-associated lung adenocarcinoma transcript 1 (MALAT1, aka nuclear-enriched abundant transcript 2; NEAT2) interfere with processes such as apoptosis, migration and invasion, proliferation, and DNA repair [45,313–315]. In conclusion, both EBV lncRNAs and EBV-induced cellular lncRNAs play indispensable roles in EBV-provoked malignancies.

5.7. KSHV and lncRNA

Approximately 16 potential KSHV lncRNAs have been described, with polyadenylated nuclear RNA (PAN) lncRNA the most important and best-characterized. This lncRNA is involved in viral gene expression, replication, and immune modulation. PAN RNA was shown to bind transcription factor interferon regulatory factor 4 (IRF4) and inhibit transcription of IRF4-responsive genes. Moreover, PAN lncRNA could interact with H3K27-specific demethylases UTX and JMJD3, and with the PRC2 components EZH2 and SUZ12. Hence, PAN lncRNA affects chromatin modification, resulting in altered host gene expression and seems to be required for efficient nuclear export of mRNA [45,128,314,316]. This viral-encoded lncRNA may play an essential role in KSHV-induced cancers.

Several cellular lncRNAs, including H19, growth arrest specific 5 (GAS5), deleted in lymphocytic leukemia 2 (DLEU2) and MALAT1 are abnormally expressed in KSHV-infected cells and aberrant expression of these lncRNAs has been associated with oncogenic processes such as proliferation, migration, invasion, cell survival and angiogenesis [314]. KSHV can also provoke changes in DNA methylation and histone modifications through induction of cellular lncRNA as shown by Yang and coworkers. The authors identified KSHV-induced KDM4A-associated transcript (KIKAT/LINC01061) lncRNA as an KSHV-induced cellular lncRNA and demonstrated that this lncRNA could interact with the histone lysine demethylase KDM4A, thereby providing an open chromatin environment and allowing activation of gene transcription. Expression of 44 genes was upregulated and some of these have been identified in cancer-related pathways. One of them was *ATOM*, which encodes angiomotin, a protein that promotes cell migration and angiogenesis. In accordance, overexpression of KIKAT/LINC01061 in SLK cells induced cell migration [317]. Hence, KSHV-induced expression of KIKAT/LINC01061 may play a role in KSHV's pathogenicity.

5.8. HBV and lncRNA

The expression of several lncRNAs are dysregulated in HBV-associated HCC compared to healthy liver tissue. Some of these lncRNAs are discussed here. For a complete overview, the reader is referred to excellent recent reviews [291,305,318,319,321,322,326].

Examples of cellular lncRNAs upregulated by HBV include highly expressed in HCC (HEIH), UCA1, HOTAIR, HULC, and LINC00152, which all were shown to interact with EZH2 and repress gene expression, thereby promoting proliferation, migration and invasion, cell survival, and tumor growth. UCA1 recruited EZH2 to the *CDKN1B* promoter, whose gene encodes cyclin dependent kinase inhibitor $p27^{KIP1}$, and repressed transcription. Another lncRNA that is upregulated is ANRIL, which binds PRC2 and represses transcription of Krüppel-like factor 2 and sequesters miR-122-5p. Knockdown of ANRIL expression induced apoptosis and suppressed proliferation, invasion and migration of HCC cells in vitro. Integration of HBV adjacent to long interspersed nuclear element 1 (LINE1) resulted in the transcription of a chimeric lncRNA, HBx-LINE1, which can be detected in ~23% of HBV-associated HCC tumors. HBx-LINE1 activated the Wnt signaling pathway by inducing nuclear localization of β-catenin, and stimulated cell proliferation and metastasis in vitro. An HBx-LINE1 transgenic mouse model revealed that the animals were more susceptible to diethylnitrosamine-induced tumor formation than wild-type mice and nuclear localization of β-catenin in hepatocytes of the transgenic animals was observed [305,319,321–323,326].

Reduced expression of the lncRNAs n346077 and downregulated expression by HBx (DREH) was observed in HBV-positive HCC included. DREH inhibited proliferation and n346077 prevented migration and invasion in vitro [305,319,321–323,326].

The mechanisms by which HBV modulates expression of lncRNAs remains largely unknown. Studies with HBx expression in liver cell lines and with an HBx transgenic mouse model showed that HBx is directly involved in regulating lncRNA expression. HBx increased expression of HULC and HEIH [319,320,324]. HBx-induced upregulation of HULC is mediated by CREB, whereas HBx-induced expression of HEIH is mediated by transcription factor Sp1 [319,324]. HBx has been shown to increase the DNA binding affinity of CREB and to induce phosphorylation of Sp1 and augment binding to DNA [336,337], suggesting that HBx promotes transcription of the HEIH gene through CREB and Sp1. The gene of lncRNA LINC00152 was shown to be hypomethylated and expression levels correlated with HBx expression levels in tumors and was induced when HBx was ectopically expressed or downregulated when HBx was silenced in vitro [325]. As mentioned in Section 2.8, HBx can cause hypomethylation through releasing DNMT3A from promoters [51]. HBx has also been found to interact with lncRNA DLEU2 and to displace EZH2 from both the viral and host genome [338]. Taken, together, HBV-induced changes in lncRNA expression may assist with the development of HCC.

6. Oncoviruses and circRNAs

6.1. Circular RNA Biogenesis and Functions

Circular RNAs (circRNAs) are single-stranded, RNAs produced from pre-mRNA by a back splicing mechanism in which the 5′ and 3′ termini are covalently joined. They vary in length from less than a hundred to several thousands of nucleotides [339,340]. CircRNAs function in transcriptional, post-transcriptional, translational and post-translational processes by acting as miRNA/protein sponges, modulators of splicing, and by recruiting proteins and affecting protein function and stability. They can also serve as mRNA that are translated into peptides [340,341]. Compiling evidence that underscores their role in cancer was the topic of recent reviews [342–344]. Some examples of circRNAs that are encoded and induced by tumor viruses are discussed below and are summarized in Table 5.

Table 5. Human tumor virus and circular RNAs with their targets and known functions in virus-induced oncogenesis. See text for details.

Virus	circRNA	Expression	Target	Effect	References
HTLV-1	ND	ND *	ND	ND	
HCV	circPSD3	Up	ND	↓ viral infectivity	[345]
MCPyV	circALTO1/2	Up	Sponge miR-M1; activation of some promoters; encodes ALTO protein	↑ LT expression	[346]
	circMCV-T		Sponge miR-M1	↑ LT expression	[347]
HR-HPV	circE7	Up	Encodes E7; sponge for several miRs	↑ proliferation; ↑ invasion; ↓ apoptosis; ↑ angiogenesis	[348,349]
	circRNA8924	Up	miR-518-d-5p and miR-519-59 sponge	↑ proliferation; ↑ invasion	[350]
	circ_0005576	Up	miR-153-3p sponge	↑ proliferation; ↑ invasion	[350]
	circ_0018239	Up	ND	↑ proliferation; ↑ invasion; ↑ immune evasion	[351]
	circ_0000263	Up	*TP53* mRNA	↑ proliferation; ↑ invasion; ↑ immune evasion	[351]
	circ_000284	Up	*SNAI2* mRNA	↑ proliferation; ↑ invasion; ↑ immune evasion	[351]
EBV	>30 EBV circRNAs	Up	miR-31, miR-203, and miR-451 sponge	↑ proliferation; ↑ EMT; ↓ apoptosis	[352,353]
	EBV circBHLF1	Up	Putative 200 aa protein	ND	[353]
KSHV	>100 KSHV circRNAs	Up	ND	Viral infection; immune modulation	[353–355]
	circ_0001400	Up	ND	Suppression viral expression	[354]
	circARFGEF1	Up	Sequesters miR-125a-3p → ↑ GLRX3	↑ proliferation; ↑ invasion; ↑ angiogenesis	[356]

Table 5. *Cont.*

Virus	circRNA	Expression	Target	Effect	References
HBV	HBV_circ_1	Up	Binds DXH9 and P0/P1/P2	Viral gene expression	[357]
	circ_100338	Up	miR-141-3p sponge	↑ proliferation; ↑ invasion; ↓ apoptosis	[351]
	circ-RNF13	Up	Sequesters miR-425-5p → ↑ TGIF2	↑ proliferation; ↑ invasion; ↓ apoptosis	[358]

↑ = increased; ↓ = decreased; * ND: not determined.

6.2. HTLV-1 and circRNA

The existence of HTLV-1 encoded circRNA and whether HTLV-1 can induce the production of cellular circRNA have not been investigated. However, after converting its ssRNA virus genome into dsDNA, the dsDNA integrates into the host chromosome and RNA polymerase II-derived viral transcripts are spliced, suggesting that viral circRNA may be produced.

6.3. HCV and circRNA

It is still unclear whether HCV encodes circRNA. The fact that this virus replicates in the cytoplasm, in the absence of the nuclear splicing machinery, may explain why no HCV circRNA is generated. A recent study examined the cellular circRNA profile in uninfected and HCV-infected liver cells. The authors found 10 circRNAs that were significantly upregulated, whereas 63 had decreased levels in the HCV-positive cells compared to control cells. The authors elaborated on the role of the upregulated circPSD3, which was generated by a backsplicing event between exon 5 and exon 8 from the pleckstrin and Sec 7 domain containing (PSD) transcript and found that depletion of circPSD3 diminished viral infectivity [345]. The mechanism by which HCV dysregulates expression of cellular circRNA and a possible role of circPSD3 and the other circRNAs in HCV-induced HCC remain to be explored.

6.4. MCPyV and circRNA

The LT and sT encoding region of MCPyV contains an alternative reading frame which encodes the ALTO protein with unknown function [359]. A recent study identified two viral circRNAs derived from ALTO mRNA, circALTO1 (762 nucleotides in length) and circALTO2 (940 nucleotides long) in MCPyV-positive MCC cell lines, whereas only circALTO2 was detected in virus-positive tumors, suggesting that the abundance of the circALTO isoforms might differ in vivo [346]. The circALTOs were stable, predominantly located in the cytoplasm, and enriched in exosomes. CircALTOs were also N6-methyladenosine (m6A) modified, which has been reported to promote cap-independent translation [360]. Indeed, circALTO but could be translated and was negatively regulated by MCPyV miR-M1. Transfection of cells with expression plasmids for circALTO1 or circALTO2 showed that ALTO stimulated the SV40 early and late promoter, and the cytomegalovirus immediate early promoter, but not the MCPyV early and late promoter, the Trichodysplasia spinulosa polyomavirus promoter, nor two cellular promoters (the elongation factor 1-alpha and the phosphoglycerate kinase 1 promoter). These findings suggest that ALTO functions as a transcriptional activator for some promoters. Accordingly, overexpression of circALTO1 resulted in significant upregulation of a large number of genes, while only a few genes were markedly downregulated. Proteins encoded by these genes included components of NFκB signaling pathway, transcription factors, and inflammatory and anti-viral cytokines, suggesting the circALTO can modulate genes and pathways implicated in MCPyV pathogenesis. As circALTOs are enriched in exosomes, it is tempting to the speculate that circALTOs could prepare recipient cells for MCPyV infection and promote tumor development [346]. MCPyV may encode additional circRNAs because potential circRNAs were predicted upstream of genes encoding the capsid protein VP2 [346]. Another study identified a 762 nucleotide long circRNA, which the authors designated circMCV-T, in MCPyV-positive MCC cell lines and tumor samples [347]. This circMCV-T was unlikely to be translated but may act as a decoy for the MCPyV-encoded microRNA miR-M1. MiR-M1 targets LT transcripts and the authors showed that circMCV-T sequestered miR-M1, thereby reversing the inhibitory effect of miR-M1 on LT expression. The authors predicted that circMCV-T may aid viral replication by sequestering miR-M1 from the viral transcripts encoding the proteins involved in replication. Indeed, they showed that exogenous expression of circMCV-T was accompanied by increase in the levels of LT and sT transcripts and stimulated viral replication. The complex interaction between viral mRNA, miRNA and circRNA is important to meticulously fine-tune viral replication. A possible role for cricMCV-T in MCPyV-induced MCC, where miR-M1 levels are undetectable or very low [265,266], remains to be addressed.

6.5. HR-HPV and circRNA

A number of HPV-encoded circRNAs have been identified in HPV positive cervical cancers. Among these viral circRNAs (v-circRNA), is circE7 the most abundant and can be translated into E7. Multiple microRNA binding sites have been identified on circE7, suggesting that it can act as a miRNA sponge. circE7 has been reported also in HPV-positive anal and head and neck cancers [348,349]. Surprisingly, circE7 levels correlated with improved survival of patients with HPV-positive cervical and anal squamous cell carcinoma [349]. The role of other HPV-encoded circRNAs is unclear, but it is assumed that by sequestering microRNA they promote tumorigenic processes such as proliferation, cell survival, invasion, migration, and angiogenesis [245,333,348,361,362].

High throughput RNA sequencing studies of HPV-positive anogenital and oropharyngeal cancers and matched adjacent non-tumor tissues discovered numerous differentially expressed cellular circRNA [114,363–365]. Ectopic expression of HPV16 E7 in the HPV negative cervical cancer cell line C33A resulted in upregulation and downregulation of numerous host cell circRNAs. Upregulated circRNAs included circRNA8924, which target miR-518-d-5p/miR-519-59, and hsa_circ_0005576, which usurps miR-153-3p [350]. These circRNAs have been shown to promote proliferation, migration or invasion [350]. Hsa_circ_0018239 is also overexpressed in cervical cancer and knockdown of this circRNA suppressed migration, proliferation and immune evasion. Other studies demonstrated enhanced levels of circRNAs that target *TP53* (circ_0000263) or *SNAI2* (circ_000284) mRNA. The latter encodes a protein involved in epithelial–mesenchymal transition (EMT) [351].

6.6. EBV and circRNA

EBV encodes more than 30 different v-circRNAs from dissimilar regions of its genome, which are stably expressed in all EBV-associated tumors [352,353,366,367]. EBV-encoded circRNAs play a role in viral replication and facilitate EBV pathogenesis and tumor development. EBV circRNAs were shown to sponge host cell microRNAs such as miR-31, miR-203, and miR-451, promote proliferation, EMT and cell survival [351,353]. Other microRNA sequestered by EBV circRNAs allowed translation of their target mRNAs, resulting in increased protein levels of, e.g., E2F3, MAPK, checkpoint kinase 1 (CHEK1), and transforming growth factor beta 1 (TGFβ1). Enhanced expression of these proteins may contribute to EBV-induced carcinogenesis [313]. Some EBV circRNAs contain open reading frames and may encode putative peptides. One example is v-circBHLF1, which may be translated in a putative 200 amino acid peptide, but the existence of this protein remains to be confirmed [353].

6.7. KSHV and circRNA

KSHV produces more than 100 circRNAs, which can be detected in Kaposi's sarcoma tumors, PEL, and multicentric Castleman's disease [316,353–355]. The functions of v-circRNAs in KSHV's pathogenesis are still largely enigmatic. Interestingly, the KSHV virion contain v-circRNAs, suggesting that they are important to establish infection and maybe exert a role as immune modulators [355]. Among virion-contained circRNAs is circ_0001400, which suppressed viral gene expression and thus may serve as an antiviral defense mechanism [354]. KSHV can also trigger the production of cellular circRNAs. Infection of endothelial cells with wild-type KSHV or KSHV with mutated vIRF1 or ectopic expression of vIRF1 demonstrated differential expression of several circRNAs. One of vIRF1-upregulated circRNAs was circARFGEF1. vIRF1 interacted with transcription factor lymphoid enhancer binding protein 1 (LEF1) and bound to the promoter region that produces the transcript from which circARFGEF1 is generated. The authors went on to show that circARFGEF1 could bind and degrade miR-125a-3p. Levels of glutaredoxin 3 (GLRX3), whose transcript is a miR-125a-3p target, were upregulated and knockdown of GLRX3 impaired motility, proliferation and angiogenesis. Accordingly, knockdown of circARFGEF1 or miR-125a-3p overexpression inhibited vIRF1-induced cell migration, proliferation and in vivo angiogenesis [356]. These results indicate that the vIRF1/circARFGEF1/miR-125a-3p/GLRX3 axis is essential for KSHV-induced invasion and angiogenesis.

6.8. HBV and circRNA

An HBV-encoded circRNA, HBV_circ_1, has been detected in HBV-infected cells and in HBV-associated HCC. HBV_circ_1 is mainly located in the cytoplasm and it was found to bind DXH9, as well as the ribosomal protein P0/P1/P2. Knockdown of DXH9 increased HBV_circ_1 levels which is in agreement with a previous study that described a role of DXH9 in repressing circRNA production [368]. Increased HBV_circ_1 levels or knockdown of DXH9 coincided with decreased levels of RNAs encoding the viral proteins. Hence, DXH9 may be an essential cellular factor in the regulation of HBV protein levels [357]. The mechanism by which HBV_circ_1 is produced and whether it may act as a decoy for microRNAs or other proteins remain to be elucidated.

There is compelling evidence that cellular circRNAs are involved in the etiology of HBV-associated HCC. By comparing the landscape of circRNA from HBV-positive HCC tissue and control tissue, differentially expressed cellular circRNAs were identified [351,358,369–371]. The role of some of these differentially expressed circRNAs in the pathogenesis of HBV was explored. For example, hsa_circRNA_100338 is upregulated and this circRNA acted as a sponge for miR-141-3p, a microRNA known to inhibit proliferation, migration and invasion and to regulate apoptosis [351]. Additionally, circ-RNF13 (=hsa_cric_0067717 or hsa_circ_103489) was upregulated in HBV-positive HCC tissue and cells compared with paired normal liver tissue or HBV-negative HCC cells. The authors showed that this circRNA sequestered miR-425-5p, which targets the TGFβ-induced homeobox 2 (TGIF2) transcript. Si-RNA mediated silencing of circ-RNF13 suppressed proliferation, migration, and invasion, and induced apoptosis in vitro, and suppressed tumor growth in vivo. Moreover, it blocked viral DNA replication and reduced the levels of hepatitis B surface and E antigens [358]. These examples show that HBV-induced circRNAs may play essential roles in HBV infection and HBV-positive HCC development.

7. Epigenetic Targeting Therapies for Treatment of Virus-Associated Tumors

Oncovirus infection has a substantial impact on the host's epigenetic landscape, which plays a crucial role in virus-driven oncogenesis. Reversing or preventing tumor virus induced epigenetic changes may therefore be a strategy for treating virus-associated tumors (for recent reviews [372–374]). A few examples will be discussed in this section.

As mentioned in Section 2, tumor viruses trigger often hypermethylation of tumor suppressor genes, resulting in silencing their expression. DNMT inhibitors can be used to reverse hypermethylation of these genes. The DNMT inhibitors 5-azacytidine and 5-aza-2′-deoxycytidine have been successfully used for treating patients with EBV-positive B cell lymphoma or HPV-positive cancers, respectively [375,376]. A recent study reported that infection of liver cells pretreated with 5-azacytidine with HBV and then challenged with IFNα, inhibited HBV replication by >50%, whereas no inhibition was measured in non-5-azacytidine treated cells [377]. This result illustrates that epigenetic reprogramming restores the antiviral activity of IFNα and suggests that demethylating drugs may have therapeutic potentials for treating HBV-infection and HBV-associated cancer.

In vitro and in vivo studies have demonstrated that HDAC inhibitors, such as the FDA approved drugs vorinostat, belinostat and panobinostat, could be a promising therapy for HPV-positive cancers [378]. A phase I/II study with the HDAC inhibitor entinostat is now recruiting patients with HPV associated malignancies (clinical trial study NCT04708470). Inhibitors of HAT are also being developed. One of them, the specific p300 inhibitor C646, reduces HR-HPV E6 and E7 expression in cervical cancer cells [379]. Several inhibitors against other histone modifying enzymes have been developed. One of the most studied is 3-deazaneplanocin (ZNep), which had a stronger anti-proliferative effect on HPV-positive oropharyngeal squamous cell carcinoma cell lines compared to virus-negative cell lines [380]. Inhibition of EZH2 with ZNep in HTLV-1 infected cells or ATL cells also reduced cell proliferation [74]. Inhibition of KDM1A with the drug GSK-LSD1 induced growth arrest and cell death of several MCPyV positive MCC cell lines and significantly reduced tumor growth in a xenograft model compared with vehicle treated animals. No synergistic effect was observed when HDAC and LSD1 inhibitors were used [381].

Anti-microRNAs have been designed to target specific microRNAs. The anti-miR-122 (Miravirsen) is used for treatment of HCV infections [382], and blocking of EBV microRNA BART17-5p, which targets the mRNA for tumor suppressor KLF2, suppressed the development of EBV associated gastric cancers [383].

How to exploit lncRNAs and circRNAs for therapeutic purposes in virus-associated cancers remains in its infancy. A recent study showed that a peptide that blocks the interaction between lncRNA HOTAIR and EZH2 decreased invasion of cancer cells in vitro and reduced tumor formation in ovarian tumor xenograft [384]. This may be relevant for virus-associated cancers because levels of HOTAIR are upregulated by several human tumor viruses (see Table 4). CRISPR/Cas9-mediated targeting of lncRNA UCA1 resulted in increased apoptosis and decreased cell proliferation, migration and invasion of bladder cancer cells in vitro and in vivo, but the application in virus-associated cancers expressing this lncRNA remains to be explored [385].

8. Conclusions

All known human tumor viruses show great diversity in their structure and genome sequence. Their oncoproteins have no similarity, yet these viruses use the same mechanisms to induce cancer. They convey the hallmarks of cancer on the host cell. One way to obtain this is by altering gene expression in the infected cell and their viral proteins may do so by functioning as transcriptional regulators, by regulating the activity of transcriptional activators and repressors, or by inducing mutations in the host genome. During recent years, it has become clear that tumor viruses also apply epigenetic mechanisms to alter cellular gene expression. Again, all human tumor viruses seem to apply the same strategies (Figure 1). They can produce their own microRNA, lncRNA and circRNA or induce these cellular non-coding RNAs. Oncoviruses can modify DNA methylation, cause PTM on histones, and induce chromatin remodeling. However, several central questions remain to be elaborated. The mechanisms by which viruses affect these processes are incompletely characterized, and the biological implications of these epigenetic changes in virus-associated cancers are not always understood. As epigenetic changes progress over time [386], and many human tumor viruses have a long incubation, it is not always easy to attribute epigenetic modification to viral infection. Tumor virus infected cells can pack microRNAs, lncRNAs, and circRNAs into extracellular vesicles which can be taken up by other cells and RNA molecules can cause epigenetic changes in the recipient cell without viral infection. Tumor virus genomes may be lost after an epigenetic pattern has been established, supporting the hit-and-run hypothesis in tumor virology [387]. N^6-methyladenosine RNA methylation adds another layer of complexity to epigenetic changes and has been shown to play a role in cancer [388]. Viral genomes and viral transcripts can be N^6-methyladenosine modified and can have an effect on the viral life cycle and pathogenicity, as was shown for HCV and HBV [389]. N^6-methyladenosine modification of circRNAs is not uncommon and plays a role in their regulation and function [390]. Two recent studies reported that the

EBV infection induces changes in N^6-methyladenosine RNA methylation of viral and host cell mRNA. These epitransciptomic changes promoted EBV infection in vitro [391,392]. Once more, viruses take advantages of cellular processes to favor their life cycle. Whether virus-mediated changes in N^6-methyladenosine RNA methylation contributes to cancer remains unknown, but it would not be a surprise. Viruses keep amazing scientists with their creativity.

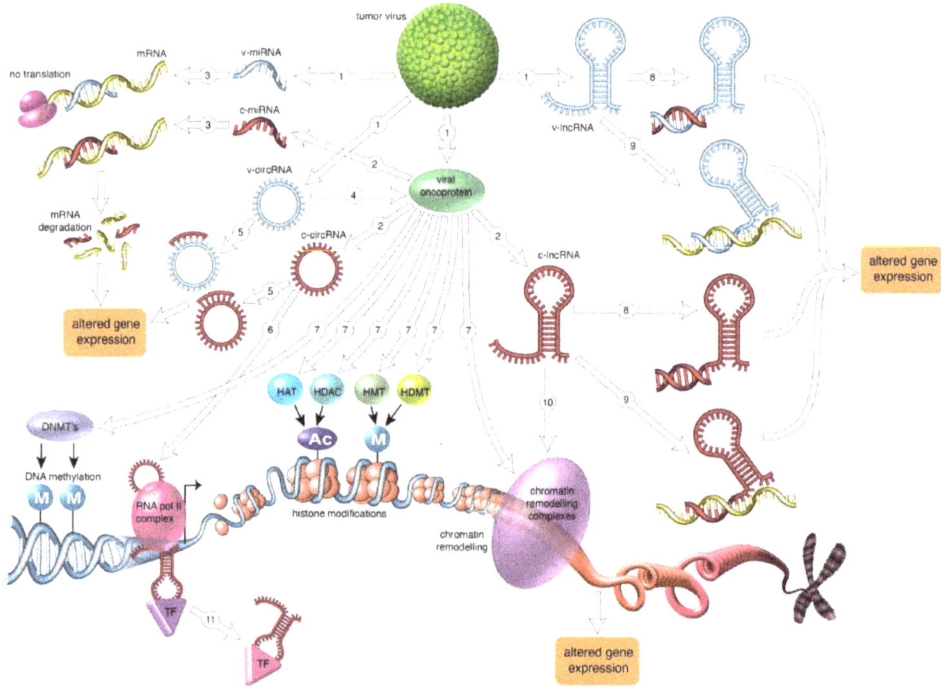

Figure 1. Epigenetic mechanisms by human tumor virus. (1) The virus encodes viral oncoproteins and its own v-microRNA, v-circRNA, and v-lncRNA. (2) Viral oncoproteins induce the expression of cellular microRNA (c-miRNA), c-circRNA, and c-lncRNA. (3) v-miRNA and c-miRNA can bind to target mRNA and induce mRNA degradation or prevent translation. (4) v-circRNA can be translated into a viral oncoprotein. (5) v-circRNA and c-circRNA act as a miRNA sponge. (6) c-circ interacts with the transcriptional machinery. (7) Viral oncoprotiens can regulate the expression of, can interact with, and can modulate the activity of DNA and histone modifying proteins and of components of chromatin remodeling complexes. (8) lncRNA sequesters miRNA. (9) lncRNA prevents translation of mRNA. (10) lncRNA recruits components of chromatin remodeling complexes. (11) lncRNA can modulate transcription by recruiting transcription factors (TF) or by sequestering TF to DNA.

Author Contributions: Conceptualization, V.P., C.P., and U.M.; writing—original draft preparation, V.P., C.P., and U.M.; writing—review and editing, V.P., C.P., and U.M. All authors have read and agreed to the published version of the manuscript.

Funding: The APC was funded by the University of Tromsø—The Arctic University of Norway.

Institutional Review Board Statement: Not applicable.

Informed Consent Statement: Not applicable.

Data Availability Statement: Not applicable.

Acknowledgments: The authors thank Roy Lyså for preparing Figure 1.

Conflicts of Interest: The authors declare no conflict of interest.

Abbreviations

aHIF	antisense to hypoxia-inducible factor 1 alpha
AICDA	activation-induced cytidine deaminase
ALTO	alternative T open reading frame
ANRIL	antisense noncoding RNA in the INK4 locus
ATG7	autophagy related 7
ATL	adult T-cell leukemia-lymphoma
ATOH1	Atonal BHLH transcription factor 1
APC	adenomatous polyposis coli protein
ATOM	angiomotin
BAF	BRG-associated factor
BART	BHRF1 cluster and the BamHI-A rightward transcript
BCL2L11	BCL2 like 11
BCLAF	Bcl-2-associated factor 1
BHLF1	BamHI leftward reading frame 1
BIM	BCL2 interacting mediator of cell death
BMI1	B lymphoma murine leukemia virus insertion region 1
BRD4	bromodomain containing 4
BRG1	BRM/SWI2-related gene
CARM1	coactivator associated arginine methyltransferase 1
Cas9	CRISPR associated protein 9
CBP	CREB-binding protein
CCDST	cervical cancer DHX9 suppressive transcript
CCL20	C-C motif chemokine ligand 20
CCND2	cyclin D2
CCR4	C-C chemokine receptor type 4
CDH1	cadherin 1
CDKN1A	cyclin-dependent kinase inhibitor 1A or p21$^{CIP1/WAF1}$
CDKN2A	cyclin-dependent kinase inhibitor 2A or p14^{ARF}
CDKN2B	cyclin-dependent kinase inhibitor 2B or p14^{INK4B}
CENP-B	centromere protein B
c-FOS	FBJ murine osteosarcoma viral oncogene homolog
CHD	chromodomain helicase DNA-binding protein
CHEK1	checkpoint kinase 1
circRNA	circular RNA
c-MYC	avian myelocytomatosis viral oncogene homolog
CRE	cAMP-response element
CREB	CRE binding protein
CRISPR	clustered regularly interspaced short palindromic sequences
CTCF	CCCTC-binding factor
CXCL	C-X-C motif chemokine ligand
CXCR	C-X-C motif chemokine receptor
DHX9	DExH-box helicase 9
DLEU2	deleted in lymphocytic leukemia 2
DNMT	DNA methyltransferase
DREH	downregulated expression by HBx
DUSP2	dual specificity phosphatase 2
EBER	EBV-encoded RNA
EBV	Epstein-Barr virus
EED	embryonic ectoderm development
EGF	epidermal growth factor
EGR3	early growth response 3
EMT	epithelial-mesenchymal transition
EpCAM	epithelial cell adhesion molecule
EPHA4	ephrin receptor 4
ETS	E-twenty six transcription factor
EVC	Ellis Van Creveld

EZH	enhancer of zeste homolog
FANCI-2	Fanconi anemia complementation group 1-2
FCER2	Fc fragment of IgE receptor II
FHIT	fragile histidine triad diadenosine triphosphate
FNDC3B	fibronectin type III domain containing 3B
FOX	Forkhead box
GAS5	growth arrest specific 5
GLRX3	glutaredoxin 3
GLUT1	glucose transporter type 1
GSTP1	glutathione S-transferase Pi 1
H3K4me3	trimethylation of histone 3 at lysine 4
H3K9ac	acetylation of histone 3 at lysine 9
H3K9me3	trimethylation of histone 3 at lysine 9
H3K27me3	trimethylation of histone 3 at lysine 27
H4K16ac	acetylation of histone 4 at lysine 16
H4R3me	methylation of histone 4 at arginine 3
H4K20me	methylation of histone 4 at lysine 20
HAT	histone acetyltransferase
HBV	hepatitis B virus
HBZ	basic zipper
HCC	hepatocellular carcinoma
HCV	hepatitis C virus
HDAC	histone deacetylases
HEIH	highly expressed in HCC
HHV4	human herpes virus-4
HHV8	human herpesvirus-8
HIF-1α	hypoxia-inducible factor 1 alpha
HLTF	helicase like transcription factor
HNGA1	head and neck squamous cell carcinoma glycolysis-associated 1
HOTAIR	HOX antisense intergenic RNA
HOX	homeobox
HR-HPV	high-risk human papillomavirus
HTLV-1	human T-lymphotropic virus 1
HULC	highly upregulated in liver cancer
IFNAR1	interferon α and β receptor subunit 1
IFNB1	interferon beta 1
IGFBP	insulin like growth factor binding protein
IL-8	interleukin 8
INO80	inositol requiring 80
IRAK1	interleukin-1 receptor-associated kinase 1
IRF4	interferon regulatory factor 4
ISWI	imitation switch
JNK	c-Jun N-terminal kinase
K	lysine
KAT2B	lysine acetyltransferase 2B
KAT7	lysine acetyltransferase 7 (or HBO1: histone acetyltransferase binding to ORC1)
KDM2B	lysine-specific demethylase 2B
KIKAT	KSHV-induced KDM4A-associated transcript
KLF4	Kruppel-like factor
KMT	lysine methyltransferase
KSHV	Kaposi's sarcoma-associated herpes virus
LANA	latency-associated nuclear antigen
LDHB	lactate dehydrogenase B
LEF1	lymphoid enhancer binding protein 1
LINE1	long interspersed nuclear element 1
LKB1	liver kinase B1
lncRNA	long non-coding RNA
LSH	lymphoid-specific helicase

LT	large tumor antigen
LTR	long terminal repeat
MALAT1	metastasis-associated lung adenocarcinoma transcript 1 (=NEAT2)
M6A	N6-methyladenosine
MAPK	mitogen-activated protein kinase
MBD	methyl-CpG-binding domain
MCC	Merkel cell carcinoma
MCIB	MHC class I polypeptide-related sequence B
MCL1	myeloid cell leukemia sequence 1
MCPyV	Merkel cell polyomavirus
MeCP2	methyl CpG binding protein 2
MHC-I	major histocompatibility complex class I
MMP9	matrix metalloproteinase 9
MYD88	myeloid differentiation primary response 88
NABP1	nucleic acid binding protein 1
NDRG2	N-myc downregulated gene 2
NEAT2	nuclear-enriched abundant transcript 2
NFκB	nuclear factor kappa B
NLRP3	NLR family pyrin domain containing 3 or cryopin
NuRD	nucleosome remodeling complex
PAN	polyadenylated nuclear RNA
PBMC	peripheral blood mononuclear cells
PCAF	p300/CBP-associated factor
PCNA	proliferating cell nuclear antigen
PDCD	programmed cell death
PEL	primary effusion lymphoma
PI3K	phosphatidylinositol 3-kinase
PLK1	polo like kinase 1
PP1R13B	protein phosphatase 1 regulatory subunit 13B
PPP2CA	protein phosphatase 2 catalytic subunit alpha
PRC	polycomb repressive complex
PRDM	PR/SET domain
PRLH1	p53-regulated lncRNA for homologous recombination repair 1
PRMT	protein arginine methyltransferase
PSD	pleckstrin and sec 7 domain containing
PTCH1	Patched 1
PTEN	phosphatase and tensin homolog
PTM	posttranslational modification
PTPRG	protein tyrosine phosphatase receptor type G
PUMA	p53 upregulated modulator of apoptosis
PVT1	plasmacytoma variant translocation 1
R	arginine
RASSF	Ras associated domain family member
RB1	retinoblastoma 1 or pRB
RbAp	retinoblastoma-binding protein
RBL2	retinoblastoma transcriptional corepressor like 2
REST	repressor element 1 silencing transcription factor
RIG-1	retinoic acid-inducible gene 1 protein
RISC	RNA-inducing silencing complex
RNF	ring finger protein 2
S	serine
SAF	Fas-Antisense
SAP30	Sin3-associated protein 30
SCL2A1	solute carrier family 2 member 1
SETD1A	SET domain containing 1A
SFRP1	secreted frizzled related protein 1
SH3BGR	SH3 domain binding glutamate-rich protein
SHP-1	Src homology-2-containing protein tyrosine phosphatase 1
SIN3A	switch independent 3A
SIRT1	sirtuin 1

SMYD3	set and mynd domain containing
SNAI	snail family transcriptional repressor
SOCS1	suppressor of cytokine signaling 1
SQSTM1	sequestosome 1
sT	small tumor antigen
STAT	Signal transducer and activator of transcription
SUV39H1	suppressor of variegation 3-9 homolog 1 (=KMT1A)
SUZ12	suppressor of zeste 12 homolog
SWI/SNF	switching defective/sucrose non-fermentable
SYK	spleen associated tyrosine kinase
T	threonine
TDG	thymine DNA glycosylase
TET	ten-eleven translocation
TERT	telomerase reverse transcriptase
TGFβ1	transforming growth factor beta 1
TGFBR2	tumor growth factor-beta type II receptor
TGIF2	TGFβ-induced homeobox 2
TLR	toll-like receptor
TMPOP2	thymopoietin pseudogene 2
TP53BP2	tumor promoter p53 binding protein 2
TP73	tumor protein p73
TRIM35	tripartite motif containing 35
TUSC7	tumor suppressor candidate 7
UCA1	urothelial carcinoma associated 1
UHRF	ubiquitin-like, containing PHD and RING finger domain
VCAM	vascular cell adhesion molecule 1
VEGF	vascular endothelial growth factor
vFLIP	viral FLICE inhibitory protein
vIL-6	viral interleukin 6
WDR5	WD repeat domain 5
WNT1	Wnt family member 1
Y	tyrosine
ZBTB	Zinc finger and BTB domain containing

References

1. De Martel, C.; Georges, D.; Bray, F.; Ferlay, J.; Clifford, G.M. Global burden of cancer attributable to infections in 2018: A worldwide incidence analysis. *Lancet Glob. Health* **2020**, *8*, e180–e190. [CrossRef]
2. Haley, C.T.; Mui, U.N.; Vangipuram, R.; Rady, P.L.; Tyring, S.K. Human oncoviruses: Mucocutaneous manifestations, pathogenesis, therapeutics, and prevention: Papillomaviruses and Merkel cell polyomavirus. *J. Am. Acad. Dermatol.* **2019**, *81*, 1–21. [CrossRef]
3. Mui, U.N.; Haley, C.T.; Vangipuram, R.; Tyring, S.K. Human oncoviruses: Mucocutaneous manifestations, pathogenesis, therapeutics, and prevention: Hepatitis viruses, human T-cell leukemia viruses, herpesviruses, and Epstein-Barr virus. *J. Am. Acad. Dermatol.* **2019**, *81*, 23–41. [CrossRef]
4. Hatano, Y.; Ideta, T.; Hirata, A.; Hatano, K.; Tomita, H.; Okada, H.; Shimizu, M.; Tanaka, T.; Hara, A. Virus-Driven Carcinogenesis. *Cancers* **2021**, *13*, 2625. [CrossRef]
5. Mesri, E.A.; Feitelson, M.A.; Munger, K. Human viral oncogenesis: A cancer hallmarks analysis. *Cell Host Microbe* **2014**, *15*, 266–282. [CrossRef] [PubMed]
6. Mui, U.N.; Haley, C.T.; Tyring, S.K. Viral Oncology: Molecular Biology and Pathogenesis. *J. Clin. Med.* **2017**, *6*, 111. [CrossRef] [PubMed]
7. Soliman, S.H.A.; Orlacchio, A.; Verginelli, F. Viral Manipulation of the Host Epigenome as a Driver of Virus-Induced Oncogenesis. *Microorganisms* **2021**, *9*, 1179. [CrossRef] [PubMed]
8. Iwanaga, M.; Watanabe, T.; Yamaguchi, K. Adult T-cell leukemia: A review of epidemiological evidence. *Front. Microbiol.* **2012**, *3*, 322. [CrossRef]
9. Becker, J.C.; Stang, A.; DeCaprio, J.A.; Cerroni, L.; Lebbé, C.; Veness, M.; Nghiem, P. Merkel cell carcinoma. *Nat. Rev. Dis. Primers.* **2017**, *3*, 17077. [CrossRef] [PubMed]
10. Mysore, K.R.; Leung, D.H. Hepatitis B and C. *Clin. Liver Dis.* **2018**, *22*, 703–722. [CrossRef]
11. Flanagan, J.M. Host epigenetic modifications by oncogenic viruses. *Br. J. Cancer* **2017**, *96*, 183–188. [CrossRef]
12. Lieberman, P.M. Chromatin organization and virus gene expression. *J. Cell. Physiol.* **2008**, *216*, 295–302. [CrossRef] [PubMed]
13. Miyazato, P.; Matsuo, M.; Katsuya, H.; Satou, Y. Transcriptional and Epigenetic Regulatory Mechanisms Affecting HTLV-1 Provirus. *Viruses* **2016**, *8*, 171. [CrossRef]

14. Burley, M.; Roberts, S.; Parish, J.L. Epigenetic regulation of human papillomavirus transcription in the productive virus life cycle. *Semin. Immunopathol.* **2020**, *42*, 159–171. [CrossRef] [PubMed]
15. Campbell, M.; Yang, W.S.; Yeh, W.W.; Kao, C.H.; Chang, P.C. Epigenetic Regulation of Kaposi's Sarcoma-Associated Herpesvirus Latency. *Front. Microbiol.* **2020**, *11*, 850. [CrossRef]
16. Chen, Z.X.; Riggs, A.D. DNA methylation and demethylation in mammals. *J. Biol. Chem.* **2011**, *286*, 18347–18353. [CrossRef]
17. Moore, L.D.; Le, T.; Fan, G. DNA methylation and its basic function. *Neuropsychopharmacology* **2013**, *38*, 23–38. [CrossRef]
18. Ng, H.H.; Zhang, Y.; Hendrich, B.; Johnson, C.A.; Turner, B.M.; Erdjument-Bromage, H.; Tempst, P.; Reinberg, D.; Bird, A. MBD2 is a transcriptional repressor belonging to the MeCP1 histone deacetylase complex. *Nat. Genet.* **1999**, *23*, 58–61. [CrossRef]
19. Feng, Q.; Zhang, Y. The MeCP1 complex represses transcription through preferential binding, remodeling, and deacetylating methylated nucleosomes. *Genes Dev.* **2001**, *15*, 827–832. [PubMed]
20. Detich, N.; Theberge, J.; Szyf, M. Promoter-specific activation and demethylation by MBD2/demethylase. *J. Biol. Chem.* **2002**, *277*, 35791–35794. [CrossRef] [PubMed]
21. Horvath, P.M.; Monteggia, L.M. MeCP2 as an Activator of Gene Expression. *Trends Neurosci.* **2018**, *41*, 72–74. [CrossRef] [PubMed]
22. Koch, A.; Joosten, S.C.; Feng, Z.; de Ruijter, T.C.; Draht, M.X.; Melotte, V.; Smits, K.M.; Veeck, J.; Herman, J.G.; Van Neste, L.; et al. Analysis of DNA methylation in cancer: Location revisited. *Nat. Rev. Clin. Oncol.* **2018**, *15*, 459–466. [CrossRef] [PubMed]
23. Greenberg, M.V.C.; Bourc'his, D. The diverse roles of DNA methylation in mammalian development and disease. *Nat. Rev. Mol. Cell Biol.* **2019**, *20*, 590–607. [CrossRef] [PubMed]
24. Oka, T.; Ouchida, M.; Koyama, M.; Ogama, Y.; Takada, S.; Nakatani, Y.; Tanaka, T.; Yoshino, T.; Hayashi, K.; Ohara, N.; et al. Gene silencing of the tyrosine phosphatase SHP1 gene by aberrant methylation in leukemias/lymphomas. *Cancer Res.* **2002**, *62*, 6390–6394. [PubMed]
25. Ego, T.; Tanaka, Y.; Shimotohno, K. Interaction of HTLV-1 Tax and methyl-CpG-binding domain 2 positively regulates the gene expression from the hypermethylated LTR. *Oncogene* **2005**, *24*, 1914–1923. [CrossRef]
26. Yan, P.; Qu, Z.; Ishikawa, C.; Mori, N.; Xiao, G. Human T-cell leukemia virus type I-mediated repression of PDZ-LIM domain-containing protein 2 involves DNA methylation but independent of the viral oncoprotein tax. *Neoplasia* **2009**, *11*, 1036–1041. [CrossRef]
27. Arora, P.; Kim, E.O.; Jung, J.K.; Jang, K.L. Hepatitis C virus core protein downregulates E-cadherin expression via activation of DNA methyltransferase 1 and 3b. *Cancer Lett.* **2008**, *261*, 244–252. [CrossRef]
28. Ripoli, M.; Barbano, R.; Balsamo, T.; Piccoli, C.; Brunetti, V.; Coco, M.; Mazzoccoli, G.; Vinciguerra, M.; Pazienza, V. Hypermethylated levels of E-cadherin promoter in Huh-7 cells expressing the HCV core protein. *Virus Res.* **2011**, *160*, 74–81. [CrossRef]
29. Park, S.H.; Lim, J.S.; Lim, S.Y.; Tiwari, I.; Jang, K.L. Hepatitis C virus Core protein stimulates cell growth by down-regulating p16 expression via DNA methylation. *Cancer Lett.* **2011**, *310*, 616–618. [CrossRef]
30. Benegiamo, G.; Vinciguerra, M.; Mazzoccoli, G.; Piepoli, A.; Andriulli, A.; Pazienza, V. DNA methyltransferases 1 and 3b expression in Huh-7 cells expressing HCV core protein of different genotypes. *Dig. Dis. Sci.* **2012**, *57*, 1598–1603. [CrossRef]
31. Lechner, M.; Fenton, T.; West, J.; Wilson, G.; Feber, A.; Henderson, S.; Thirlwell, C.; Dibra, H.K.; Jay, A.; Butcher, L.; et al. Identification and functional validation of HPV-mediated hypermethylation in head and neck squamous cell carcinoma. *Genome Med.* **2013**, *5*, 15. [CrossRef] [PubMed]
32. Ekanayake Weeramange, C.; Tang, K.D.; Vasani, S.; Langton-Lockton, J.; Kenny, L.; Punyadeera, C. DNA Methylation Changes in Human Papillomavirus-Driven Head and Neck Cancers. *Cells* **2020**, *9*, 1359. [CrossRef]
33. Tsai, C.N.; Tsai, C.L.; Tse, K.P.; Chang, H.Y.; Chang, Y.S. The Epstein-Barr virus oncogene product, latent membrane protein 1, induces the downregulation of E-cadherin gene expression via activation of DNA methyltransferases. *Proc. Natl. Acad. Sci. USA* **2002**, *99*, 10084–10089. [CrossRef] [PubMed]
34. Tsai, C.L.; Li, H.P.; Lu, Y.J.; Hsueh, C.; Liang, Y.; Chen, C.L.; Tsao, S.W.; Tse, K.P.; Yu, J.S.; Chang, Y.S. Activation of DNA methyltransferase 1 by EBV LMP1 Involves c-Jun NH(2)-terminal kinase signaling. *Cancer Res.* **2006**, *66*, 11668–11676. [CrossRef]
35. Peng, H.; Chen, Y.; Gong, P.; Cai, L.; Lyu, X.; Jiang, Q.; Wang, J.; Lu, J.; Yao, K.; Liu, K.; et al. Higher methylation intensity induced by EBV LMP1 via NF-κB/DNMT3b signaling contributes to silencing of PTEN gene. *Oncotarget* **2016**, *7*, 40025–40037. [CrossRef]
36. Luo, X.; Hong, L.; Cheng, C.; Li, N.; Zhao, X.; Shi, F.; Liu, J.; Fan, J.; Zhou, J.; Bode, A.M.; et al. DNMT1 mediates metabolic reprogramming induced by Epstein-Barr virus latent membrane protein 1 and reversed by grifolin in nasopharyngeal carcinoma. *Cell Death Dis.* **2018**, *9*, 619. [CrossRef]
37. Vargas-Ayala, R.C.; Jay, A.; Manara, F.; Maroui, M.A.; Hernandez-Vargas, H.; Diederichs, A.; Robitaille, A.; Sirand, C.; Ceraolo, M.G.; Romero-Medina, M.C.; et al. Interplay between the Epigenetic Enzyme Lysine (K)-Specific Demethylase 2B and Epstein-Barr Virus Infection. *J. Virol.* **2019**, *93*, e00273. [CrossRef] [PubMed]
38. Leonard, S.; Wei, W.; Anderton, J.; Vockerodt, M.; Rowe, M.; Murray, P.G.; Woodman, C.B. Epigenetic and transcriptional changes which follow Epstein-Barr virus infection of germinal center B cells and their relevance to the pathogenesis of Hodgkin's lymphoma. *J. Virol.* **2011**, *85*, 9568–9577. [CrossRef]
39. Hino, R.; Uozaki, H.; Murakami, N.; Ushiku, T.; Shinozaki, A.; Ishikawa, S.; Morikawa, T.; Nakaya, T.; Sakatani, T.; Takada, K.; et al. Activation of DNA methyltransferase 1 by EBV latent membrane protein 2A leads to promoter hypermethylation of PTEN gene in gastric carcinoma. *Cancer Res.* **2009**, *69*, 2766–2774. [CrossRef]

40. Wang, J.; Liu, W.; Zhang, X.; Zhang, Y.; Xiao, H.; Luo, B. LMP2A induces DNA methylation and expression repression of AQP3 in EBV-associated gastric carcinoma. *Virology* **2019**, *534*, 87–95. [CrossRef] [PubMed]
41. Namba-Fukuyo, H.; Funata, S.; Matsusaka, K.; Fukuyo, M.; Rahmutulla, B.; Mano, Y.; Fukayama, M.; Aburatani, H.; Kaneda, A. TET2 functions as a resistance factor against DNA methylation acquisition during Epstein-Barr virus infection. *Oncotarget* **2016**, *7*, 81512–81526. [CrossRef] [PubMed]
42. Zhang, S.; Pei, Y.; Lang, F.; Sun, K.; Singh, R.K.; Lamplugh, Z.L.; Saha, A.; Robertson, E.S. EBNA3C facilitates RASSF1A downregulation through ubiquitin-mediated degradation and promoter hypermethylation to drive B-cell proliferation. *PLoS Pathog.* **2019**, *15*, e1007514. [CrossRef] [PubMed]
43. Platt, G.; Carbone, A.; Mittnacht, S. p16INK4a loss and sensitivity in KSHV associated primary effusion lymphoma. *Oncogene* **2002**, *21*, 1823–1831. [CrossRef]
44. Shamay, M.; Krithivas, A.; Zhang, J.; Hayward, S.D. Recruitment of the de novo DNA methyltransferase Dnmt3a by Kaposi's sarcoma-associated herpesvirus LANA. *Proc. Natl. Acad. Sci. USA* **2006**, *103*, 14554–14559. [CrossRef] [PubMed]
45. Pei, Y.; Wong, J.H.; Robertson, E.S. Herpesvirus Epigenetic Reprogramming and Oncogenesis. *Annu. Rev. Virol.* **2020**, *7*, 309–331. [CrossRef]
46. Krithivas, A.; Fujimuro, M.; Weidner, M.; Young, D.B.; Hayward, S.D. Protein interactions targeting the latency-associated nuclear antigen of Kaposi's sarcoma-associated herpesvirus to cell chromosomes. *J. Virol.* **2002**, *76*, 11596–11604. [CrossRef]
47. Wu, J.; Xu, Y.; Mo, D.; Huang, P.; Sun, R.; Huang, L.; Pan, S.; Xu, J. Kaposi's sarcoma-associated herpesvirus (KSHV) vIL-6 promotes cell proliferation and migration by upregulating DNMT1 via STAT3 activation. *PLoS ONE* **2014**, *9*, e93478. [CrossRef]
48. Li, W.; Wang, Q.; Feng, Q.; Wang, F.; Yan, Q.; Gao, S.J.; Lu, C. Oncogenic KSHV-encoded interferon regulatory factor upregulates HMGB2 and CMPK1 expression to promote cell invasion by disrupting a complex lncRNA-OIP5-AS1/miR-218-5p network. *PLoS Pathog.* **2019**, *15*, e1007578. [CrossRef]
49. Li, W.; Wang, Q.; Qi, X.; Guo, Y.; Lu, H.; Chen, Y.; Lu, Z.; Yan, Q.; Zhu, X.; Jung, J.U.; et al. Viral interleukin-6 encoded by an oncogenic virus promotes angiogenesis and cellular transformation by enhancing STAT3-mediated epigenetic silencing of caveolin 1. *Oncogene* **2020**, *39*, 4603–4618. [CrossRef]
50. Lee, S.M.; Lee, Y.G.; Bae, J.B.; Choi, J.K.; Tayama, C.; Hata, K.; Yun, Y.; Seong, J.K.; Kim, Y.J. HBx induces hypomethylation of distal intragenic CpG islands required for active expression of developmental regulators. *Proc. Natl. Acad. Sci. USA* **2014**, *111*, 9555–9560. [CrossRef]
51. Zheng, D.L.; Zhang, L.; Cheng, N.; Xu, X.; Deng, Q.; Teng, X.M.; Wang, K.S.; Zhang, X.; Huang, J.; Han, Z.G. Epigenetic modification induced by hepatitis B virus X protein via interaction with de novo DNA methyltransferase DNMT3A. *J. Hepatol.* **2009**, *50*, 377–387. [CrossRef]
52. Park, I.Y.; Sohn, B.H.; Yu, E.; Suh, D.J.; Chung, Y.H.; Lee, J.H.; Surzycki, S.J.; Lee, Y.I. Aberrant epigenetic modifications in hepatocarcinogenesis induced by hepatitis B virus X protein. *Gastroenterology* **2007**, *132*, 1476–1494. [CrossRef] [PubMed]
53. Jung, J.K.; Arora, P.; Pagano, J.S.; Jang, K.L. Expression of DNA methyltransferase 1 is activated by hepatitis B virus X protein via a regulatory circuit involving the p16INK4a-cyclin D1-CDK 4/6-pRb-E2F1 pathway. *Cancer Res.* **2007**, *67*, 5771–5778. [CrossRef] [PubMed]
54. Zhao, J.; Wu, G.; Bu, F.; Lu, B.; Liang, A.; Cao, L.; Tong, X.; Lu, X.; Wu, M.; Guo, Y. Epigenetic silence of ankyrin-repeat-containing, SH3-domain-containing, and proline-rich-region-containing protein 1 (ASPP1) and ASPP2 genes promotes tumor growth in hepatitis B virus-positive hepatocellular carcinoma. *Hepatology* **2010**, *51*, 142–153. [CrossRef] [PubMed]
55. Huang, J.; Wang, Y.; Guo, Y.; Sun, S. Down-regulated microRNA-152 induces aberrant DNA methylation in hepatitis B virus-related hepatocellular carcinoma by targeting DNA methyltransferase 1. *Hepatology* **2010**, *52*, 60–70. [CrossRef]
56. Wei, X.; Xiang, T.; Ren, G.; Tan, C.; Liu, R.; Xu, X.; Wu, Z. miR-101 is down-regulated by the hepatitis B virus x protein and induces aberrant DNA methylation by targeting DNA methyltransferase 3A. *Cell Signal.* **2013**, *25*, 439–446. [CrossRef] [PubMed]
57. Ishitsuka, K.; Tamura, K. Human T-cell leukaemia virus type I and adult T-cell leukaemia-lymphoma. *Lancet Oncol.* **2014**, *15*, e517–e526. [CrossRef]
58. Giam, C.Z.; Semmes, O.J. HTLV-1 Infection and Adult T-Cell Leukemia/Lymphoma-A Tale of Two Proteins: Tax and HBZ. *Viruses* **2016**, *8*, 161. [CrossRef] [PubMed]
59. Bangham, C.R.; Araujo, A.; Yamano, Y.; Taylor, G.P. HTLV-1-associated myelopathy/tropical spastic paraparesis. *Nat. Rev. Dis. Primers* **2015**, *1*, 15012. [CrossRef]
60. Matsuoka, M.; Jeang, K.T. Human T-cell leukaemia virus type 1 (HTLV-1) infectivity and cellular transformation. *Nat. Rev. Cancer* **2007**, *7*, 270–280. [CrossRef]
61. Bangham, C.R.; Ratner, L. How does HTLV-1 cause adult T-cell leukaemia/lymphoma (ATL)? *Curr. Opin. Virol.* **2015**, *14*, 93–100. [CrossRef] [PubMed]
62. Zhao, T. The Role of HBZ in HTLV-1-Induced Oncogenesis. *Viruses.* **2016**, *8*, 34. [CrossRef] [PubMed]
63. Takeda, S.; Maeda, M.; Morikawa, S.; Taniguchi, Y.; Yasunaga, J.; Nosaka, K.; Tanaka, Y.; Matsuoka, M. Genetic and epigenetic inactivation of tax gene in adult T-cell leukemia cells. *Int. J. Cancer* **2004**, *109*, 559–567. [CrossRef] [PubMed]
64. Satou, Y.; Yasunaga, J.; Yoshida, M.; Matsuoka, M. HTLV-I basic leucine zipper factor gene mRNA supports proliferation of adult T cell leukemia cells. *Proc. Natl. Acad. Sci. USA* **2006**, *103*, 720–725. [CrossRef] [PubMed]

65. Zhang, X.; Odom, D.T.; Koo, S.H.; Conkright, M.D.; Canettieri, G.; Best, J.; Chen, H.; Jenner, R.; Herbolsheimer, E.; Jacobsen, E.; et al. Genome-wide analysis of cAMP-response element binding protein occupancy, phosphorylation, and target gene activation in human tissues. *Proc. Natl. Acad. Sci. USA* **2005**, *102*, 4459–4464. [CrossRef] [PubMed]
66. Yoshida, M.; Nosaka, K.; Yasunaga, J.; Nishikata, I.; Morishita, K.; Matsuoka, M. Aberrant expression of the MEL1S gene identified in association with hypomethylation in adult T-cell leukemia cells. *Blood* **2004**, *103*, 2753–2760. [CrossRef]
67. Morishita, K. Leukemogenesis of the EVI1/MEL1 gene family. *Int. J. Hematol.* **2007**, *85*, 279–286. [CrossRef]
68. Sato, H.; Oka, T.; Shinnou, Y.; Kondo, T.; Washio, K.; Takano, M.; Takata, K.; Morito, T.; Huang, X.; Tamura, M.; et al. Multi-step aberrant CpG island hyper-methylation is associated with the progression of adult T-cell leukemia/lymphoma. *Am. J. Pathol.* **2010**, *176*, 402–415. [CrossRef]
69. Kataoka, K.; Nagata, Y.; Kitanaka, A.; Shiraishi, Y.; Shimamura, T.; Yasunaga, J.; Totoki, Y.; Chiba, K.; Sato-Otsubo, A.; Nagae, G.; et al. Integrated molecular analysis of adult T cell leukemia/lymphoma. *Nat. Genet.* **2015**, *47*, 1304–1315. [CrossRef]
70. Yasunaga, J.; Taniguchi, Y.; Nosaka, K.; Yoshida, M.; Satou, Y.; Sakai, T.; Mitsuya, H.; Matsuoka, M. Identification of aberrantly methylated genes in association with adult T-cell leukemia. *Cancer Res.* **2004**, *64*, 6002–6009. [CrossRef]
71. Shimazu, Y.; Shimazu, Y.; Hishizawa, M.; Hamaguchi, M.; Nagai, Y.; Sugino, N.; Fujii, S.; Kawahara, M.; Kadowaki, N.; Nishikawa, H.; et al. Hypomethylation of the Treg-Specific Demethylated Region in FOXP3 Is a Hallmark of the Regulatory T-cell Subtype in Adult T-cell Leukemia. *Cancer Immunol. Res.* **2016**, *4*, 136–145. [CrossRef]
72. Yamano, Y.; Takenouchi, N.; Li, H.C.; Tomaru, U.; Yao, K.; Grant, C.W.; Maric, D.A.; Jacobson, S. Virus-induced dysfunction of CD4+CD25+ T cells in patients with HTLV-I-associated neuroimmunological disease. *J. Clin. Investig.* **2005**, *115*, 1361–1368. [CrossRef] [PubMed]
73. Satou, Y.; Yasunaga, J.; Zhao, T.; Yoshida, M.; Miyazato, P.; Takai, K.; Shimizu, K.; Ohshima, K.; Green, P.L.; Ohkura, N.; et al. HTLV-1 bZIP factor induces T-cell lymphoma and systemic inflammation in vivo. *PLoS Pathog.* **2011**, *7*, e1001274. [CrossRef] [PubMed]
74. Ichikawa, T.; Nakahata, S.; Fujii, M.; Iha, H.; Shimoda, K.; Morishita, K. The regulation of NDRG2 expression during ATLL development after HTLV-1 infection. *Biochim. Biophys. Acta Mol. Basis Dis.* **2019**, *1865*, 2633–2646. [CrossRef] [PubMed]
75. Jeong, S.J.; Lu, H.; Cho, W.K.; Park, H.U.; Pise-Masison, C.; Brady, J.N. Coactivator-associated arginine methyltransferase 1 enhances transcriptional activity of the human T-cell lymphotropic virus type 1 long terminal repeat through direct interaction with Tax. *J. Virol.* **2006**, *80*, 10036–10044. [CrossRef]
76. Issa, J.P. Age-related epigenetic changes and the immune system. *Clin. Immunol.* **2003**, *109*, 103–108. [CrossRef]
77. Klutstein, M.; Nejman, D.; Greenfield, R.; Cedar, H. DNA Methylation in Cancer and Aging. *Cancer Res.* **2016**, *76*, 3446–3450. [CrossRef]
78. Arzumanyan, A.; Reis, H.M.; Feitelson, M.A. Pathogenic mechanisms in HBV- and HCV-associated hepatocellular carcinoma. *Nat. Rev. Cancer* **2013**, *13*, 123–135. [CrossRef]
79. Rusyn, I.; Lemon, S.M. Mechanisms of HCV-induced liver cancer: What did we learn from in vitro and animal studies? *Cancer Lett.* **2014**, *345*, 210–215. [CrossRef]
80. McGivern, D.R.; Lemon, S.M. Virus-specific mechanisms of carcinogenesis in hepatitis C virus associated liver cancer. *Oncogene* **2011**, *30*, 1969–1983. [CrossRef]
81. Rongrui, L.; Na, H.; Zongfang, L.; Fanpu, J.; Shiwen, J. Epigenetic mechanism involved in the HBV/HCV-related hepatocellular carcinoma tumorigenesis. *Curr. Pharm. Des.* **2014**, *20*, 1715–1725. [CrossRef] [PubMed]
82. Pazienza, V.; Panebianco, C.; Andriulli, A. Hepatitis viruses exploitation of host DNA methyltransferases functions. *Clin. Exp. Med.* **2016**, *16*, 265–272. [CrossRef]
83. Mekky, M.A.; Salama, R.H.; Abdel-Aal, M.F.; Ghaliony, M.A.; Zaky, S. Studying the frequency of aberrant DNA methylation of APC, P14, and E-cadherin genes in HCV-related hepatocarcinogenesis. *Cancer Biomark.* **2018**, *22*, 503–509. [CrossRef] [PubMed]
84. Zhou, Y.; Wang, X.B.; Qiu, X.P.; Shuai, Z.; Wang, C.; Zheng, F. CDKN2A promoter methylation and hepatocellular carcinoma risk: A meta-analysis. *Clin. Res. Hepatol. Gastroenterol.* **2018**, *42*, 529–541. [CrossRef] [PubMed]
85. Pietropaolo, V.; Prezioso, C.; Moens, U. Merkel Cell Polyomavirus and Merkel Cell Carcinoma. *Cancers* **2020**, *12*, 1774. [CrossRef]
86. Calvignac-Spencer, S.; Feltkamp, M.C.; Daugherty, M.D.; Moens, U.; Ramqvist, T.; Johne, R.; Ehlers, B. A taxonomy update for the family Polyomaviridae. *Arch. Virol.* **2016**, *161*, 1739–1750. [CrossRef]
87. Angermeyer, S.; Hesbacher, S.; Becker, J.C.; Schrama, D.; Houben, R. Merkel cell polyomavirus-positive Merkel cell carcinoma cells do not require expression of the viral small T antigen. *J. Investig. Dermatol.* **2013**, *133*, 2059–2064. [CrossRef]
88. Helmbold, P.; Lahtz, C.; Enk, A.; Herrmann-Trost, P.; Marsch, W.; Kutzner, H.; Dammann, R.H. Frequent occurrence of RASSF1A promoter hypermethylation and Merkel cell polyomavirus in Merkel cell carcinoma. *Mol. Carcinog.* **2009**, *48*, 903–909. [CrossRef]
89. Richter, A.M.; Haag, T.; Walesch, S.; Herrmann-Trost, P.; Marsch, W.C.; Kutzner, H.; Helmbold, P.; Dammann, R.H. Aberrant Promoter Hypermethylation of RASSF Family Members in Merkel Cell Carcinoma. *Cancers* **2013**, *5*, 1566–1576. [CrossRef]
90. Ricci, C.; Morandi, L.; Ambrosi, F.; Righi, A.; Gibertoni, D.; Maletta, F.; Agostinelli, C.; Corradini, A.G.; Uccella, S.; Asioli, S.; et al. Intron 4–5 hTERT DNA Hypermethylation in Merkel Cell Carcinoma: Frequency, Association with Other Clinico-pathological Features and Prognostic Relevance. *Endocr. Pathol.* **2021**, 1–11. [CrossRef]
91. Sahi, H.; Savola, S.; Sihto, H.; Koljonen, V.; Bohling, T.; Knuutila, S. RB1 gene in Merkel cell carcinoma: Hypermethylation in all tumors and concurrent heterozygous deletions in the polyomavirus-negative subgroup. *APMIS* **2014**, *122*, 1157–1166. [CrossRef] [PubMed]

92. Amara, K.; Trimeche, M.; Ziadi, S.; Laatiri, A.; Hachana, M.; Sriha, B.; Mokni, M.; Korbi, S. Presence of simian virus 40 DNA sequences in diffuse large B-cell lymphomas in Tunisia correlates with aberrant promoter hypermethylation of multiple tumor suppressor genes. *Int. J. Cancer* **2007**, *121*, 2693–2702. [CrossRef]
93. Lassacher, A.; Heitzer, E.; Kerl, H.; Wolf, P. p14ARF hypermethylation is common but INK4a-ARF locus or p53 mutations are rare in Merkel cell carcinoma. *J. Investig. Dermatol.* **2008**, *128*, 1788–1796. [CrossRef] [PubMed]
94. Haag, T.; Richter, A.M.; Schneider, M.B.; Jiménez, A.P.; Dammann, R.H. The dual specificity phosphatase 2 gene is hypermethylated in human cancer and regulated by epigenetic mechanisms. *BMC Cancer* **2016**, *16*, 49. [CrossRef] [PubMed]
95. Higaki-Mori, H.; Kuwamoto, S.; Iwasaki, T.; Kato, M.; Murakami, I.; Nagata, K.; Sano, H.; Horie, Y.; Yoshida, Y.; Yamamoto, O.; et al. Association of Merkel cell polyomavirus infection with clinicopathological differences in Merkel cell carcinoma. *Hum. Pathol.* **2012**, *43*, 2282–2291. [CrossRef] [PubMed]
96. Fan, K.; Gravemeyer, J.; Ritter, C.; Rasheed, K.; Gambichler, T.; Moens, U.; Shuda, M.; Schrama, D.; Becker, J.C. MCPyV Large T Antigen-Induced Atonal Homolog 1 Is a Lineage-Dependency Oncogene in Merkel Cell Carcinoma. *J. Investig. Dermatol.* **2020**, *140*, 56–65.e3. [CrossRef]
97. Gambichler, T.; Dreißigacker, M.; Kasakovski, D.; Skrygan, M.; Wieland, U.; Silling, S.; Gravemeyer, J.; Melior, A.; Cherouny, A.; Stücker, M.; et al. Patched 1 expression in Merkel cell carcinoma. *J. Dermatol.* **2021**, *48*, 64–74. [CrossRef] [PubMed]
98. Chteinberg, E.; Sauer, C.M.; Rennspiess, D.; Beumers, L.; Schiffelers, L.; Eben, J.; Haugg, A.; Winnepenninckx, V.; Kurz, A.K.; Speel, E.J.; et al. Neuroendocrine Key Regulator Gene Expression in Merkel Cell Carcinoma. *Neoplasia* **2018**, *20*, 1227–1235. [CrossRef]
99. Improta, G.; Ritter, C.; Pettinato, A.; Vasta, V.; Schrama, D.; Fraggetta, F.; Becker, J.C. MGMT promoter methylation status in Merkel cell carcinoma: In vitro versus in vivo. *J. Cancer Res. Clin. Oncol.* **2017**, *143*, 1489–1497. [CrossRef]
100. Chteinberg, E.; Vogt, J.; Kolarova, J.; Bormann, F.; van den Oord, J.; Speel, E.J.; Winnepenninckx, V.; Kurz, A.K.; Zenke, M.; Siebert, R.; et al. The curious case of Merkel cell carcinoma: Epigenetic youth and lack of pluripotency. *Epigenetics* **2020**, *15*, 1319–1324. [CrossRef]
101. Soejima, K.; Fang, W.; Rollins, B.J. DNA methyltransferase 3b contributes to oncogenic transformation induced by SV40T antigen and activated Ras. *Oncogene* **2003**, *22*, 4723–4733. [CrossRef] [PubMed]
102. Ricci, C.; Morandi, L.; Righi, A.; Gibertoni, D.; Maletta, F.; Ambrosi, F.; Agostinelli, C.; Uccella, S.; Asioli, S.; Sessa, F.; et al. PD-1 (PDCD1) promoter methylation in Merkel cell carcinoma: Prognostic relevance and relationship with clinico-pathological parameters. *Mod. Pathol.* **2019**, *32*, 1359–1372. [CrossRef] [PubMed]
103. Van Doorslaer, K.; Chen, Z.; Bernard, H.U.; Chan, P.K.S.; DeSalle, R.; Dillner, J.; Forslund, O.; Haga, T.; McBride, A.A.; Villa, L.L.; et al. ICTV Report. *J. Gen. Virol.* **2018**, *99*, 989–990. [CrossRef] [PubMed]
104. Forman, D.; de Martel, C.; Lacey, C.J.; Soerjomataram, I.; Lortet-Tieulent, J.; Bruni, L.; Vignat, J.; Ferlay, J.; Bray, F.; Plummer, M.; et al. Global burden of human papillomavirus and related diseases. *Vaccine* **2012**, *30* (Suppl. 5), F12–F23. [CrossRef] [PubMed]
105. Doorbar, J.; Egawa, N.; Griffin, H.; Kranjec, C.; Murakami, I. Human papillomavirus molecular biology and disease association. *Rev. Med. Virol.* **2015**, *25* (Suppl. 1), 2–23. [CrossRef]
106. Marur, S.; D'Souza, G.; Westra, W.H.; Forastiere, A.A. HPV-associated head and neck cancer: A virus-related cancer epidemic. *Lancet Oncol.* **2010**, *11*, 781–789. [CrossRef]
107. Scarth, J.A.; Patterson, M.R.; Morgan, E.L.; Macdonald, A. The human papillomavirus oncoproteins: A review of the host pathways targeted on the road to transformation. *J. Gen. Virol.* **2021**, *102*, 001540. [CrossRef]
108. Durzynska, J.; Lesniewicz, K.; Poreba, E. Human papillomaviruses in epigenetic regulations. *Mutat. Res. Mutat. Res.* **2017**, *772*, 36–50. [CrossRef] [PubMed]
109. Boscolo-Rizzo, P.; Furlan, C.; Lupato, V.; Polesel, J.; Fratta, E. Novel insights into epigenetic drivers of oropharyngeal squamous cell carcinoma: Role of HPV and lifestyle factors. *Clin. Epigenet.* **2017**, *9*, 124. [CrossRef]
110. Sen, P.; Ganguly, P.; Ganguly, N. Modulation of DNA methylation by human papillomavirus E6 and E7 oncoproteins in cervical cancer. *Oncol. Lett.* **2018**, *15*, 11–22. [CrossRef]
111. Kuss-Duerkop, S.K.; Westrich, J.A.; Pyeon, D. DNA Tumor Virus Regulation of Host DNA Methylation and Its Implications for Immune Evasion and Oncogenesis. *Viruses* **2018**, *10*, 82. [CrossRef] [PubMed]
112. Feng, C.; Dong, J.; Chang, W.; Cui, M.; Xu, T. The Progress of Methylation Regulation in Gene Expression of Cervical Cancer. *Int. J. Genom.* **2018**, *2018*, 8260652. [CrossRef] [PubMed]
113. Nakagawa, T.; Kurokawa, T.; Mima, M.; Imamoto, S.; Mizokami, H.; Kondo, S.; Okamoto, Y.; Misawa, K.; Hanazawa, T.; Kaneda, A. DNA Methylation and HPV-Associated Head and Neck Cancer. *Microorganisms* **2021**, *9*, 801. [CrossRef]
114. Da Silva, M.L.R.; De Albuquerque, B.; Allyrio, T.; De Almeida, V.D.; Cobucci, R.N.O.; Bezerra, F.L.; Andrade, V.S.; Lanza, D.C.F.; De Azevedo, J.C.V.; De Araújo, J.M.G.; et al. The role of HPV-induced epigenetic changes in cervical carcinogenesis. *Biomed. Rep.* **2012**, *15*, 60. [CrossRef] [PubMed]
115. Lin, T.S.; Lee, H.; Chen, R.A.; Ho, M.L.; Lin, C.Y.; Chen, Y.H.; Tsai, Y.Y.; Chou, M.C.; Cheng, Y.W. An association of DNMT3b protein expression with P16INK4a promoter hypermethylation in non-smoking female lung cancer with human papillomavirus infection. *Cancer Lett.* **2005**, *226*, 77–84. [CrossRef]
116. Shannon-Lowe, C.; Rickinson, A. The Global Landscape of EBV-Associated Tumors. *Front. Oncol.* **2019**, *9*, 713. [CrossRef]
117. El-Sharkawy, A.; Al Zaidan, L.; Malki, A. Epstein-Barr Virus-Associated Malignancies: Roles of Viral Oncoproteins in Carcinogenesis. *Front. Ooncol.* **2018**, *8*, 265. [CrossRef]

118. Chang, M.S.; Uozaki, H.; Chong, J.M.; Ushiku, T.; Sakuma, K.; Ishikawa, S.; Hino, R.; Barua, R.R.; Iwasaki, Y.; Arai, K.; et al. CpG island methylation status in gastric carcinoma with and without infection of Epstein-Barr virus. *Clin. Cancer Res.* **2006**, *12*, 2995–3002. [CrossRef] [PubMed]
119. Saha, A.; Jha, H.C.; Upadhyay, S.K.; Robertson, E.S. Epigenetic silencing of tumor suppressor genes during in vitro Epstein-Barr virus infection. *Proc. Natl. Acad. Sci. USA* **2015**, *112*, E5199–E5207. [CrossRef]
120. Dai, W.; Zheng, H.; Cheung, A.K.; Lung, M.L. Genetic and epigenetic landscape of nasopharyngeal carcinoma. *Chin. Clin. Oncol.* **2016**, *5*, 16. [CrossRef]
121. Hong, S.; Liu, D.; Luo, S.; Fang, W.; Zhan, J.; Fu, S.; Zhang, Y.; Wu, X.; Zhou, H.; Chen, X.; et al. The genomic landscape of Epstein-Barr virus-associated pulmonary lymphoepithelioma-like carcinoma. *Nat. Commun.* **2019**, *10*, 3108. [CrossRef]
122. Stanland, L.J.; Luftig, M.A. The Role of EBV-Induced Hypermethylation in Gastric Cancer Tumorigenesis. *Viruses* **2020**, *12*, 1222. [CrossRef]
123. Paschos, K.; Smith, P.; Anderton, E.; Middeldorp, J.M.; White, R.E.; Allday, M.J. Epstein-barr virus latency in B cells leads to epigenetic repression and CpG methylation of the tumour suppressor gene Bim. *PLoS Pathog.* **2009**, *5*, e1000492. [CrossRef]
124. van Roy, F. Beyond E-cadherin: Roles of other cadherin superfamily members in cancer. *Nat. Rev. Cancer* **2014**, *14*, 121–134. [CrossRef]
125. Gao, Y.; Fu, Y.; Wang, J.; Zheng, X.; Zhou, J.; Ma, J. EBV as a high infection risk factor promotes RASSF10 methylation and induces cell proliferation in EBV-associated gastric cancer. *Biochem. Biophys. Res. Commun.* **2021**, *547*, 1–8. [CrossRef]
126. Malnati, M.S.; Dagna, L.; Ponzoni, M.; Lusso, P. Human herpesvirus 8 (HHV-8/KSHV) and hematologic malignancies. *Rev. Clin. Exp. Hematol.* **2003**, *7*, 375–405. [PubMed]
127. Mesri, E.A.; Cesarman, E.; Boshoff, C. Kaposi's sarcoma and its associated herpesvirus. *Nat. Rev. Cancer* **2010**, *10*, 707–719. [CrossRef]
128. Fröhlich, J.; Grundhoff, A. Epigenetic control in Kaposi sarcoma-associated herpesvirus infection and associated disease. *Semin. Immunopathol.* **2020**, *42*, 143–157. [CrossRef] [PubMed]
129. Journo, G.; Tushinsky, C.; Shterngas, A.; Avital, N.; Eran, Y.; Karpuj, M.V.; Frenkel-Morgenstern, M.; Shamay, M. Modulation of Cellular CpG DNA Methylation by Kaposi's Sarcoma-Associated Herpesvirus. *J. Virol.* **2018**, *92*, e00008. [CrossRef] [PubMed]
130. Di Bartolo, D.L.; Cannon, M.; Liu, Y.F.; Renne, R.; Chadburn, A.; Boshoff, C.; Cesarman, E. KSHV LANA inhibits TGF-beta signaling through epigenetic silencing of the TGF-beta type II receptor. *Blood* **2008**, *111*, 4731–4740. [CrossRef]
131. Sung, H.; Ferlay, J.; Siegel, R.L.; Laversanne, M.; Soerjomataram, I.; Jemal, A.; Bray, F. Global Cancer Statistics 2020: GLOBOCAN Estimates of Incidence and Mortality Worldwide for 36 Cancers in 185 Countries. *CA Cancer J. Clin.* **2021**, *71*, 209–249. [CrossRef]
132. Levrero, M.; Zucman-Rossi, J. Mechanisms of HBV-induced hepatocellular carcinoma. *J. Hepatol.* **2016**, *64*, S84–S101. [CrossRef]
133. Lee, J.O.; Kwun, H.J.; Jung, J.K.; Choi, K.H.; Min, D.S.; Jang, K.L. Hepatitis B virus X protein represses E-cadherin expression via activation of DNA methyltransferase 1. *Oncogene* **2005**, *24*, 6617–6625. [CrossRef] [PubMed]
134. Herceg, Z.; Paliwal, A. Epigenetic mechanisms in hepatocellular carcinoma: How environmental factors influence the epigenome. *Mutat. Res.* **2011**, *727*, 55–61. [CrossRef] [PubMed]
135. Tian, Y.; Ou, J.H. Genetic and epigenetic alterations in hepatitis B virus-associated hepatocellular carcinoma. *Virol. Sin.* **2015**, *30*, 85–91. [CrossRef] [PubMed]
136. Wahid, B.; Ali, A.; Rafique, S.; Idrees, M. New Insights into the Epigenetics of Hepatocellular Carcinoma. *BioMed Res. Int.* **2017**, *2017*, 1609575. [CrossRef] [PubMed]
137. Lv, X.; Ye, G.; Zhang, X.; Huang, T. p16 Methylation was associated with the development, age, hepatic viruses infection of hepatocellular carcinoma, and p16 expression had a poor survival: A systematic meta-analysis (PRISMA). *Medicine* **2017**, *96*, e8106. [CrossRef] [PubMed]
138. Peng, J.L.; Wu, J.Z.; Li, G.J.; Wu, J.L.; Xi, Y.M.; Li, X.Q.; Wang, L. Association of RASSF1A hypermethylation with risk of HBV/HCV-induced hepatocellular carcinoma: A meta-analysis. *Pathol. Res. Pract.* **2020**, *216*, 153099. [CrossRef]
139. Li, B.; Carey, M.; Workman, J.L. The role of chromatin during transcription. *Cell* **2007**, *128*, 707–719. [CrossRef]
140. Bannister, A.J.; Kouzarides, T. Regulation of chromatin by histone modifications. *Cell Res.* **2011**, *21*, 381–395. [CrossRef]
141. Venkatesh, S.; Workman, J.L. Histone exchange, chromatin structure and the regulation of transcription. *Nat. Rev. Mol. Cell Biol.* **2015**, *16*, 178–189. [CrossRef] [PubMed]
142. Tessarz, P.; Kouzarides, T. Histone core modifications regulating nucleosome structure and dynamics. *Nat. Rev. Mol. Cell Biol.* **2014**, *15*, 703–708. [CrossRef]
143. Chen, T.; Dent, S.Y. Chromatin modifiers and remodellers: Regulators of cellular differentiation. *Nat. Rev. Genet.* **2014**, *15*, 93–106. [CrossRef] [PubMed]
144. Clapier, C.R.; Cairns, B.R. The biology of chromatin remodeling complexes. *Annu. Rev. Biochem.* **2009**, *78*, 273–304. [CrossRef] [PubMed]
145. Clapier, C.R.; Iwasa, J.; Cairns, B.R.; Peterson, C.L. Mechanisms of action and regulation of ATP-dependent chromatin-remodelling complexes. *Nat. Rev. Mol. Cell Biol.* **2017**, *18*, 407–422. [CrossRef] [PubMed]
146. Zhao, S.; Allis, C.D.; Wang, G.G. The language of chromatin modification in human cancers. *Nat. Rev. Cancer* **2021**, *21*, 413–430. [CrossRef]

147. Clerc, I.; Polakowski, N.; André-Arpin, C.; Cook, P.; Barbeau, B.; Mesnard, J.M.; Lemasson, I. An interaction between the human T cell leukemia virus type 1 basic leucine zipper factor (HBZ) and the KIX domain of p300/CBP contributes to the down-regulation of tax-dependent viral transcription by HBZ. *J. Biol. Chem.* **2008**, *283*, 23903–23913. [CrossRef] [PubMed]
148. Kwok, R.P.; Laurance, M.E.; Lundblad, J.R.; Goldman, P.S.; Shih, H.; Connor, L.M.; Marriott, S.J.; Goodman, R.H. Control of cAMP-regulated enhancers by the viral transactivator Tax through CREB and the co-activator CBP. *Nature* **1996**, *380*, 642–646. [CrossRef] [PubMed]
149. Ego, T.; Ariumi, Y.; Shimotohno, K. The interaction of HTLV-1 Tax with HDAC1 negatively regulates the viral gene expression. *Oncogene* **2002**, *21*, 7241–7246. [CrossRef]
150. Tang, H.M.; Gao, W.W.; Chan, C.P.; Cheng, Y.; Deng, J.J.; Yuen, K.S.; Iha, H.; Jin, D.Y. SIRT1 Suppresses Human T-Cell Leukemia Virus Type 1 Transcription. *J. Virol.* **2015**, *89*, 8623–8631. [CrossRef]
151. Fujikawa, D.; Nakagawa, S.; Hori, M.; Kurokawa, N.; Soejima, A.; Nakano, K.; Yamochi, T.; Nakashima, M.; Kobayashi, S.; Tanaka, Y.; et al. Polycomb-dependent epigenetic landscape in adult T-cell leukemia. *Blood* **2016**, *127*, 1790–1802. [CrossRef] [PubMed]
152. Wu, K.; Bottazzi, M.E.; de la Fuente, C.; Deng, L.; Gitlin, S.D.; Maddukuri, A.; Dadgar, S.; Li, H.; Vertes, A.; Pumfery, A.; et al. Protein profile of tax-associated complexes. *J. Biol. Chem.* **2004**, *279*, 495–508. [CrossRef] [PubMed]
153. Lan, R.; Wang, Q. Deciphering structure, function and mechanism of lysine acetyltransferase HBO1 in protein acetylation, transcription regulation, DNA replication and its oncogenic properties in cancer. *Cell. Mol. Life Sci.* **2020**, *77*, 637–649. [CrossRef]
154. Mukai, R.; Ohshima, T. HTLV-1 bZIP factor suppresses the centromere protein B (CENP-B)-mediated trimethylation of histone H3K9 through the abrogation of DNA-binding ability of CENP-B. *J. Gen. Virol.* **2015**, *96*, 159–164. [CrossRef]
155. Satou, Y.; Miyazato, P.; Ishihara, K.; Yaguchi, H.; Melamed, A.; Miura, M.; Fukuda, A.; Nosaka, K.; Watanabe, T.; Rowan, A.G.; et al. The retrovirus HTLV-1 inserts an ectopic CTCF-binding site into the human genome. *Proc. Natl. Acad. Sci. USA* **2016**, *113*, 3054–3059. [CrossRef]
156. Baharudin, R.; Tieng, F.Y.F.; Lee, L.H.; Ab Mutalib, N.S. Epigenetics of SFRP1: The Dual Roles in Human Cancers. *Cancers* **2020**, *12*, 445. [CrossRef]
157. Kasai, H.; Mochizuki, K.; Tanaka, T.; Yamashita, A.; Matsuura, Y.; Moriishi, K. Induction of HOX Genes by Hepatitis C Virus Infection via Impairment of Histone H2A Monoubiquitination. *J. Virol.* **2021**, *95*, e01784-20. [CrossRef]
158. Duong, F.H.; Christen, V.; Lin, S.; Heim, M.H. Hepatitis C virus-induced up-regulation of protein phosphatase 2A inhibits histone modification and DNA damage repair. *Hepatology* **2010**, *51*, 741–751.
159. Cheng, J.; Park, D.E.; Berrios, C.; White, E.A.; Arora, R.; Yoon, R.; Branigan, T.; Xiao, T.; Westerling, T.; Federation, A.; et al. Merkel cell polyomavirus recruits MYCL to the EP400 complex to promote oncogenesis. *PLoS Pathog.* **2017**, *13*, e1006668. [CrossRef]
160. Patel, D.; Huang, S.M.; Baglia, L.A.; McCance, D.J. The E6 protein of human papillomavirus type 16 binds to and inhibits co-activation by CBP and p300. *EMBO J.* **1999**, *18*, 5061–5072. [CrossRef]
161. Fraga, M.F.; Ballestar, E.; Villar-Garea, A.; Boix-Chornet, M.; Espada, J.; Schotta, G.; Bonaldi, T.; Haydon, C.; Ropero, S.; Petrie, K.; et al. Loss of acetylation at Lys16 and trimethylation at Lys20 of histone H4 is a common hallmark of human cancer. *Nat. Genet.* **2005**, *37*, 391–400. [CrossRef] [PubMed]
162. Zhang, L.; Tian, S.; Pei, M.; Zhao, M.; Wang, L.; Jiang, Y.; Yang, T.; Zhao, J.; Song, L.; Yang, X. Crosstalk between histone modification and DNA methylation orchestrates the epigenetic regulation of the costimulatory factors, Tim-3 and galectin-9, in cervical cancer. *Oncol. Rep.* **2019**, *42*, 2655–2669. [CrossRef] [PubMed]
163. Zhang, Y.; Dakic, A.; Chen, R.; Dai, Y.; Schlegel, R.; Liu, X. Direct HPV E6/Myc interactions induce histone modifications, Pol II phosphorylation, and hTERT promoter activation. *Oncotarget* **2017**, *8*, 96323–96339. [CrossRef] [PubMed]
164. Hsu, C.H.; Peng, K.L.; Jhang, H.C.; Lin, C.H.; Wu, S.Y.; Chiang, C.M.; Lee, S.C.; Yu, W.C.; Juan, L.J. The HPV E6 oncoprotein targets histone methyltransferases for modulating specific gene transcription. *Oncogene* **2012**, *31*, 2335–2349. [CrossRef] [PubMed]
165. Hyland, P.L.; McDade, S.S.; McCloskey, R.; Dickson, G.J.; Arthur, K.; McCance, D.J.; Patel, D. Evidence for alteration of EZH2, BMI1, and KDM6A and epigenetic reprogramming in human papillomavirus type 16 E6/E7-expressing keratinocytes. *J. Virol.* **2011**, *85*, 10999–11006. [CrossRef] [PubMed]
166. Jansma, A.L.; Martinez-Yamout, M.A.; Liao, R.; Sun, P.; Dyson, H.J.; Wright, P.E. The high-risk HPV16 E7 oncoprotein mediates interaction between the transcriptional coactivator CBP and the retinoblastoma protein pRb. *J. Mol. Biol.* **2014**, *426*, 4030–4048. [CrossRef]
167. Huang, S.M.; McCance, D.J. Down regulation of the interleukin-8 promoter by human papillomavirus type 16 E6 and E7 through effects on CREB binding protein/p300 and P/CAF. *J. Virol.* **2002**, *76*, 8710–8721. [CrossRef]
168. Brehm, A.; Nielsen, S.J.; Miska, E.A.; McCance, D.J.; Reid, J.L.; Bannister, A.J.; Kouzarides, T. The E7 oncoprotein associates with Mi2 and histone deacetylase activity to promote cell growth. *EMBO J.* **1999**, *18*, 2449–2458. [CrossRef]
169. Lee, D.; Lim, C.; Seo, T.; Kwon, H.; Min, H.; Choe, J. The viral oncogene human papillomavirus E7 deregulates transcriptional silencing by Brm-related gene 1 via molecular interactions. *J. Biol. Chem.* **2002**, *277*, 48842–48848. [CrossRef] [PubMed]
170. McLaughlin-Drubin, M.E.; Crum, C.P.; Münger, K. Human papillomavirus E7 oncoprotein induces KDM6A and KDM6B histone demethylase expression and causes epigenetic reprogramming. *Proc. Natl. Acad. Sci. USA* **2011**, *108*, 2130–2135. [CrossRef]
171. Huang, B.H.; Laban, M.; Leung, C.H.; Lee, L.; Lee, C.K.; Salto-Tellez, M.; Raju, G.C.; Hooi, S.C. Inhibition of histone deacetylase 2 increases apoptosis and p21Cip1/WAF1 expression, independent of histone deacetylase 1. *Cell Death Differ.* **2005**, *12*, 395–404. [CrossRef]

172. Anderton, J.A.; Bose, S.; Vockerodt, M.; Vrzalikova, K.; Wei, W.; Kuo, M.; Helin, K.; Christensen, J.; Rowe, M.; Murray, P.G.; et al. The H3K27me3 demethylase, KDM6B, is induced by Epstein-Barr virus and over-expressed in Hodgkin's Lymphoma. *Oncogene* **2011**, *30*, 2037–2043. [CrossRef] [PubMed]
173. Wang, L.; Grossman, S.R.; Kieff, E. Epstein-Barr virus nuclear protein 2 interacts with p300, CBP, and PCAF histone acetyltransferases in activation of the LMP1 promoter. *Proc. Natl. Acad. Sci. USA* **2000**, *97*, 43043–43045. [CrossRef] [PubMed]
174. Portal, D.; Zhou, H.; Zhao, B.; Kharchenko, P.V.; Lowry, E.; Wong, L.; Quackenbush, J.; Holloway, D.; Jiang, S.; Lu, Y.; et al. Epstein-Barr virus nuclear antigen leader protein localizes to promoters and enhancers with cell transcription factors and EBNA2. *Proc. Natl. Acad. Sci. USA* **2013**, *110*, 18537–18542. [CrossRef] [PubMed]
175. Cotter, M.A., 2nd; Robertson, E.S. Modulation of histone acetyltransferase activity through interaction of epstein-barr nuclear antigen 3C with prothymosin alpha. *Mol. Cell. Biol.* **2000**, *20*, 5722–5735. [CrossRef]
176. Knight, J.S.; Lan, K.; Subramanian, C.; Robertson, E.S. Epstein-Barr virus nuclear antigen 3C recruits histone deacetylase activity and associates with the corepressors mSin3A and NCoR in human B-cell lines. *J. Virol.* **2003**, *77*, 4261–4272. [CrossRef] [PubMed]
177. Adamson, A.L.; Kenney, S. The Epstein-Barr virus BZLF1 protein interacts physically and functionally with the histone acetylase CREB-binding protein. *J. Virol.* **1999**, *73*, 6551–6558. [CrossRef] [PubMed]
178. Schaeffner, M.; Mrozek-Gorska, P.; Buschle, A.; Woellmer, A.; Tagawa, T.; Cernilogar, F.M.; Schotta, G.; Krietenstein, N.; Lieleg, C.; Korber, P.; et al. BZLF1 interacts with chromatin remodelers promoting escape from latent infections with EBV. *Life Sci. Alliance* **2019**, *2*, e201800108. [CrossRef]
179. Swenson, J.J.; Holley-Guthrie, E.; Kenney, S.C. Epstein-Barr virus immediate-early protein BRLF1 interacts with CBP, promoting enhanced BRLF1 transactivation. *J. Virol.* **2001**, *75*, 6228–6234. [CrossRef] [PubMed]
180. He, X.; Yan, B.; Liu, S.; Jia, J.; Lai, W.; Xin, X.; Tang, C.E.; Luo, D.; Tan, T.; Jiang, Y.; et al. Chromatin Remodeling Factor LSH Drives Cancer Progression by Suppressing the Activity of Fumarate Hydratase. *Cancer Res.* **2016**, *76*, 5743–5755. [CrossRef]
181. Krithivas, A.; Young, D.B.; Liao, G.; Greene, D.; Hayward, S.D. Human herpesvirus 8 LANA interacts with proteins of the mSin3 corepressor complex and negatively regulates Epstein-Barr virus gene expression in dually infected PEL cells. *J. Virol.* **2000**, *74*, 9637–9645. [CrossRef]
182. Sakakibara, S.; Ueda, K.; Nishimura, K.; Do, E.; Ohsaki, E.; Okuno, T.; Yamanishi, K. Accumulation of heterochromatin components on the terminal repeat sequence of Kaposi's sarcoma-associated herpesvirus mediated by the latency-associated nuclear antigen. *J. Virol.* **2004**, *78*, 7299–7310. [CrossRef]
183. He, M.; Zhang, W.; Bakken, T.; Schutten, M.; Toth, Z.; Jung, J.U.; Gill, P.; Cannon, M.; Gao, S.J. Cancer angiogenesis induced by Kaposi sarcoma-associated herpesvirus is mediated by EZH2. *Cancer Res.* **2012**, *72*, 3582–3592. [CrossRef]
184. Hu, J.; Yang, Y.; Turner, P.C.; Jain, V.; McIntyre, L.M.; Renne, R. LANA binds to multiple active viral and cellular promoters and associates with the H3K4methyltransferase hSET1 complex. *PLoS Pathog.* **2014**, *10*, e1004240. [CrossRef]
185. Hellert, J.; Weidner-Glunde, M.; Krausze, J.; Richter, U.; Adler, H.; Fedorov, R.; Pietrek, M.; Rückert, J.; Ritter, C.; Schulz, T.F.; et al. A structural basis for BRD2/4-mediated host chromatin interaction and oligomer assembly of Kaposi sarcoma-associated herpesvirus and murine gammaherpesvirus LANA proteins. *PLoS Pathog.* **2013**, *9*, e1003640. [CrossRef] [PubMed]
186. Li, M.; Damania, B.; Alvarez, X.; Ogryzko, V.; Ozato, K.; Jung, J.U. Inhibition of p300 histone acetyltransferase by viral interferon regulatory factor. *Mol. Cell. Biol.* **2000**, *20*, 8254–8263. [CrossRef]
187. Lee, H.R.; Li, F.; Choi, U.Y.; Yu, H.R.; Aldrovandi, G.M.; Feng, P.; Gao, S.J.; Hong, Y.K.; Jung, J.U. Deregulation of HDAC5 by Viral Interferon Regulatory Factor 3 Plays an Essential Role in Kaposi's Sarcoma-Associated Herpesvirus-Induced Lymphangiogenesis. *mBio* **2018**, *9*, e02217-17. [CrossRef] [PubMed]
188. Gwack, Y.; Baek, H.J.; Nakamura, H.; Lee, S.H.; Meisterernst, M.; Roeder, R.G.; Jung, J.U. Principal role of TRAP/mediator and SWI/SNF complexes in Kaposi's sarcoma-associated herpesvirus RTA-mediated lytic reactivation. *Mol. Cell. Biol.* **2003**, *23*, 2055–2067. [CrossRef]
189. Cougot, D.; Wu, Y.; Cairo, S.; Caramel, J.; Renard, C.A.; Lévy, L.; Buendia, M.A.; Neuveut, C. The hepatitis B virus X protein functionally interacts with CREB-binding protein/p300 in the regulation of CREB-mediated transcription. *J. Biol. Chem.* **2007**, *282*, 4277–4287. [CrossRef] [PubMed]
190. Tian, Y.; Yang, W.; Song, J.; Wu, Y.; Ni, B. Hepatitis B virus X protein-induced aberrant epigenetic modifications contributing to human hepatocellular carcinoma pathogenesis. *Mol. Cell. Biol.* **2013**, *33*, 2810–2816. [CrossRef]
191. Yoo, Y.G.; Na, T.Y.; Seo, H.W.; Seong, J.K.; Park, C.K.; Shin, Y.K.; Lee, M.O. Hepatitis B virus X protein induces the expression of MTA1 and HDAC1, which enhances hypoxia signaling in hepatocellular carcinoma cells. *Oncogene* **2008**, *27*, 3405–3413. [CrossRef]
192. Zhang, H.; Diab, A.; Fan, H.; Mani, S.K.; Hullinger, R.; Merle, P.; Andrisani, O. PLK1 and HOTAIR Accelerate Proteasomal Degradation of SUZ12 and ZNF198 during Hepatitis B Virus-Induced Liver Carcinogenesis. *Cancer Res.* **2015**, *75*, 2363–2374. [CrossRef]
193. Rivière, L.; Gerossier, L.; Ducroux, A.; Dion, S.; Deng, Q.; Michel, M.L.; Buendia, M.A.; Hantz, O.; Neuveut, C. HBx relieves chromatin-mediated transcriptional repression of hepatitis B viral cccDNA involving SETDB1 histone methyltransferase. *J. Hepatol.* **2015**, *63*, 1093–1102. [CrossRef] [PubMed]
194. Shi, X.Y.; Zhang, Y.Y.; Zhou, X.W.; Lu, J.S.; Guo, Z.K.; Huang, P.T. Hepatitis B virus X protein regulates the mEZH2 promoter via the E2F1-binding site in AML12 cells. *Chin. J. Cancer* **2011**, *30*, 273–279. [CrossRef] [PubMed]

195. Chen, S.L.; Liu, L.L.; Lu, S.X.; Luo, R.Z.; Wang, C.H.; Wang, H.; Cai, S.H.; Yang, X.; Xie, D.; Zhang, C.Z.; et al. HBx-mediated decrease of AIM2 contributes to hepatocellular carcinoma metastasis. *Mol. Oncol.* **2017**, *11*, 1225–1240. [CrossRef]
196. Yang, L.; He, J.; Chen, L.; Wang, G. Hepatitis B virus X protein upregulates expression of SMYD3 and C-MYC in HepG2 cells. *Med. Oncol.* **2009**, *26*, 445–451. [CrossRef]
197. Fan, H.; Zhang, H.; Pascuzzi, P.E.; Andrisani, O. Hepatitis B virus X protein induces EpCAM expression via active DNA demethylation directed by RelA in complex with EZH2 and TET2. *Oncogene* **2016**, *35*, 715–726. [CrossRef]
198. Gao, W.; Jia, Z.; Tian, Y.; Yang, P.; Sun, H.; Wang, C.; Ding, Y.; Zhang, M.; Zhang, Y.; Yang, D.; et al. HBx Protein Contributes to Liver Carcinogenesis by H3K4me3 Modification Through Stabilizing WD Repeat Domain 5 Protein. *Hepatology* **2020**, *71*, 1678–1695. [CrossRef]
199. de La Fuente, C.; Deng, L.; Santiago, F.; Arce, L.; Wang, L.; Kashanchi, F. Gene expression array of HTLV type 1-infected T cells: Up-regulation of transcription factors and cell cycle genes. *AIDS Res. Hum. Retrovir.* **2000**, *16*, 1695–1700. [CrossRef] [PubMed]
200. Wurm, T.; Wright, D.G.; Polakowski, N.; Mesnard, J.M.; Lemasson, I. The HTLV-1-encoded protein HBZ directly inhibits the acetyl transferase activity of p300/CBP. *Nucleic Acids Res.* **2012**, *40*, 5910–5925. [CrossRef]
201. Wang, F.; Marshall, C.B.; Ikura, M. Transcriptional/epigenetic regulator CBP/p300 in tumorigenesis: Structural and functional versatility in target recognition. *Cell. Mol. Life Sci.* **2013**, *70*, 3989–4008. [CrossRef]
202. Wright, D.G.; Marchal, C.; Hoang, K.; Ankney, J.A.; Nguyen, S.T.; Rushing, A.W.; Polakowski, N.; Miotto, B.; Lemasson, I. Human T-cell leukemia virus type-1-encoded protein HBZ represses p53 function by inhibiting the acetyltransferase activity of p300/CBP and HBO1. *Oncotarget* **2016**, *7*, 1687–1706. [CrossRef]
203. Kozako, T.; Suzuki, T.; Yoshimitsu, M.; Uchida, Y.; Kuroki, A.; Aikawa, A.; Honda, S.; Arima, N.; Soeda, S. Novel small-molecule SIRT1 inhibitors induce cell death in adult T-cell leukaemia cells. *Sci. Rep.* **2015**, *5*, 11345. [CrossRef]
204. Yamagishi, M.; Hori, M.; Fujikawa, D.; Ohsugi, T.; Honma, D.; Adachi, N.; Katano, H.; Hishima, T.; Kobayashi, S.; Nakano, K.; et al. Targeting Excessive EZH1 and EZH2 Activities for Abnormal Histone Methylation and Transcription Network in Malignant Lymphomas. *Cell Rep.* **2019**, *29*, 2321–2337.e7. [CrossRef] [PubMed]
205. Takahashi, R.; Yamagishi, M.; Nakano, K.; Yamochi, T.; Yamochi, T.; Fujikawa, D.; Nakashima, M.; Tanaka, Y.; Uchimaru, K.; Utsunomiya, A.; et al. Epigenetic deregulation of Ellis Van Creveld confers robust Hedgehog signaling in adult T-cell leukemia. *Cancer Sci.* **2014**, *105*, 1160–1169. [CrossRef]
206. Briscoe, J.; Thérond, P.P. The mechanisms of Hedgehog signalling and its roles in development and disease. *Nat. Rev. Mol. Cell Biol.* **2013**, *14*, 416–429. [CrossRef] [PubMed]
207. Tanaka-Nakanishi, A.; Yasunaga, J.; Takai, K.; Matsuoka, M. HTLV-1 bZIP factor suppresses apoptosis by attenuating the function of FoxO3a and altering its localization. *Cancer Res.* **2014**, *74*, 188–200. [CrossRef] [PubMed]
208. Alasiri, A.; Abboud Guerr, J.; Hall, W.W.; Sheehy, N. Novel Interactions between the Human T-Cell Leukemia Virus Type 1 Antisense Protein HBZ and the SWI/SNF Chromatin Remodeling Family: Implications for Viral Life Cycle. *J. Virol.* **2019**, *93*, e00412-19. [CrossRef]
209. Perez, S.; Kaspi, A.; Domovitz, T.; Davidovich, A.; Lavi-Itzkovitz, A.; Meirson, T.; Alison Holmes, J.; Dai, C.Y.; Huang, C.F.; Chung, R.T.; et al. Hepatitis C virus leaves an epigenetic signature post cure of infection by direct-acting antivirals. *PLoS Genet.* **2019**, *15*, e1008181. [CrossRef]
210. Eckner, R.; Ludlow, J.W.; Lill, N.L.; Oldread, E.; Arany, Z.; Modjtahedi, N.; DeCaprio, J.A.; Livingston, D.M.; Morgan, J.A. Association of p300 and CBP with simian virus 40 large T antigen. *Mol. Cell. Biol.* **1996**, *16*, 3454–3464. [CrossRef] [PubMed]
211. Cho, S.; Tian, Y.; Benjamin, T.L. Binding of p300/CBP co-activators by polyoma large T antigen. *J. Biol. Chem.* **2001**, *276*, 33533–33539. [CrossRef]
212. Valls, E.; de la Cruz, X.; Martínez-Balbás, M.A. The SV40 T antigen modulates CBP histone acetyltransferase activity. *Nucleic Acids Res.* **2003**, *31*, 3114–3122. [CrossRef]
213. Sáenz Robles, M.T.; Shivalila, C.; Wano, J.; Sorrells, S.; Roos, A.; Pipas, J.M. Two independent regions of simian virus 40 T antigen increase CBP/p300 levels, alter patterns of cellular histone acetylation, and immortalize primary cells. *J. Virol.* **2013**, *87*, 13499–13509. [CrossRef] [PubMed]
214. Busam, K.J.; Pulitzer, M.P.; Coit, D.C.; Arcila, M.; Leng, D.; Jungbluth, A.A.; Wiesner, T. Reduced H3K27me3 expression in Merkel cell polyoma virus-positive tumors. *Mod. Pathol.* **2017**, *30*, 877–883. [CrossRef] [PubMed]
215. Jha, S.; Vande Pol, S.; Banerjee, N.S.; Dutta, A.B.; Chow, L.T.; Dutta, A. Destabilization of TIP60 by human papillomavirus E6 results in attenuation of TIP60-dependent transcriptional regulation and apoptotic pathway. *Mol. Cell* **2010**, *38*, 700–711. [CrossRef] [PubMed]
216. Longworth, M.S.; Laimins, L.A. The binding of histone deacetylases and the integrity of zinc finger-like motifs of the E7 protein are essential for the life cycle of human papillomavirus type 31. *J. Virol.* **2004**, *78*, 3533–3541. [CrossRef] [PubMed]
217. Nichol, J.N.; Dupéré-Richer, D.; Ezponda, T.; Licht, J.D.; Miller, W.H., Jr. H3K27 Methylation: A Focal Point of Epigenetic Deregulation in Cancer. *Adv. Cancer Res.* **2016**, *131*, 59–95. [PubMed]
218. Wiles, E.T.; Selker, E.U. H3K27 methylation: A promiscuous repressive chromatin mark. *Curr. Opin. Genet. Dev.* **2017**, *43*, 31–37. [CrossRef]
219. Gameiro, S.F.; Kolendowski, B.; Zhang, A.; Barrett, J.W.; Nichols, A.C.; Torchia, J.; Mymryk, J.S. Human papillomavirus dysregulates the cellular apparatus controlling the methylation status of H3K27 in different human cancers to consistently alter gene expression regardless of tissue of origin. *Oncotarget* **2017**, *8*, 72564–72576. [CrossRef]

220. Chen, X.; Loo, J.X.; Shi, X.; Xiong, W.; Guo, Y.; Ke, H.; Yang, M.; Jiang, Y.; Xia, S.; Zhao, M.; et al. E6 Protein Expressed by High-Risk HPV Activates Super-Enhancers of the EGFR and c-MET Oncogenes by Destabilizing the Histone Demethylase KDM5C. *Cancer Res.* **2018**, *78*, 1418–1430. [CrossRef]
221. Buschle, A.; Hammerschmidt, W. Epigenetic lifestyle of Epstein-Barr virus. *Semin. Immunopathol.* **2020**, *42*, 131–142. [CrossRef]
222. Leong, M.M.L.; Cheung, A.K.L.; Dai, W.; Tsao, S.W.; Tsang, C.M.; Dawson, C.W.; Mun Yee Ko, J.; Lung, M.L. EBV infection is associated with histone bivalent switch modifications in squamous epithelial cells. *Proc. Natl. Acad. Sci. USA* **2019**, *116*, 14144–14153. [CrossRef]
223. Hernando, H.; Islam, A.B.; Rodríguez-Ubreva, J.; Forné, I.; Ciudad, L.; Imhof, A.; Shannon-Lowe, C.; Ballestar, E. Epstein-Barr virus-mediated transformation of B cells induces global chromatin changes independent to the acquisition of proliferation. *Nucleic Acids Res.* **2014**, *42*, 249–263. [CrossRef]
224. Wu, D.Y.; Krumm, A.; Schubach, W.H. Promoter-specific targeting of human SWI-SNF complex by Epstein-Barr virus nuclear protein 2. *J. Virol.* **2000**, *74*, 8893–8903. [CrossRef] [PubMed]
225. Cordier, M.; Calender, A.; Billaud, M.; Zimber, U.; Rousselet, G.; Pavlish, O.; Banchereau, J.; Tursz, T.; Bornkamm, G.; Lenoir, G.M. Stable transfection of Epstein-Barr virus (EBV) nuclear antigen 2 in lymphoma cells containing the EBV P3HR1 genome induces expression of B-cell activation molecules CD21 and CD23. *J. Virol.* **1990**, *64*, 1002–1013. [CrossRef]
226. Naipauer, J.; Rosario, S.; Gupta, S.; Premer, C.; Méndez-Solís, O.; Schlesinger, M.; Ponzinibbio, V.; Jain, V.; Gay, L.; Renne, R.; et al. PDGFRA defines the mesenchymal stem cell Kaposi's sarcoma progenitors by enabling KSHV oncogenesis in an angiogenic environment. *PLoS Pathog.* **2019**, *15*, e1008221. [CrossRef] [PubMed]
227. Urbich, C.; Rössig, L.; Kaluza, D.; Potente, M.; Boeckel, J.N.; Knau, A.; Diehl, F.; Geng, J.G.; Hofmann, W.K.; Zeiher, A.M.; et al. HDAC5 is a repressor of angiogenesis and determines the angiogenic gene expression pattern of endothelial cells. *Blood* **2009**, *113*, 5669–5679. [CrossRef]
228. Günther, T.; Schreiner, S.; Dobner, T.; Tessmer, U.; Grundhoff, A. Influence of ND10 components on epigenetic determinants of early KSHV latency establishment. *PLoS Pathog.* **2014**, *10*, e1004274. [CrossRef] [PubMed]
229. Luo, D.; Wang, Z.; Wu, J.; Jiang, C.; Wu, J. The role of hypoxia inducible factor-1 in hepatocellular carcinoma. *BioMed Res. Int.* **2014**, *2014*, 409272. [CrossRef] [PubMed]
230. Wong, C.M.; Wei, L.; Law, C.T.; Ho, D.W.; Tsang, F.H.; Au, S.L.; Sze, K.M.; Lee, J.M.; Wong, C.C.; Ng, I.O. Up-regulation of histone methyltransferase SETDB1 by multiple mechanisms in hepatocellular carcinoma promotes cancer metastasis. *Hepatology* **2016**, *63*, 474–487. [CrossRef] [PubMed]
231. Laugesen, A.; Helin, K. Chromatin repressive complexes in stem cells, development, and cancer. *Cell Stem Cell* **2014**, *14*, 735–751. [CrossRef]
232. Sanna, L.; Marchesi, I.; Melone, M.A.B.; Bagella, L. The role of enhancer of zeste homolog 2: From viral epigenetics to the carcinogenesis of hepatocellular carcinoma. *J. Cell. Physiol.* **2018**, *233*, 6508–6517. [CrossRef] [PubMed]
233. Ha, M.; Kim, V.N. Regulation of microRNA biogenesis. *Nat. Rev. Mol. Cell Biol.* **2014**, *15*, 509–524. [CrossRef]
234. Gebert, L.F.R.; MacRae, I.J. Regulation of microRNA function in animals. *Nat. Rev. Mol. Cell Biol.* **2019**, *20*, 21–37. [CrossRef]
235. Moles, R.; Bellon, M.; Nicot, C. STAT1: A Novel Target of miR-150 and miR-223 Is Involved in the Proliferation of HTLV-I-Transformed and ATL Cells. *Neoplasia* **2015**, *17*, 449–462. [CrossRef] [PubMed]
236. Christova, R.; Jones, T.; Wu, P.J.; Bolzer, A.; Costa-Pereira, A.P.; Watling, D.; Kerr, I.M.; Sheer, D. P-STAT1 mediates higher-order chromatin remodelling of the human MHC in response to IFNgamma. *J. Cell Sci.* **2007**, *120*, 3262–3270. [CrossRef]
237. Vernin, C.; Thenoz, M.; Pinatel, C.; Gessain, A.; Gout, O.; Delfau-Larue, M.H.; Nazaret, N.; Legras-Lachuer, C.; Wattel, E.; Mortreux, F. HTLV-1 bZIP factor HBZ promotes cell proliferation and genetic instability by activating OncomiRs. *Cancer Res.* **2014**, *74*, 6082–6093. [CrossRef]
238. Pascut, D.; Hoang, M.; Nguyen, N.N.Q.; Pratama, M.Y.; Tiribelli, C. HCV Proteins Modulate the Host Cell miRNA Expression Contributing to Hepatitis C Pathogenesis and Hepatocellular Carcinoma Development. *Cancers* **2021**, *13*, 2485. [CrossRef]
239. Zheng, F.; Liao, Y.J.; Cai, M.Y.; Liu, Y.H.; Liu, T.H.; Chen, S.P.; Bian, X.W.; Guan, X.Y.; Lin, M.C.; Zeng, Y.X.; et al. The putative tumour suppressor microRNA-124 modulates hepatocellular carcinoma cell aggressiveness by repressing ROCK2 and EZH2. *Gut* **2012**, *61*, 278–289. [CrossRef] [PubMed]
240. Seo, G.J.; Chen, C.J.; Sullivan, C.S. Merkel cell polyomavirus encodes a microRNA with the ability to autoregulate viral gene expression. *Virology* **2009**, *383*, 183–187. [CrossRef]
241. Akhbari, P.; Tobin, D.; Poterlowicz, K.; Roberts, W.; Boyne, J.R. MCV-miR-M1 Targets the Host-Cell Immune Response Resulting in the Attenuation of Neutrophil Chemotaxis. *J. Investig. Dermatol.* **2018**, *138*, 2343–2354. [CrossRef]
242. Kumar, S.; Xie, H.; Shi, H.; Gao, J.; Juhlin, C.C.; Björnhagen, V.; Höög, A.; Lee, L.; Larsson, C.; Lui, W.O. Merkel cell polyomavirus oncoproteins induce microRNAs that suppress multiple autophagy genes. *Int. J. Cancer* **2020**, *146*, 1652–1666. [CrossRef]
243. Veija, T.; Sahi, H.; Koljonen, V.; Bohling, T.; Knuutila, S.; Mosakhani, N. miRNA-34a underexpressed in Merkel cell polyomavirus-negative Merkel cell carcinoma. *Virch. Arch.* **2015**, *466*, 289–295. [CrossRef]
244. Kumar, S.; Xie, H.; Scicluna, P.; Lee, L.; Björnhagen, V.; Höög, A.; Larsson, C.; Lui, W.O. MiR-375 Regulation of LDHB Plays Distinct Roles in Polyomavirus-Positive and -Negative Merkel Cell Carcinoma. *Cancers* **2018**, *10*, 443. [CrossRef] [PubMed]
245. Tornesello, M.L.; Faraonio, R.; Buonaguro, L.; Annunziata, C.; Starita, N.; Cerasuolo, A.; Pezzuto, F.; Tornesello, A.L.; Buonaguro, F.M. The Role of microRNAs, Long Non-coding RNAs, and Circular RNAs in Cervical Cancer. *Front. Oncol.* **2020**, *10*, 150. [CrossRef] [PubMed]

246. Jiménez-Wences, H.; Peralta-Zaragoza, O.; Fernández-Tilapa, G. Human papilloma virus, DNA methylation and microRNA expression in cervical cancer. *Oncol. Rep.* **2014**, *31*, 2467–2476. [CrossRef] [PubMed]
247. Zhu, Y.; Haecker, I.; Yang, Y.; Gao, S.J.; Renne, R. γ-Herpesvirus-encoded miRNAs and their roles in viral biology and pathogenesis. *Curr. Opin. Virol.* **2013**, *3*, 266–275. [CrossRef]
248. Piedade, D.; Azevedo-Pereira, J.M. The Role of microRNAs in the Pathogenesis of Herpesvirus Infection. *Viruses* **2016**, *8*, 156. [CrossRef]
249. Li, W.; Yi, W.; Yang, D.; Li, G. Epstein-Barr virus-encoded microRNAs involve in tumorigenesis and development. *J. Cent. South Univ. Med. Sci.* **2021**, *46*, 300–308.
250. Choi, Y.B.; Cousins, E.; Nicholas, J. Novel Functions and Virus-Host Interactions Implicated in Pathogenesis and Replication of Human Herpesvirus 8. *Recent Results Cancer Res.* **2021**, *217*, 245–301. [PubMed]
251. Hussein, H.A.M.; Alfhili, M.A.; Pakala, P.; Simon, S.; Hussain, J.; McCubrey, J.A.; Akula, S.M. miRNAs and their roles in KSHV pathogenesis. *Virus Res.* **2019**, *266*, 15–24. [CrossRef]
252. Liang, D.; Lin, X.; Lan, K. Looking at Kaposi's Sarcoma-Associated Herpesvirus-Host Interactions from a microRNA Viewpoint. *Front. Microbiol.* **2011**, *2*, 271. [CrossRef]
253. Lu, F.; Stedman, W.; Yousef, M.; Renne, R.; Lieberman, P.M. Epigenetic regulation of Kaposi's sarcoma-associated herpesvirus latency by virus-encoded microRNAs that target Rta and the cellular Rbl2-DNMT pathway. *J. Virol.* **2010**, *84*, 2697–2706. [CrossRef] [PubMed]
254. Lagos, D.; Pollara, G.; Henderson, S.; Gratrix, F.; Fabani, M.; Milne, R.S.; Gotch, F.; Boshoff, C. miR-132 regulates antiviral innate immunity through suppression of the p300 transcriptional co-activator. *Nat. Cell Biol.* **2010**, *12*, 513–519. [CrossRef]
255. Loureiro, D.; Tout, I.; Narguet, S.; Benazzouz, S.M.; Mansouri, A.; Asselah, T. miRNAs as Potential Biomarkers for Viral Hepatitis B and C. *Viruses.* **2020**, *12*, 1440. [CrossRef]
256. Tang, J.; Xiao, X.; Jiang, Y.; Tian, Y.; Peng, Z.; Yang, M.; Xu, Z.; Gong, G. miR-3 Encoded by Hepatitis B Virus Downregulates PTEN Protein Expression and Promotes Cell Proliferation. *J. Hepatocell. Carcinoma* **2020**, *7*, 257–269. [CrossRef] [PubMed]
257. Sartorius, K.; Makarova, J.; Sartorius, B.; An, P.; Winkler, C.; Chuturgoon, A.; Kramvis, A. The Regulatory Role of MicroRNA in Hepatitis-B Virus-Associated Hepatocellular Carcinoma (HBV-HCC) Pathogenesis. *Cells* **2019**, *8*, 1504. [CrossRef] [PubMed]
258. Ohtani, K.; Iwanaga, R.; Arai, M.; Huang, Y.; Matsumura, Y.; Nakamura, M. Cell type-specific E2F activation and cell cycle progression induced by the oncogene product Tax of human T-cell leukemia virus type I. *J. Biol. Chem.* **2000**, *275*, 11154–11163. [CrossRef] [PubMed]
259. Pulikkan, J.A.; Dengler, V.; Peramangalam, P.S.; Peer Zada, A.A.; Müller-Tidow, C.; Bohlander, S.K.; Tenen, D.G.; Behre, G. Cell-cycle regulator E2F1 and microRNA-223 comprise an autoregulatory negative feedback loop in acute myeloid leukemia. *Blood* **2010**, *115*, 1768–1778. [CrossRef]
260. Kawatsuki, A.; Yasunaga, J.I.; Mitobe, Y.; Green, P.L.; Matsuoka, M. HTLV-1 bZIP factor protein targets the Rb/E2F-1 pathway to promote proliferation and apoptosis of primary CD4(+) T cells. *Oncogene* **2016**, *35*, 4509–4517. [CrossRef] [PubMed]
261. Ura, S.; Honda, M.; Yamashita, T.; Ueda, T.; Takatori, H.; Nishino, R.; Sunakozaka, H.; Sakai, Y.; Horimoto, K.; Kaneko, S. Differential microRNA expression between hepatitis B and hepatitis C leading disease progression to hepatocellular carcinoma. *Hepatology* **2009**, *49*, 1098–1112. [CrossRef]
262. Zeng, B.; Li, Z.; Chen, R.; Guo, N.; Zhou, J.; Zhou, Q.; Lin, Q.; Cheng, D.; Liao, Q.; Zheng, L.; et al. Epigenetic regulation of miR-124 by hepatitis C virus core protein promotes migration and invasion of intrahepatic cholangiocarcinoma cells by targeting SMYD3. *FEBS Lett.* **2012**, *586*, 3271–3278. [CrossRef]
263. Oura, K.; Morishita, A.; Masaki, T. Molecular and Functional Roles of MicroRNAs in the Progression of Hepatocellular Carcinoma-A Review. *Int. J. Mol. Sci.* **2020**, *21*, 8362. [CrossRef] [PubMed]
264. Wang, Y.; Kato, N.; Jazag, A.; Dharel, N.; Otsuka, M.; Taniguchi, H.; Kawabe, T.; Omata, M. Hepatitis C virus core protein is a potent inhibitor of RNA silencing-based antiviral response. *Gastroenterology* **2006**, *130*, 883–892. [CrossRef]
265. Lee, S.; Paulson, K.G.; Murchison, E.P.; Afanasiev, O.K.; Alkan, C.; Leonard, J.H.; Byrd, D.R.; Hannon, G.J.; Nghiem, P. Identification and validation of a novel mature microRNA encoded by the Merkel cell polyomavirus in human Merkel cell carcinomas. *J. Clin. Virol.* **2011**, *52*, 272–275. [CrossRef] [PubMed]
266. Chen, C.J.; Cox, J.E.; Azarm, K.D.; Wylie, K.N.; Woolard, K.D.; Pesavento, P.A.; Sullivan, C.S. Identification of a polyomavirus microRNA highly expressed in tumors. *Virology* **2015**, *476*, 43–53. [CrossRef]
267. Konstatinell, A.; Coucheron, D.H.; Sveinbjørnsson, B.; Moens, U. MicroRNAs as Potential Biomarkers in Merkel Cell Carcinoma. *Int. J. Mol. Sci.* **2018**, *19*, 1873. [CrossRef]
268. Xie, H.; Lee, L.; Caramuta, S.; Höög, A.; Browaldh, N.; Björnhagen, V.; Larsson, C.; Lui, W.O. MicroRNA expression patterns related to merkel cell polyomavirus infection in human merkel cell carcinoma. *J. Investig. Dermatol.* **2014**, *134*, 507–517. [CrossRef] [PubMed]
269. Fan, K.; Ritter, C.; Nghiem, P.; Blom, A.; Verhaegen, M.E.; Dlugosz, A.; Ødum, N.; Woetmann, A.; Tothill, R.W.; Hicks, R.J.; et al. Circulating Cell-Free miR-375 as Surrogate Marker of Tumor Burden in Merkel Cell Carcinoma. *Clin. Cancer Res.* **2018**, *24*, 5873–5882. [CrossRef]
270. Abraham, K.J.; Zhang, X.; Vidal, R.; Paré, G.C.; Feilotter, H.E.; Tron, V.A. Roles for miR-375 in Neuroendocrine Differentiation and Tumor Suppression via Notch Pathway Suppression in Merkel Cell Carcinoma. *Am. J. Pathol.* **2016**, *186*, 1025–1035. [CrossRef] [PubMed]

271. McCleland, M.L.; Adler, A.S.; Shang, Y.; Hunsaker, T.; Truong, T.; Peterson, D.; Torres, E.; Li, L.; Haley, B.; Stephan, J.P.; et al. An integrated genomic screen identifies LDHB as an essential gene for triple-negative breast cancer. *Cancer Res.* **2012**, *72*, 5812–5823. [CrossRef] [PubMed]
272. Brisson, L.; Bański, P.; Sboarina, M.; Dethier, C.; Danhier, P.; Fontenille, M.J.; Van Hée, V.F.; Vazeille, T.; Tardy, M.; Falces, J.; et al. Lactate Dehydrogenase B Controls Lysosome Activity and Autophagy in Cancer. *Cancer Cell* **2016**, *30*, 418–431. [CrossRef]
273. Ostrowski, S.M.; Wright, M.C.; Bolock, A.M.; Geng, X.; Maricich, S.M. Ectopic Atoh1 expression drives Merkel cell production in embryonic, postnatal and adult mouse epidermis. *Development* **2015**, *142*, 2533–2544. [CrossRef] [PubMed]
274. Gambichler, T.; Mohtezebsade, S.; Wieland, U.; Silling, S.; Höh, A.K.; Dreißigacker, M.; Schaller, J.; Schulze, H.J.; Oellig, F.; Kreuter, A.; et al. Prognostic relevance of high atonal homolog-1 expression in Merkel cell carcinoma. *J. Cancer Res. Clin. Oncol.* **2017**, *143*, 43–49. [CrossRef] [PubMed]
275. Fan, K.; Zebisch, A.; Horny, K.; Schrama, D.; Becker, J.C. Highly Expressed miR-375 is not an Intracellular Oncogene in Merkel Cell Polyomavirus-Associated Merkel Cell Carcinoma. *Cancers* **2020**, *12*, 529. [CrossRef]
276. Fan, K.; Spassova, I.; Gravemeyer, J.; Ritter, C.; Horny, K.; Lange, A.; Gambichler, T.; Ødum, N.; Schrama, D.; Schadendorf, D.; et al. Merkel cell carcinoma-derived exosome-shuttle miR-375 induces fibroblast polarization by inhibition of RBPJ and p53. *Oncogene* **2021**, *40*, 980–996. [CrossRef]
277. Csoboz, B.; Rasheed, K.; Sveinbjørnsson, B.; Moens, U. Merkel cell polyomavirus and non-Merkel cell carcinomas: Guilty or circumstantial evidence? *APMIS* **2020**, *128*, 104–120. [CrossRef] [PubMed]
278. Lasithiotaki, I.; Tsitoura, E.; Koutsopoulos, A.; Lagoudaki, E.; Koutoulaki, C.; Pitsidianakis, G.; Spandidos, D.A.; Siafakas, N.M.; Sourvinos, G.; Antoniou, K.M. Aberrant expression of miR-21, miR-376c and miR-145 and their target host genes in Merkel cell polyomavirus-positive non-small cell lung cancer. *Oncotarget* **2017**, *8*, 112371–112383. [CrossRef]
279. Gu, W.; An, J.; Ye, P.; Zhao, K.N.; Antonsson, A. Prediction of conserved microRNAs from skin and mucosal human papillomaviruses. *Arch. Virol.* **2011**, *156*, 1161–1171. [CrossRef]
280. Qian, K.; Pietilä, T.; Rönty, M.; Michon, F.; Frilander, M.J.; Ritari, J.; Tarkkanen, J.; Paulín, L.; Auvinen, P.; Auvinen, E. Identification and validation of human papillomavirus encoded microRNAs. *PLoS ONE* **2013**, *8*, e70202. [CrossRef]
281. Virtanen, E.; Pietilä, T.; Nieminen, P.; Qian, K.; Auvinen, E. Low expression levels of putative HPV encoded microRNAs in cervical samples. *SpringerPlus* **2016**, *5*, 1856. [CrossRef] [PubMed]
282. Emmett, S.; Whiteman, D.C.; Panizza, B.J.; Antonsson, A. An Update on Cellular MicroRNA Expression in Human Papillomavirus-Associated Head and Neck Squamous Cell Carcinoma. *Oncology* **2018**, *95*, 193–201. [CrossRef] [PubMed]
283. Harden, M.E.; Munger, K. Perturbation of DROSHA and DICER expression by human papillomavirus 16 oncoproteins. *Virology* **2017**, *507*, 192–198. [CrossRef] [PubMed]
284. Snoek, B.C.; Babion, I.; Koppers-Lalic, D.; Pegtel, D.M.; Steenbergen, R.D. Altered microRNA processing proteins in HPV-induced cancers. *Curr. Opin. Virol.* **2019**, *39*, 23–32. [CrossRef]
285. Dong, M.; Chen, J.N.; Huang, J.T.; Gong, L.P.; Shao, C.K. The roles of EBV-encoded microRNAs in EBV-associated tumors. *Crit. Rev. Oncol. Hematol.* **2019**, *135*, 30–38. [CrossRef]
286. Zebardast, A.; Tehrani, S.S.; Latifi, T.; Sadeghi, F. Critical review of Epstein-Barr virus microRNAs relation with EBV-associated gastric cancer. *J. Cell. Physiol.* **2012**, *236*, 6136–6153. [CrossRef]
287. Gottwein, E.; Corcoran, D.L.; Mukherjee, N.; Skalsky, R.L.; Hafner, M.; Nusbaum, J.D.; Shamulailatpam, P.; Love, C.L.; Dave, S.S.; Tuschl, T.; et al. Viral microRNA targetome of KSHV-infected primary effusion lymphoma cell lines. *Cell Host Microbe* **2011**, *10*, 515–526. [CrossRef]
288. Qin, J.; Li, W.; Gao, S.J.; Lu, C. KSHV microRNAs: Tricks of the Devil. *Trends Microbiol.* **2017**, *25*, 648–661. [CrossRef]
289. Li, W.; Hu, M.; Wang, C.; Lu, H.; Chen, F.; Xu, J.; Shang, Y.; Wang, F.; Qin, J.; Yan, Q.; et al. A viral microRNA downregulates metastasis suppressor CD82 and induces cell invasion and angiogenesis by activating the c-Met signaling. *Oncogene* **2017**, *36*, 5407–5420. [CrossRef]
290. Dahlke, C.; Maul, K.; Christalla, T.; Walz, N.; Schult, P.; Stocking, C.; Grundhoff, A. A microRNA encoded by Kaposi sarcoma-associated herpesvirus promotes B-cell expansion in vivo. *PLoS ONE* **2012**, *7*, e49435. [CrossRef]
291. Wong, C.M.; Tsang, F.H.; Ng, I.O. Non-coding RNAs in hepatocellular carcinoma: Molecular functions and pathological implications. *Nat. Rev. Gastroenterol. Hepatol.* **2018**, *15*, 137–151. [CrossRef] [PubMed]
292. Sartorius, K.; An, P.; Winkler, C.; Chuturgoon, A.; Li, X.; Makarova, J.; Kramvis, A. The Epigenetic Modulation of Cancer and Immune Pathways in Hepatitis B Virus-Associated Hepatocellular Carcinoma: The Influence of HBx and miRNA Dysregulation. *Front. Immunol.* **2021**, *12*, 661204. [CrossRef]
293. Lee, S.; Kim, W.; Ko, C.; Ryu, W.S. Hepatitis B virus X protein enhances Myc stability by inhibiting SCF(Skp2) ubiquitin E3 ligase-mediated Myc ubiquitination and contributes to oncogenesis. *Oncogene* **2016**, *35*, 1857–1867. [CrossRef] [PubMed]
294. Chirillo, P.; Falco, M.; Puri, P.L.; Artini, M.; Balsano, C.; Levrero, M.; Natoli, G. Hepatitis B virus pX activates NF-kappa B-dependent transcription through a Raf-independent pathway. *J. Virol.* **1996**, *70*, 641–646. [CrossRef]
295. Bui, T.V.; Mendell, J.T. Myc: Maestro of MicroRNAs. *Genes Cancer* **2010**, *1*, 568–575. [CrossRef]
296. Feng, Z.; Zhang, C.; Wu, R.; Hu, W. Tumor suppressor p53 meets microRNAs. *J. Mol. Cell Biol.* **2011**, *3*, 44–50. [CrossRef]
297. Markopoulos, G.S.; Roupakia, E.; Tokamani, M.; Alabasi, G.; Sandaltzopoulos, R.; Marcu, K.B.; Kolettas, E. Roles of NF-κB Signaling in the Regulation of miRNAs Impacting on Inflammation in Cancer. *Biomedicines* **2018**, *6*, 40. [CrossRef] [PubMed]

298. Ren, M.; Qin, D.; Li, K.; Qu, J.; Wang, L.; Wang, Z.; Huang, A.; Tang, H. Correlation between hepatitis B virus protein and microRNA processor Drosha in cells expressing HBV. *Antiviral Res.* **2012**, *94*, 225–231. [CrossRef] [PubMed]
299. Frankish, A.; Diekhans, M.; Jungreis, I.; Lagarde, J.; Loveland, J.E.; Mudge, J.M.; Sisu, C.; Wright, J.C.; Armstrong, J.; Barnes, I.; et al. GENCODE 2021. *Nucleic Acids Res.* **2021**, *49*, D916–D923. [CrossRef] [PubMed]
300. Quinn, J.J.; Chang, H.Y. Unique features of long non-coding RNA biogenesis and function. *Nat. Rev. Genet.* **2016**, *17*, 47–62. [CrossRef]
301. Statello, L.; Guo, C.J.; Chen, L.L.; Huarte, M. Gene regulation by long non-coding RNAs and its biological functions. *Nat. Rev. Mol. Cell Biol.* **2021**, *22*, 96–118. [CrossRef] [PubMed]
302. Ma, G.; Yasunaga, J.I.; Shimura, K.; Takemoto, K.; Watanabe, M.; Amano, M.; Nakata, H.; Liu, B.; Zuo, X.; Matsuoka, M. Human retroviral antisense mRNAs are retained in the nuclei of infected cells for viral persistence. *Proc. Natl. Acad. Sci. USA* **2021**, *118*, e2014783118. [CrossRef]
303. Song, Z.; Wu, W.; Chen, M.; Cheng, W.; Yu, J.; Fang, J.; Xu, L.; Yasunaga, J.I.; Matsuoka, M.; Zhao, T. Long Noncoding RNA ANRIL Supports Proliferation of Adult T-Cell Leukemia Cells through Cooperation with EZH2. *J. Virol.* **2018**, *92*, e00909-18. [CrossRef]
304. Plissonnier, M.L.; Herzog, K.; Levrero, M.; Zeisel, M.B. Non-Coding RNAs and Hepatitis C Virus-Induced Hepatocellular Carcinoma. *Viruses* **2018**, *10*, 591. [CrossRef]
305. Unfried, J.P.; Sangro, P.; Prats-Mari, L.; Sangro, B.; Fortes, P. The Landscape of lncRNAs in Hepatocellular Carcinoma: A Translational Perspective. *Cancers* **2021**, *13*, 2651. [CrossRef]
306. Wang, Y. Upregulated lncRNA-HNGA1, a target of miR-375, contributes to aerobic glycolysis of head and neck squamous cell carci-noma through increasing levels of the glucose transporter protein SCL2A1. *Eur. J. Cancer* **2016**, *1*, S14–S15. [CrossRef]
307. He, H.; Liu, X.; Liu, Y.; Zhang, M.; Lai, Y.; Hao, Y.; Wang, Q.; Shi, D.; Wang, N.; Luo, X.G.; et al. Human Papillomavirus E6/E7 and Long Noncoding RNA TMPOP2 Mutually Upregulated Gene Expression in Cervical Cancer Cells. *J. Virol.* **2019**, *93*, e01808-18. [CrossRef]
308. Ding, X.; Jia, X.; Wang, C.; Xu, J.; Gao, S.J.; Lu, C. A DHX9-lncRNA-MDM2 interaction regulates cell invasion and angiogenesis of cervical cancer. *Cell Death Differ.* **2019**, *26*, 1750–1765. [CrossRef] [PubMed]
309. Liu, H.; Xu, J.; Yang, Y.; Wang, X.; Wu, E.; Majerciak, V.; Zhang, T.; Steenbergen, R.D.M.; Wang, H.K.; Banerjee, N.S.; et al. Oncogenic HPV promotes the expression of the long noncoding RNA lnc-FANCI-2 through E7 and YY1. *Proc. Natl. Acad. Sci. USA* **2021**, *118*, e2014195118. [CrossRef] [PubMed]
310. Herbert, K.M.; Pimienta, G. Consideration of Epstein-Barr Virus-Encoded Noncoding RNAs EBER1 and EBER2 as a Functional Backup of Viral Oncoprotein Latent Membrane Protein 1. *mBio* **2016**, *7*, e01926-15. [CrossRef]
311. Lee, N. The many ways Epstein-Barr virus takes advantage of the RNA tool kit. *RNA Biol.* **2021**, *18*, 759–766. [CrossRef]
312. Verhoeven, R.J.A.; Tong, S.; Mok, B.W.; Liu, J.; He, S.; Zong, J.; Chen, Y.; Tsao, S.W.; Lung, M.L.; Chen, H. Epstein-Barr Virus BART Long Non-coding RNAs Function as Epigenetic Modulators in Nasopharyngeal Carcinoma. *Front. Oncol.* **2019**, *9*, 1120. [CrossRef]
313. Wang, H.; Liu, W.; Luo, B. The roles of miRNAs and lncRNAs in Epstein-Barr virus associated epithelial cell tumors. *Virus Res.* **2021**, *291*, 198217. [CrossRef]
314. Liu, W.; Zhang, Y.; Luo, B. Long Non-coding RNAs in Gammaherpesvirus Infections: Their Roles in Tumorigenic Mechanisms. *Front. Microbiol.* **2020**, *11*, 604536. [CrossRef] [PubMed]
315. Zhang, J.; Li, X.; Hu, J.; Cao, P.; Yan, Q.; Zhang, S.; Dang, W.; Lu, J. Long noncoding RNAs involvement in Epstein-Barr virus infection and tumorigenesis. *Virol. J.* **2020**, *17*, 51. [CrossRef]
316. Withers, J.B.; Li, E.S.; Vallery, T.K.; Yario, T.A.; Steitz, J.A. Two herpesviral noncoding PAN RNAs are functionally homologous but do not associate with common chromatin loci. *PLoS Pathog.* **2018**, *14*, e1007389. [CrossRef]
317. Yang, W.S.; Yeh, W.W.; Campbell, M.; Chang, L.; Chang, P.C. Long non-coding RNA KIKAT/LINC01061 as a novel epigenetic regulator that relocates KDM4A on chromatin and modulates viral reactivation. *PLoS Pathog.* **2021**, *17*, e1009670. [CrossRef] [PubMed]
318. Shi, L.; Peng, F.; Tao, Y.; Fan, X.; Li, N. Roles of long noncoding RNAs in hepatocellular carcinoma. *Virus Res.* **2016**, *223*, 131–139. [CrossRef] [PubMed]
319. Zhang, H.; Chen, X.; Zhang, J.; Wang, X.; Chen, H.; Liu, L.; Liu, S. Long non-coding RNAs in HBV-related hepatocellular carcinoma. *Int. J. Oncol.* **2020**, *56*, 18–32. [CrossRef]
320. Zhang, B.; Han, S.; Feng, B.; Chu, X.; Chen, L.; Wang, R. Hepatitis B virus X protein-mediated non-coding RNA aberrations in the development of human hepatocellular carcinoma. *Exp. Mol. Med.* **2017**, *49*, e293. [CrossRef]
321. Qiu, L.; Wang, T.; Xu, X.; Wu, Y.; Tang, Q.; Chen, K. Long Non-Coding RNAs in Hepatitis B Virus-Related Hepatocellular Carcinoma: Regulation, Functions, and Underlying Mechanisms. *Int. J. Mol. Sci.* **2017**, *18*, 2505. [CrossRef]
322. Han, T.S.; Hur, K.; Cho, H.S.; Ban, H.S. Epigenetic Associations between lncRNA/circRNA and miRNA in Hepatocellular Carcinoma. *Cancers* **2020**, *12*, 2622. [CrossRef] [PubMed]
323. Hu, J.J.; Song, W.; Zhang, S.D.; Shen, X.H.; Qiu, X.M.; Wu, H.Z.; Gong, P.H.; Lu, S.; Zhao, Z.J.; He, M.L.; et al. HBx-upregulated lncRNA UCA1 promotes cell growth and tumorigenesis by recruiting EZH2 and repressing p27Kip1/CDK2 signaling. *Sci. Rep.* **2016**, *6*, 23521. [CrossRef]

324. Du, Y.; Kong, G.; You, X.; Zhang, S.; Zhang, T.; Gao, Y.; Ye, L.; Zhang, X. Elevation of highly up-regulated in liver cancer (HULC) by hepatitis B virus X protein promotes hepatoma cell proliferation via down-regulating p18. *J. Biol. Chem.* **2012**, *287*, 26302–26311. [CrossRef]
325. Deng, X.; Zhao, X.F.; Liang, X.Q.; Chen, R.; Pan, Y.F.; Liang, J. Linc00152 promotes cancer progression in hepatitis B virus-associated hepatocellular carcinoma. *Biomed. Pharmacother.* **2017**, *90*, 100–108. [CrossRef]
326. Moyo, B.; Nicholson, S.A.; Arbuthnot, P.B. The role of long non-coding RNAs in hepatitis B virus-related hepatocellular carcinoma. *Virus Res.* **2016**, *212*, 103–113. [CrossRef] [PubMed]
327. Lemasson, I.; Thébault, S.; Sardet, C.; Devaux, C.; Mesnard, J.M. Activation of E2F-mediated transcription by human T-cell leukemia virus type I Tax protein in a p16(INK4A)-negative T-cell line. *J. Biol. Chem.* **1998**, *273*, 23598–23604. [CrossRef] [PubMed]
328. Moon, S.L.; Blackinton, J.G.; Anderson, J.R.; Dozier, M.K.; Dodd, B.J.; Keene, J.D.; Wilusz, C.J.; Bradrick, S.S.; Wilusz, J. XRN1 stalling in the 5′ UTR of Hepatitis C virus and Bovine Viral Diarrhea virus is associated with dysregulated host mRNA stability. *PLoS Pathog.* **2015**, *11*, e1004708. [CrossRef] [PubMed]
329. Carnero, E.; Barriocanal, M.; Prior, C.; Pablo Unfried, J.; Segura, V.; Guruceaga, E.; Enguita, M.; Smerdou, C.; Gastaminza, P.; Fortes, P. Long noncoding RNA EGOT negatively affects the antiviral response and favors HCV replication. *EMBO Rep.* **2016**, *17*, 1013–1028. [CrossRef] [PubMed]
330. Berrios, C.; Padi, M.; Keibler, M.A.; Park, D.E.; Molla, V.; Cheng, J.; Lee, S.M.; Stephanopoulos, G.; Quackenbush, J.; DeCaprio, J.A. Merkel Cell Polyomavirus Small T Antigen Promotes Pro-Glycolytic Metabolic Perturbations Required for Transformation. *PLoS Pathog.* **2016**, *12*, e1006020. [CrossRef]
331. Kitamura, S.; Yanagi, T.; Inamura-Takashima, Y.; Imafuku, K.; Hata, H.; Uehara, J.; Ishida, Y.; Otsuka, A.; Hirata, K.; Shimizu, H. Retrospective study on the correlation between 18-fluorodeoxyglucose uptake in positron emission tomography-computer tomography and tumour volume, cytological activity as assessed with Ki-67 and GLUT-1 staining in 10 cases of Merkel cell carcinoma. *J. Eur. Acad. Dermatol. Venereol.* **2018**, *32*, e285–e287. [CrossRef] [PubMed]
332. Toberer, F.; Haenssle, H.A.; Heinzel-Gutenbrunner, M.; Enk, A.; Hartschuh, W.; Helmbold, P.; Kutzner, H. Metabolic reprogramming and angiogenesis in primary cutaneous Merkel cell carcinoma: Expression of hypoxia-inducible factor-1α and its central downstream factors. *J. Eur. Acad. Dermatol. Venereol.* **2021**, *35*, 88–94. [CrossRef]
333. Casarotto, M.; Fanetti, G.; Guerrieri, R.; Palazzari, E.; Lupato, V.; Steffan, A.; Polesel, J.; Boscolo-Rizzo, P.; Fratta, E. Beyond MicroRNAs: Emerging Role of Other Non-Coding RNAs in HPV-Driven Cancers. *Cancers* **2020**, *12*, 1246. [CrossRef] [PubMed]
334. He, J.; Huang, B.; Zhang, K.; Liu, M.; Xu, T. Long non-coding RNA in cervical cancer: From biology to therapeutic opportunity. *Biomed. Pharmacother.* **2020**, *127*, 110209. [CrossRef] [PubMed]
335. Cheng, S.; Li, Z.; He, J.; Fu, S.; Duan, Y.; Zhou, Q.; Yan, Y.; Liu, X.; Liu, L.; Feng, C.; et al. Epstein-Barr virus noncoding RNAs from the extracellular vesicles of nasopharyngeal carcinoma (NPC) cells promote angiogenesis via TLR3/RIG-I-mediated VCAM-1 expression. *Biochim. Biophys. Acta Mol. Basis Dis.* **2019**, *1865*, 1201–1213. [CrossRef] [PubMed]
336. Maguire, H.F.; Hoeffler, J.P.; Siddiqui, A. HBV X protein alters the DNA binding specificity of CREB and ATF-2 by protein-protein interactions. *Science* **1991**, *252*, 842–844. [CrossRef]
337. Lee, Y.I.; Lee, S.; Lee, Y.; Bong, Y.S.; Hyun, S.W.; Yoo, Y.D.; Kim, S.J.; Kim, Y.W.; Poo, H.R. The human hepatitis B virus transactivator X gene product regulates Sp1 mediated transcription of an insulin-like growth factor II promoter 4. *Oncogene* **1998**, *16*, 2367–2380. [CrossRef] [PubMed]
338. Salerno, D.; Chiodo, L.; Alfano, V.; Floriot, O.; Cottone, G.; Paturel, A.; Pallocca, M.; Plissonnier, M.L.; Jeddari, S.; Belloni, L.; et al. Hepatitis B protein HBx binds the DLEU2 lncRNA to sustain cccDNA and host cancer-related gene transcription. *Gut* **2020**, *69*, 2016–2024. [CrossRef]
339. Kristensen, L.S.; Andersen, M.S.; Stagsted, L.V.W.; Ebbesen, K.K.; Hansen, T.B.; Kjems, J. The biogenesis, biology and characterization of circular RNAs. *Nat. Rev. Genet.* **2019**, *20*, 675–691. [CrossRef]
340. Xiao, M.S.; Ai, Y.; Wilusz, J.E. Biogenesis and Functions of Circular RNAs Come into Focus. *Trends Cell Biol.* **2020**, *30*, 226–240. [CrossRef]
341. Chen, L.L. The expanding regulatory mechanisms and cellular functions of circular RNAs. *Nat. Rev. Mol. Cell Biol.* **2020**, *21*, 475–490. [CrossRef]
342. Liu, J.; Zhang, X.; Yan, M.; Li, H. Emerging Role of Circular RNAs in Cancer. *Front. Oncol.* **2020**, *10*, 663. [CrossRef]
343. Wang, X.; Li, H.; Lu, Y.; Cheng, L. Circular RNAs in Human Cancer. *Front. Oncol.* **2020**, *10*, 577118. [CrossRef] [PubMed]
344. Tang, X.; Ren, H.; Guo, M.; Qian, J.; Yang, Y.; Gu, C. Review on circular RNAs and new insights into their roles in cancer. *Comput. Struct. Biotechnol. J.* **2021**, *19*, 910–928. [CrossRef]
345. Chen, T.C.; Tallo-Parra, M.; Cao, Q.M.; Kadener, S.; Böttcher, R.; Pérez-Vilaró, G.; Boonchuen, P.; Somboonwiwat, K.; Díez, J.; Sarnow, P. Host-derived circular RNAs display proviral activities in Hepatitis C virus-infected cells. *PLoS Pathog.* **2020**, *16*, e1008346. [CrossRef] [PubMed]
346. Yang, R.; Lee, E.E.; Kim, J.; Choi, J.H.; Kolitz, E.; Chen, Y.; Crewe, C.; Salisbury, N.J.H.; Scherer, P.E.; Cockerell, C.; et al. Characterization of ALTO-encoding circular RNAs expressed by Merkel cell polyomavirus and trichodysplasia spinulosa polyomavirus. *PLoS Pathog.* **2021**, *17*, e1009582. [CrossRef] [PubMed]
347. Abere, B.; Zhou, H.; Li, J.; Cao, S.; Toptan, T.; Grundhoff, A.; Fischer, N.; Moore, P.S.; Chang, Y. Merkel Cell Polyomavirus Encodes Circular RNAs (circRNAs) Enabling a Dynamic circRNA/microRNA/mRNA Regulatory Network. *mBio* **2020**, *11*, e03059-20. [CrossRef] [PubMed]

348. Zhao, J.; Lee, E.E.; Kim, J.; Yang, R.; Chamseddin, B.; Ni, C.; Gusho, E.; Xie, Y.; Chiang, C.M.; Buszczak, M.; et al. Transforming activity of an oncoprotein-encoding circular RNA from human papillomavirus. *Nat. Commun.* **2019**, *10*, 2300. [CrossRef]
349. Chamseddin, B.H.; Lee, E.E.; Kim, J.; Zhan, X.; Yang, R.; Murphy, K.M.; Lewis, C.; Hosler, G.A.; Hammer, S.T.; Wang, R.C. Assessment of circularized E7 RNA, GLUT1, and PD-L1 in anal squamous cell carcinoma. *Oncotarget* **2019**, *10*, 5958–5969. [CrossRef]
350. Ma, H.; Tian, T.; Liu, X.; Xia, M.; Chen, C.; Mai, L.; Xie, S.; Yu, L. Upregulated circ_0005576 facilitates cervical cancer progression via the miR-153/KIF20A axis. *Biomed. Pharmacother.* **2019**, *118*, 109311. [CrossRef]
351. Nahand, J.S.; Jamshidi, S.; Hamblin, M.R.; Mahjoubin-Tehran, M.; Vosough, M.; Jamali, M.; Khatami, A.; Moghoofei, M.; Baghi, H.B.; Mirzaei, H. Circular RNAs: New Epigenetic Signatures in Viral Infections. *Front. Microbiol.* **2020**, *11*, 1853. [CrossRef]
352. Ungerleider, N.; Concha, M.; Lin, Z.; Roberts, C.; Wang, X.; Cao, S.; Baddoo, M.; Moss, W.N.; Yu, Y.; Seddon, M.; et al. The Epstein Barr virus circRNAome. *PLoS Pathog.* **2018**, *14*, e1007206. [CrossRef] [PubMed]
353. Avilala, J.; Becnel, D.; Abdelghani, R.; Nanbo, A.; Kahn, J.; Li, L.; Lin, Z. Role of Virally Encoded Circular RNAs in the Pathogenicity of Human Oncogenic Viruses. *Front. Microbiol.* **2021**, *12*, 657036. [CrossRef] [PubMed]
354. Tagawa, T.; Gao, S.; Koparde, V.N.; Gonzalez, M.; Spouge, J.L.; Serquiña, A.P.; Lurain, K.; Ramaswami, R.; Uldrick, T.S.; Yarchoan, R.; et al. Discovery of Kaposi's sarcoma herpesvirus-encoded circular RNAs and a human antiviral circular RNA. *Proc. Natl. Acad. Sci. USA* **2018**, *115*, 12805–12810. [CrossRef] [PubMed]
355. Abere, B.; Li, J.; Zhou, H.; Toptan, T.; Moore, P.S.; Chang, Y. Kaposi's Sarcoma-Associated Herpesvirus-Encoded circRNAs Are Expressed in Infected Tumor Tissues and Are Incorporated into Virions. *mBio* **2020**, *11*, e03027-19. [CrossRef]
356. Yao, S.; Jia, X.; Wang, F.; Sheng, L.; Song, P.; Cao, Y.; Shi, H.; Fan, W.; Ding, X.; Gao, S.J.; et al. CircRNA ARFGEF1 functions as a ceRNA to promote oncogenic KSHV-encoded viral interferon regulatory factor induction of cell invasion and angiogenesis by upregulating glutaredoxin 3. *PLoS Pathog.* **2021**, *17*, e1009294. [CrossRef]
357. Sekiba, K.; Otsuka, M.; Ohno, M.; Kishikawa, T.; Yamagami, M.; Suzuki, T.; Ishibashi, R.; Seimiya, T.; Tanaka, E.; Koike, K. DHX9 regulates production of hepatitis B virus-derived circular RNA and viral protein levels. *Oncotarget* **2018**, *9*, 20953–20964. [CrossRef]
358. Chen, Y.; Li, S.; Wei, Y.; Xu, Z.; Wu, X. Circ-RNF13, as an oncogene, regulates malignant progression of HBV-associated hepatocellular carcinoma cells and HBV expression and replication through circ-RNF13/miR-424-5p/TGIF2 ceRNA pathway. *Bosn. J. Basic Med. Sci.* **2021**. Online ahead of print. [CrossRef]
359. Carter, J.J.; Daugherty, M.D.; Qi, X.; Bheda-Malge, A.; Wipf, G.C.; Robinson, K.; Roman, A.; Malik, H.S.; Galloway, D.A. Identification of an overprinting gene in Merkel cell polyomavirus provides evolutionary insight into the birth of viral genes. *Proc. Natl. Acad. Sci. USA* **2013**, *110*, 12744–12749. [CrossRef]
360. Yang, Y.; Fan, X.; Mao, M.; Song, X.; Wu, P.; Zhang, Y.; Jin, Y.; Yang, Y.; Chen, L.L.; Wang, Y.; et al. Extensive translation of circular RNAs driven by N(6)-methyladenosine. *Cell Res.* **2017**, *27*, 626–641. [CrossRef] [PubMed]
361. Wang, H.; Zhao, Y.; Chen, M.; Cui, J. Identification of Novel Long Non-coding and Circular RNAs in Human Papillomavirus-Mediated Cervical Cancer. *Front. Microbiol.* **2017**, *8*, 1720. [CrossRef]
362. Chaichian, S.; Shafabakhsh, R.; Mirhashemi, S.M.; Moazzami, B.; Asemi, Z. Circular RNAs: A novel biomarker for cervical cancer. *J. Cell. Physiol.* **2020**, *235*, 718–724. [CrossRef] [PubMed]
363. Hu, C.; Wang, Y.; Li, A.; Zhang, J.; Xue, F.; Zhu, L. Overexpressed circ_0067934 acts as an oncogene to facilitate cervical cancer progression via the miR-545/EIF3C axis. *J. Cell. Physiol.* **2019**, *234*, 9225–9232. [CrossRef] [PubMed]
364. Bonelli, P.; Borrelli, A.; Tuccillo, F.M.; Buonaguro, F.M.; Tornesello, M.L. The Role of circRNAs in Human Papillomavirus (HPV)-Associated Cancers. *Cancers* **2021**, *13*, 1173. [CrossRef]
365. Xie, J.; Chen, Q.; Zhou, P.; Fan, W. Circular RNA hsa_circ_0000511 Improves Epithelial Mesenchymal Transition of Cervical Cancer by Regulating hsa-mir-296-5p/HMGA1. *J. Immunol. Res.* **2021**, *2021*, 9964538. [CrossRef] [PubMed]
366. Toptan, T.; Abere, B.; Nalesnik, M.A.; Swerdlow, S.H.; Ranganathan, S.; Lee, N.; Shair, K.H.; Moore, P.S.; Chang, Y. Circular DNA tumor viruses make circular RNAs. *Proc. Natl. Acad. Sci. USA* **2018**, *115*, E8737–E8745. [CrossRef]
367. Ungerleider, N.A.; Tibbetts, S.A.; Renne, R.; Flemington, E.K. Gammaherpesvirus RNAs Come Full Circle. *mBio* **2019**, *10*, e00071-19. [CrossRef]
368. Aktaş, T.; Avşar Ilık, İ.; Maticzka, D.; Bhardwaj, V.; Pessoa Rodrigues, C.; Mittler, G.; Manke, T.; Backofen, R.; Akhtar, A. DHX9 suppresses RNA processing defects originating from the Alu invasion of the human genome. *Nature* **2017**, *544*, 115–119. [CrossRef]
369. Cui, S.; Qian, Z.; Chen, Y.; Li, L.; Li, P.; Ding, H. Screening of up- and downregulation of circRNAs in HBV-related hepatocellular carcinoma by microarray. *Oncol. Lett.* **2018**, *15*, 423–432. [CrossRef]
370. Wang, S.; Cui, S.; Zhao, W.; Qian, Z.; Liu, H.; Chen, Y.; Lv, F.; Ding, H.G. Screening and bioinformatics analysis of circular RNA expression profiles in hepatitis B-related hepatocellular carcinoma. *Cancer Biomark.* **2018**, *22*, 631–640. [CrossRef] [PubMed]
371. Wu, C.; Deng, L.; Zhuo, H.; Chen, X.; Tan, Z.; Han, S.; Tang, J.; Qian, X.; Yao, A. Circulating circRNA predicting the occurrence of hepatocellular carcinoma in patients with HBV infection. *J. Cell. Mol. Med.* **2020**, *24*, 10216–10222. [CrossRef] [PubMed]
372. Nehme, Z.; Pasquereau, S.; Herbein, G. Control of viral infections by epigenetic-targeted therapy. *Clin. Epigenet.* **2019**, *11*, 55. [CrossRef] [PubMed]
373. Cheng, Y.; He, C.; Wang, M.; Ma, X.; Mo, F.; Yang, S.; Han, J.; Wei, X. Targeting epigenetic regulators for cancer therapy: Mechanisms and advances in clinical trials. *Signal Transduct. Target. Ther.* **2019**, *4*, 62. [CrossRef]

374. Chen, Z.; Xie, H.; Hu, M.; Huang, T.; Hu, Y.; Sang, N.; Zhao, Y. Recent progress in treatment of hepatocellular carcinoma. *Am. J. Cancer Res.* **2020**, *10*, 2993–3036.
375. Saillard, C.; Guermouche, H.; Derrieux, C.; Bruneau, J.; Frenzel, L.; Couronne, L.; Asnafi, V.; Macintyre, E.; Trinquand, A.; Lhermitte, L.; et al. Response to 5-azacytidine in a patient with TET2-mutated angioimmunoblastic T-cell lymphoma and chronic myelomonocytic leukaemia preceded by an EBV-positive large B-cell lymphoma. *Hematol. Oncol.* **2017**, *35*, 864–868. [CrossRef]
376. Biktasova, A.; Hajek, M.; Sewell, A.; Gary, C.; Bellinger, G.; Deshpande, H.A.; Bhatia, A.; Burtness, B.; Judson, B.; Mehra, S.; et al. Demethylation Therapy as a Targeted Treatment for Human Papillomavirus-Associated Head and Neck Cancer. *Clin. Cancer Res.* **2017**, *23*, 7276–7287. [CrossRef]
377. Gailhouste, L.; Sudoh, M.; Qin, X.Y.; Watashi, K.; Wakita, T.; Ochiya, T.; Matsuura, T.; Kojima, S.; Furutani, Y. Epigenetic reprogramming promotes the antiviral action of IFNα in HBV-infected cells. *Cell Death Discov.* **2021**, *7*, 130. [CrossRef]
378. Lourenço de Freitas, N.; Deberaldini, M.G.; Gomes, D.; Pavan, A.R.; Sousa, Â.; Dos Santos, J.L.; Soares, C.P. Histone Deacetylase Inhibitors as Therapeutic Interventions on Cervical Cancer Induced by Human Papillomavirus. *Front. Cell Dev. Biol.* **2020**, *8*, 592868. [CrossRef]
379. He, H.; Lai, Y.; Hao, Y.; Liu, Y.; Zhang, Z.; Liu, X.; Guo, C.; Zhang, M.; Zhou, H.; Wang, N.; et al. Selective p300 inhibitor C646 inhibited HPV E6-E7 genes, altered glucose metabolism and induced apoptosis in cervical cancer cells. *Eur. J. Pharmacol.* **2017**, *812*, 206–215. [CrossRef] [PubMed]
380. Lindsay, C.D.; Kostiuk, M.A.; Harris, J.; O'Connell, D.A.; Seikaly, H.; Biron, V.L. Efficacy of EZH2 inhibitory drugs in human papillomavirus-positive and human papillomavirus-negative oropharyngeal squamous cell carcinomas. *Clin. Epigenet.* **2017**, *9*, 95. [CrossRef]
381. Leiendecker, L.; Jung, P.S.; Krecioch, I.; Neumann, T.; Schleiffer, A.; Mechtler, K.; Wiesner, T.; Obenauf, A.C. LSD1 inhibition induces differentiation and cell death in Merkel cell carcinoma. *EMBO Mol. Med.* **2020**, *12*, e12525. [CrossRef] [PubMed]
382. van der Ree, M.H.; van der Meer, A.J.; de Bruijne, J.; Maan, R.; van Vliet, A.; Welzel, T.M.; Zeuzem, S.; Lawitz, E.J.; Rodriguez-Torres, M.; Kupcova, V.; et al. Long-term safety and efficacy of microRNA-targeted therapy in chronic hepatitis C patients. *Antiviral Res.* **2014**, *111*, 53–59. [CrossRef] [PubMed]
383. Yoon, J.H.; Min, K.; Lee, S.K. Epstein-Barr Virus miR-BART17-5p Promotes Migration and Anchorage-Independent Growth by Targeting Kruppel-Like Factor 2 in Gastric Cancer. *Microorganisms* **2020**, *8*, 258. [CrossRef]
384. Özeş, A.R.; Wang, Y.; Zong, X.; Fang, F.; Pilrose, J.; Nephew, K.P. Therapeutic targeting using tumor specific peptides inhibits long non-coding RNA HOTAIR activity in ovarian and breast cancer. *Sci. Rep.* **2017**, *7*, 894. [CrossRef]
385. Zhen, S.; Hua, L.; Liu, Y.H.; Sun, X.M.; Jiang, M.M.; Chen, W.; Zhao, L.; Li, X. Inhibition of long non-coding RNA UCA1 by CRISPR/Cas9 attenuated malignant phenotypes of bladder cancer. *Oncotarget* **2017**, *8*, 9634–9646. [CrossRef]
386. Pal, S.; Tyler, J.K. Epigenetics and aging. *Sci. Adv.* **2016**, *2*, e1600584. [CrossRef]
387. Ferreira, D.A.; Tayyar, Y.; Idris, A.; McMillan, N.A.J. A "hit-and-run" affair—A possible link for cancer progression in virally driven cancers. *Biochim. Biophys. Acta Rev. Cancer* **2021**, *1875*, 188476. [CrossRef] [PubMed]
388. Chen, X.Y.; Zhang, J.; Zhu, J.S. The role of m(6)A RNA methylation in human cancer. *Mol. Cancer* **2019**, *18*, 103. [CrossRef]
389. Kim, G.W.; Imam, H.; Siddiqui, A. The RNA Binding Proteins YTHDC1 and FMRP Regulate the Nuclear Export of N(6)-Methyladenosine-Modified Hepatitis B Virus Transcripts and Affect the Viral Life Cycle. *J. Virol.* **2021**, *95*, e0009721. [CrossRef]
390. Zhang, L.; Hou, C.; Chen, C.; Guo, Y.; Yuan, W.; Yin, D.; Liu, J.; Sun, Z. The role of N(6)-methyladenosine (m(6)A) modification in the regulation of circRNAs. *Mol. Cancer* **2020**, *19*, 105. [CrossRef]
391. Zheng, X.; Wang, J.; Zhang, X.; Fu, Y.; Peng, Q.; Lu, J.; Wei, L.; Li, Z.; Liu, C.; Wu, Y.; et al. RNA m(6) A methylation regulates virus-host interaction and EBNA2 expression during Epstein-Barr virus infection. *Immun. Inflamm. Dis.* **2021**, *9*, 351–362. [CrossRef] [PubMed]
392. Dai, D.L.; Li, X.; Wang, L.; Xie, C.; Jin, Y.; Zeng, M.S.; Zuo, Z.; Xia, T.L. Identification of an N6-methyladenosine-mediated positive feedback loop that promotes Epstein-Barr virus infection. *J. Biol. Chem.* **2021**, *296*, 100547. [CrossRef] [PubMed]

MDPI
St. Alban-Anlage 66
4052 Basel
Switzerland
Tel. +41 61 683 77 34
Fax +41 61 302 89 18
www.mdpi.com

International Journal of Molecular Sciences Editorial Office
E-mail: ijms@mdpi.com
www.mdpi.com/journal/ijms

www.ingramcontent.com/pod-product-compliance
Lightning Source LLC
Chambersburg PA
CBHW040829250926
51850CB00007BA/1292
9783036578576